COMPREHENSIVE CHEMICAL KINETICS

COMPREHENSIVE

Section 1. THE PRACTICE AND THEORY OF KINETICS

Volume 1   The Practice of Kinetics

Volume 2   The Theory of Kinetics

Volume 3   The Formation and Decay of Excited Species

Section 2. HOMOGENEOUS DECOMPOSITION AND ISOMERISATION REACTIONS

Volume 4   Decomposition of Inorganic and Organometallic Compounds

Volume 5   Decomposition and Isomerisation of Organic Compounds

Section 3. INORGANIC REACTIONS

Volume 6   Reactions of Non-metallic Inorganic Compounds

Volume 7   Reactions of Metallic Salts and Complexes, and Organometallic Compounds

Section 4. ORGANIC REACTIONS (6 volumes)

Volume 9   Addition and Elimination Reactions of Aliphatic Compounds

Volume 10 Ester Formation and Hydrolysis and Related Reactions

Volume 12 Electrophilic Substitution at a Saturated Carbon Atom

Volume 13 Reactions of Aromatic Compounds

Section 5. POLYMERISATION REACTIONS (2 volumes)

Section 6. OXIDATION AND COMBUSTION REACTIONS (2 volumes)

Section 7. SELECTED ELEMENTARY REACTIONS (2 volumes)

Additional Sections

HETEROGENEOUS REACTIONS

SOLID STATE REACTIONS

KINETICS AND TECHNOLOGICAL PROCESSES

CHEMICAL KINETICS

EDITED BY

C. H. BAMFORD

M.A., Ph.D., Sc.D. (Cantab.), F.R.I.C., F.R.S.

Campbell-Brown Professor of Industrial Chemistry,
University of Liverpool

AND

C. F. H. TIPPER

Ph.D. (Bristol), D.Sc. (Edinburgh)

Senior Lecturer in Physical Chemistry,
University of Liverpool

VOLUME 12

ELECTROPHILIC SUBSTITUTION AT A SATURATED

CARBON ATOM

ELSEVIER SCIENTIFIC PUBLISHING COMPANY

AMSTERDAM - LONDON - NEW YORK

1973

ELSEVIER SCIENTIFIC PUBLISHING COMPANY
335 JAN VAN GALENSTRAAT
P.O. BOX 211, AMSTERDAM, THE NETHERLANDS

AMERICAN ELSEVIER PUBLISHING COMPANY, INC.
52 VANDERBILT AVENUE
NEW YORK, NEW YORK 10017

LIBRARY OF CONGRESS CARD NUMBER 72-83196
ISBN 0-444-41052-X

WITH 5 ILLUSTRATIONS AND 135 TABLES

PRINTED IN THE NETHERLANDS

COMPREHENSIVE CHEMICAL KINETICS

Contributor to Volume 12

All chapters in this Volume have been written by

M. H. ABRAHAM   Department of Chemistry,
University of Surrey,
Guildford, Surrey, England

Preface

Section 4 deals almost exclusively with reactions recognised as organic in a traditional sense, but excluding (unless very relevant) those already considered in Sections 2 and 3 and biochemical systems. Also oxidations, *e.g.* of hydrocarbons, by molecular oxygen, polymerization reactions and fully heterogeneous processes are considered later. The relationships of mechanism and kinetics, *e.g.* effects of structure of reactants and solvent, isotope effects, are fully discussed. Rate parameters of individual elementary steps, as well as of overall processes, are given if available. We have endeavoured, in conformity with our earlier policy, to organise this section according to the types of chemical transformation and with the minimum of recourse to mechanistic classification. Nevertheless it seemed desirable to divide up certain general processes on the basis of their nucleophilic or electrophilic character.

Volume 12 is devoted entirely to the subject of electrophilic substitution at a saturated carbon atom, with particular reference to kinetic studies involving metal alkyls as substrates. In Chapter 2 the generation of carbanions is discussed generally, and in Chapter 3 possible mechanisms of the electrophilic substitution at saturated carbon are considered. Chapters 5–10 are arranged on the basis of reaction type; substitution reactions, mercury for mercury exchanges, other metal for metal exchanges and acidolysis and halogenolysis of organometallic compounds. Finally in Chapter 11 observations on structural, salt and solvent effects are correlated.

The editors are very grateful for the continuing support and advice from members of the editorial board, and also for much invaluable assistance from Dr. D. Bethell of the Department of Organic Chemistry.

Liverpool　　　　　　　　　　　　　　　　　　　　　C. H. Bamford
September, 1972　　　　　　　　　　　　　　　　　　C. H. F. Tipper

Contents

Preface VII

Chapter 1

Introduction 1

REFERENCES. 3

Chapter 2

The formation and reactions of carbanions 5

  1. GENERAL THERMODYNAMIC CONSIDERATIONS 5

  2. FORMATION OF CARBANIONS FROM HYDROCARBONS 7

  3. REACTIONS AND PROPERTIES OF CARBANIONS 8

REFERENCES. 9

Chapter 3

Mechanisms of electrophilic substitution at saturated carbon 11

  1. THE FUNDAMENTAL MECHANISMS 11
    1.1 Unimolecular electrophilic substitution; S_E1 11
    1.2 Bimolecular electrophilic substitution *via* an open transition state; S_E2, S_E2(open) 12
    1.3 Bimolecular electrophilic substitution *via* a cyclic transition state; S_E2, S_Ei, S_F2, S_E2(cyclic) 12
    1.4 Electrophilic substitution *via* co-ordination 13
    1.5 Summary of the nomenclature of the fundamental mechanisms 15

  2. SUBSTITUTION WITH REARRANGEMENT. 16
    2.1 Unimolecular electrophilic substitution with rearrangement; S_E1', $[S_E1']$. . . . 16
    2.2 Bimolecular electrophilic substitution with rearrangement; S_E2', $[S_E2'$(open)] 16
    2.3 Bimolecular electrophilic substitution, *via* a cyclic transition state, with rearrangement; $S_Ei'[S_E2'$(cyclic)]. 17
    2.4 Electrophilic substitution *via* co-ordination, with rearrangement; $[S_E2'$(co-ord)] 17

  3. MECHANISMS INVOLVING NUCLEOPHILIC OR ANIONIC CATALYSIS 17
    3.1 The solvent as a nucleophilic catalyst. 18
    3.2 The unimolecular mechanism, S_E1, and nucleophilic catalysis 19
    3.3 The bimolecular mechanism, S_E2(open), and nucleophilic catalysis 19
    3.4 The bimolecular mechanism, S_E2(cyclic), and nucleophilic catalysis. 20
    3.5 Substitution *via* co-ordination, S_E2(co-ord), and nucleophilic catalysis 20
    3.6 Mechanisms involving substitution with rearrangement and nucleophilic catalysis 21

REFERENCES 21

Chapter 4

The unimolecular mechanism, S_E1 23

  1. SUBSTITUTIONS IN ORGANOMERCURY COMPOUNDS 23

1.1 The fundamental mechanism, S_E1 23
1.2 The unimolecular mechanism, S_E1, and nucleophilic catalysis 25

2. SUBSTITUTIONS IN OTHER ORGANOMETALLIC COMPOUNDS 31
2.1 The substitution of 1-cyano-1-carbethoxypentyl(triphenyl-phosphine)-gold(I) by alkylmercuric salts 31
2.2 The hydrolysis of 4-pyridiomethyl-manganese, -molybdenum, and -tungsten compounds 31
2.3 Alkyl–metal compounds studied by proton magnetic resonance spectroscopy (PMR) 32
2.4 The cleavage of organometallic compounds of Group IVA by alkali 34

REFERENCES 36

Chapter 5

Mercury-for-mercury exchanges 39

1. THE ONE-ALKYL MERCURY-FOR-MERCURY EXCHANGE 39
1.1 Stereochemical studies 39
1.2 Exchange between simple alkylmercuric salts and mercuric salts 40
1.3 Exchange between benzylmercuric bromide and mercuric bromide 44
1.4 Exchange between α-carbethoxybenzylmercuric bromide and mercuric bromide 47

2. THE TWO-ALKYL (*syn*-PROPORTIONATION) MERCURY-FOR-MERCURY EXCHANGE 48
2.1 Stereochemical studies 48
2.2 *syn*-Proportionation of dialkylmercurys and mercuric salts 49

3. THE TWO-ALKYL (SYMMETRISATION) MERCURY-FOR-MERCURY EXCHANGE 53
3.1 Symmetrisation of α-carbalkoxybenzylmercuric bromides by ammonia in chloroform 53
3.2 Co-symmetrisation of α-carbethoxybenzylmercuric bromides by ammonia in chloroform 56
3.3 Symmetrisation of α-carbethoxybenzylmeruric bromide by diphenylmercury . . 57

4. THE THREE-ALKYL MERCURY-FOR-MERCURY EXCHANGE 58

REFERENCES 60

Chapter 6

Other metal-for-metal exchanges 63

1. SUBSTITUTION OF ALKYL–METAL COMPOUNDS BY METAL SALTS 63
1.1 Substitution of methylmagnesium halides by trialkylsilicon halides 63
1.2 Substitution of dialkylzincs by phenylmercuric chloride 64
1.3 Substitution of alkylboronic acids by mercuric chloride 65
1.4 Substitution of alkylthallium compounds by alkylmercuric compounds 69
1.5 Substitution of tetraalkyltins by mercuric salts 70
1.6 Substitution of tetraalkyltins by dimethyltin dichloride 87
1.7 Substitutions of trimethylantimony by antimony trichloride 90
1.8 Substitution of pentaaquopyridiomethylchromium(III) ions by mercuric(II) salts 91
1.9 Substitution of pentaaquopyridiomethylchromium(III) ions by thallium(III) salts 95
1.10 Substitution of pyridiomethyldicarbonyl-π-cyclopentadienyliron by mercury(II) and thallium(III) salts 96
1.11 Substitution of pyridiomethylpentacyanocobaltate(III) ions by mercury(II), thallium(III), indium(III) and gallium(III) salts 97

1.12 Substitution of pentaaquopyridiomethylchromium(III) ions by mercury(I) . .   99
1.13 Substitution of alkylgold(I) and alkylgold(III) complexes by mercuric salts and
     by alkylmercuric salts   100
     1.13.1 Substitution of alkyl(triphenylphosphine)-gold(I) complexes by mercuric
            salts   100
     1.13.2 Substitution of alkyl (triphenylphosphine)gold(I) complexes by methyl-
            mercuric salts   102
     1.13.3 Substitution of trimethyl(triphenylphosphine)gold(III) by mercuric
            salts   104

2. ALKYL EXCHANGES BETWEEN COMPOUNDS R_nM AND R_mM'   105

REFERENCES   105

Chapter 7

Acidolysis of organometallic compounds   107

1. STEREOCHEMICAL STUDIES   107

2. ACIDOLYSIS OF ORGANOMAGNESIUM COMPOUNDS   108
   2.1 The cleavage of Grignard reagents by 1-alkynes   108
   2.2 The cleavage of dialkylmagnesiums by hexyne-1   110

3. ACIDOLYSIS OF DIALKYLZINCS AND OF DIALKYLCADMIUMS   111

4. ACIDOLYSES OF ORGANOMERCURY COMPOUNDS   114
   4.1 Acidolysis of alkylmercuric iodides by aqueous sulphuric and perchloric acids   114
   4.2 Acidolysis of dialkylmercurys by acetic acid and by perchloric acid in solvent
       acetic acid   115
   4.3 Acidolysis of dialkylmercurys by hydrogen chloride   117
   4.4 A note on the Kharasch "electronegativity" series   119
   4.5 Acidolysis of benzylmercuric compounds   120
   4.6 The anion-catalysed acidolysis of α-carbethoxybenzylmercuric chloride   121

5. ACIDOLYSIS OF TRIETHYLBORON BY CARBOXYLIC ACIDS   122

6. ACIDOLYSIS OF THE ALKYLS OF METALS IN GROUP IVA   124

7. ACIDOLYSIS OF THE ALKYLS OF METALS IN GROUP VIII AND GROUP VIB   129
   7.1 Acidolysis of *trans*-bis(triethylphosphine)halo-(methyl)platinum(II)   129
   7.2 Acidolysis of the benzylchromium ion   132

REFERENCES   133

Chapter 8

Halogenolysis (halogenodemetallation) of organometallic compounds   135

1. STEREOCHEMICAL STUDIES   135

2. HALOGENOLYSIS OF ORGANOMERCURY COMPOUNDS   137
   2.1 Iodinolysis of alkylmercuric iodides by I_2/I^-   137
   2.2 General kinetic expressions for iodinolyses in the presence of iodide ion   138
   2.3 Iodinolysis of benzylmercuric chloride by I_2/I^-   139
   2.4 Brominolysis of benzylmercuric chloride by Br_2/Br^-   143
   2.5 Halogenolysis of benzylmercuric chloride by halogens in solvent carbon tetra-
       chloride   144
   2.6 Brominolysis of *sec.*-butylmercuric bromide by bromine in solvent carbon
       tetrachloride   144
   2.7 Iodinolysis of dialkylmercurys by iodine   145

3. HALOGENOLYSIS OF TETRAALKYLTINS 145
   3.1 Iodinolysis of tetraalkyltins by I_2/I^- in solvent methanol 146
   3.2 Iodinolysis of tetraalkyltins by iodine in solvent acetic acid 151
   3.3 Iodinolysis of tetraalkyltins by iodine in solvent dimethylsulphoxide (DMSO) 152
   3.4 Brominolysis of tetraalkyltins by Br_2/Br^- in solvent dimethylformamide (DMF) 153
   3.5 Brominolysis of tetraalkyltins and trialkyltin bromides by bromine, and bromino-
       lysis of tetraalkyltins by Br_2/Br^-, in solvent acetic acid 154
   3.6 Iodinolysis of tetraalkyltins by iodine in solvent chlorobenzene. 155
   3.7 Brominolysis of tetraalkyltins by bromine in solvent chlorobenzene. 156
      3.7.1 The second-order term in equation (40) 158
      3.7.2 The third-order term in equation (40) 161
   3.8 Brominolysis of tetraalkyltins by bromine in solvent carbon tetrachloride . . . 163
   3.9 Iodinolysis of tetra-n-butyltin by iodine in some non-polar solvents. 165

4. HALOGENOLYSIS OF TETRAALKYLLEADS 166
   4.1 Iodinolysis of tetraalkylleads by I_2/I^- in polar solvents 166
   4.2 Iodinolysis of tetraalkylleads by iodine in solvents benzene and carbon tetra-
       chloride 170
   4.3 The brominolysis of tetraalkylleads by Br_2/Br^- in solvent methanol 171

5. IODINE BROMIDE CLEAVAGE OF TETRAALKYLTINS 172

6. SOLVENT EFFECTS ON THE IODINOLYSIS OF TETRAALKYTINS AND TETRAALKYLLEADS . . . 173

REFERENCES 175

Chapter 9

Other electrophilic substitutions of organometallic compounds 179

1. THE STEREOCHEMICAL COURSE OF CARBONATION OF ALKYLLITHIUM AND ALKYLMAGNESIUM
   COMPOUNDS 179

2. ADDITION REACTIONS OF ORGANOMAGNESIUM COMPOUNDS 179

3. THE REACTION BETWEEN DIALKYLZINCS AND BENZALDEHYDE 180

4. ALKYLATION OF TRIARYLBROMOMETHANE BY ALKYLMERCURIC HALIDES 180

5. CLEAVAGES OF ALKYLBORONIC ACIDS 182
   5.1 The action of hydrogen peroxide on alkylboronic acids 182
   5.2 The oxidation of alkylboronic acids by chromic acid in aqueous solution. . . . 184

6. ADDITION OF TRIALKYLALUMINIUMS TO NITRILES, ESTERS, AND KETONES 185
   6.1 Addition of trimethyl- and triethylaluminium to benzonitrile in solvent xylene,
       and to esters and ketones 185
   6.2 Addition of trimethylaluminium to benzophenone 186

7. DISPROPORTIONATION OF TETRAALKYLSILANES CATALYSED BY ALUMINIUM BROMIDE . . . 188

8. THE OXIDATION OF TETRAALKYLTINS BY CHROMIUM TRIOXIDE IN SOLVENT ACETIC ACID . 189

9. THE DISPLACEMENT OF COBALT(III) AND CHROMIUM(III) FROM PYRIDIOMETHYLCOBALTATE
   IONS AND -CHROMIUM(III) IONS BY NITROSATING AGENTS 190

REFERENCES 192

Chapter 10

Electrophilic substitutions of allyl–metal compounds 195

1. BASE CATALYSED S_E1 CLEAVAGE OF 3-PHENALLYLSILICON AND -TIN COMPOUNDS 195

2. METAL-FOR-METAL EXCHANGES OF ALLYL–METAL COMPOUNDS   197
    2.1 Substitution of allylic derivatives of Group IVA elements by mercuric salts . .   197

3. ACIDOLYSIS OF ALLYL–METAL COMPOUNDS   201
    3.1 Acidolysis of allylmercuric iodide by aqueous acid   201
    3.2 Acidolysis of allylmercuric chloride by HCl in 90 % aqueous dioxan   203
    3.3 Acidolyses of crotylmercuric bromide and cinnamylmercuric bromide.   203
    3.4 Acidolysis of some allyl- and allenyltins.   204

4. HALOGENOLYSIS (HALOGENODEMETALLATION) OF ALLYL–METAL COMPOUNDS.   206
    4.1 Iodinolysis of tetraallyltin by I_2/I^- in solvent acetone.   206
    4.2 Iodinolyses of allyltin compounds by I_2/I^- in various solvents   208

REFERENCES.   210

Chapter 11

Constitutional effects, salt effects, and solvent effects in electrophilic substitution at saturated
carbon.   211

1. ENTROPIES OF ACTIVATION AS A CRITERION OF MECHANISM. SUBSTITUTION BY MECHANISM
   S_E2   211

2. CONSTITUTIONAL INFLUENCES IN THE ELECTROPHILIC REAGENT   212
    2.1 Substitution by mechanism S_E2   212
    2.2 Substitution by mechanism S_E1   215

3. CONSTITUTIONAL INFLUENCES IN THE SUBSTRATE   215
    3.1 The effect of p-substituents on the electrophilic cleavage of benzyl–metal and
         α-carbethoxybenzyl–metal bonds   215
    3.2 More general influences in the substrate.   218
         3.2.1 Substitution by mechanism S_E2   218
         3.2.2 Substitution by mechanism S_E1   218
    3.3 Reactivities of substrates RMX_n with respect to the metal atom M   218
         3.3.1 Substitution by mechanism S_E2   218
         3.3.2 Substitution by mechanism S_E1   220

4. SALT EFFECTS AND CO-SOLVENT EFFECTS IN ELECTROPHILIC SUBSTITUTION AT SATURATED
   CARBON   221
    4.1 Substitution by mechanism S_E2   221
    4.2 Substitution by mechanism S_E1   224

5. SOLVENT EFFECTS AND CONSTITUTIONAL EFFECTS OF THE SUBSTRATE, IN ELECTROPHILIC
   SUBSTITUTION AT SATURATED CARBON   224
    5.1 Relative reactivities of substrates RMX_n with respect to the alkyl group R . . .   224
         5.1.1 Substitution by mechanism S_E2   224
    5.2 Solvent effects on the rates of electrophilic substitutions   234
         5.2.1 Substitution by mechanism S_E2   234
         5.2.2 Substitution by mechanism S_E1   238

6. METAL–METAL BONDING IN S_E2 TRANSITION STATES   239

REFERENCES.   240

Index   243

Chapter 1

Introduction

The heterolytic fission of a bond between a carbon atom and a leaving group (L) can in principle occur in either of two ways.

$$R\overset{\frown}{L} \longrightarrow R^+ + L^- \tag{1}$$

$$\overset{\frown}{R\!-\!L} \longrightarrow R^- + L^+ \tag{2}$$

If R is an alkyl group, reaction (1) leads to the familiar mechanism of nucleophilic substitution at saturated carbon whilst reaction (2) leads to an electrophilic substitution of saturated carbon. Of course for these mechanisms to be followed it is not necessary for a completely developed carbonium ion or carbanion to be formed, and both nucleophilic and electrophilic substitution at saturated carbon may proceed by mechanisms in which the carbon atom undergoing substitution has a "carbonium ion character" or a "carbanion character" respectively.

Nevertheless, it is clear from (1) and (2) that nucleophilic substitutions of a substrate R–L will be favoured if the leaving group is an electronegative group such as –Cl, –Br, –I, $-OH_2^+$, etc. Electrophilic substitution, on the other hand, will normally require an electropositive group, L, to be attached to the alkyl group, R. Most of the common functional groups in aliphatic chemistry are, however, electronegative and the resultant lack of suitable substrates is no doubt the major reason for the late development of electrophilic substitution at saturated carbon as a field for kinetic and mechanistic studies. Metals, and metal-containing groups, are prime examples of electropositive groups and hence reactions of metal alkyls have provided nearly all of the examples of kinetic investigations in this field (apart from the well-known[1] proton transfers, reaction (2) (L = H), which are normally better considered as Lowry-Brönsted acid–base reactions).

The possibility of electrophilic substitution at saturated carbon as an independent mechanism was considered by Hughes and Ingold[2] in 1935, but this mechanism was not kinetically demonstrated with metal alkyls as substrates until 1955, when Winstein and Traylor[3] published their results on the acetolysis of dialkylmercurys. At about the same time, stereochemical studies on electrophilic substitutions at saturated carbon were commenced by Winstein and by Reutov, again using alkylmercury compounds as substrates. Notable studies on the kinetics and stereochemistry of substitution at saturated carbon have been carried out by Ingold and his co-workers and by Reutov and his co-workers. Ingold[4] has published reviews of his own work on organomercury compounds as substrates, and Reutov[5] has published numerous reviews dealing also mainly with organomercury compounds. In addition, a number of general reviews on electrophilic substitution at saturated carbon are available[6] and Cram[7] has detailed his own work on electrophilic substitutions involving leaving groups such as hydrogen, carbon, nitrogen, and oxygen. Reutov and Beletskaya[8] have written a book on general reaction mechanisms in organometallic chemistry dealing with electrophilic substitution at saturated carbon as well as with topics such as aromatic substitution, homolytic reactions etc., and Jensen and Rickborn[9] have published a detailed and critical account of electrophilic substitution in organomercury compounds. Data on redistribution reactions have been collected by Lockhart[10].

The present work is concerned with electrophilic substitution at saturated carbon with particular reference to kinetic studies involving metal alkyls as substrates. In Chapter 2, a more general view is taken in order to include results on the generation of carbanions from hydrocarbons. Chapter 3 provides an outline of the possible mechanisms of electrophilic substitution at saturated carbon, and also serves as an introduction to the nomenclature used in the description of these mechanisms. The main section of the work, Chapters 5–9, is arranged on the basis of reaction type, although I have found it convenient to collect together in Chapter 4 the few examples of reactions proceeding by mechanism S_E1, and to deal with substitutions involving allyl groups separately in Chapter 10. Finally, in Chapter 11, I have collated the numerous observations that have been made on structural, salt, and solvent effects in electrophilic substitution at saturated carbon.

Throughout this volume, the formula RMX_n is used to denote an organometallic substrate in which R is the group undergoing substitution, M is a metal atom, and X_n represents any other atoms or groups (including alkyl groups) attached to the metal. The entire entity MX_n is thus the leaving group. Very frequently the formula E–N will be used to describe an electrophilic reagent in which E is the electrophilic pole and N is the nucleophilic pole (*e.g.* H–Cl, H–OH, I–I, etc.).

Rate coefficients are normally given in units of sec^{-1} or $l.mole^{-1}.sec^{-1}$, and where necessary literature values given in other units have been converted into values based on these units. First-order rate coefficients are denoted as k_1 and second-order rate coefficients as k_2. Where it has been necessary to refer a rate coefficient to a given reaction, then the subscript in parentheses refers to that reaction and not to any particular order. For example $k_{(14)}$ is the rate coefficient for reaction (14). Temperatures are normally given in °C except where specified; in the Arrhenius equation, of course, all temperatures refer to °K. The sign = has been used for stoichiometric equations, and the sign → for reactions presumed to be elementary ones.

Thermodynamic activation parameters calculated from second-order rate coefficients, unless specifically indicated to the contrary, all refer to the molar scale (standard state 1 mole/l and unit activity).

REFERENCES

1 R. P. BELL, *The Proton in Chemistry*, Methuen, London, 1959.
2 E. D. HUGHES AND C. K. INGOLD, *J. Chem. Soc.*, (1935) 244.
3 S. WINSTEIN AND T. G. TRAYLOR, *J. Am. Chem. Soc.*, 77 (1955) 3747.
4 C. K. INGOLD, *Helv. Chim. Acta*, 47 (1964) 1191; *Record Chem. Progr.*, 25 (1964) 145.
5 O. A. REUTOV, *Angew. Chem.*, 72 (1960) 198; *Record Chem. Progr.*, 22 (1961) 1; *Zh. Vsesoyuz. Khim. Obshchestva im. D. I. Mendeleeva*, 7 (1962) 290; *Rev. Chim. Acad. Rep. Populaire Roumamine*, 12 (1967) 313; *Usp. Khim.*, 36 (1967) 414; *Russ. Chem. Revs., Engl. Transl.*, 36 (1967) 163; *Fortschr. Chem. Forsch.*, 8 (1967) 61.
6 M. GIELEN AND J. NASIELSKI, *Ind. Chim. Belge*, 26 (1961) 1393; G. KOBRICH, *Angew. Chem.*, 74 (1962) 453; *Angew. Chem. Intern. Ed. Engl.*, 1 (1962) 382; M. H. ABRAHAM AND J. A. HILL, *J. Organometal. Chem.*, 7 (1967) 11; I. P. BELETSKAYA, *Zh. Vsesoyuz. Khim. Obshchestra im D.I. Mendeleeva*, 12 (1967) 3; D. S. MATTESON, *Organometal. Chem. Rev.*, 4A (1969) 263; W. KITCHING, *Rev. Pure Appl. Chem.*, 19 (1969) 1; I. P. BELETSKAYA, K. P. BUTIN AND O. A. REUTOV, *Organometal. Chem. Rev.*, 7A (1971) 51.
7 D. J. CRAM, *Fundamentals of Carbanion Chemistry*, Academic Press, New York, 1965; *Surv. Progr. Chem.*, 4 (1968) 45.
8 O. A. REUTOV AND I. P. BELETSKAYA, *Reaction Mechanisms of Organometallic Compounds*, North Holland Press, Amsterdam, 1968.
9 F. R. JENSEN AND B. RICKBORN, *Electrophilic Substitution of Organomercurials*, McGraw-Hill, New York, 1968.
10 J. C. LOCKHART, *Redistribution Reactions*, Academic Press, New York, 1971.

Chapter 2

The Formation and Reactions of Carbanions

1. General thermodynamic considerations

When an electrophilic substitution at saturated carbon occurs, either a carbanion is liberated as such or, if no carbanion is actually formed, the carbon atom undergoing substitution has a certain amount of "carbanion character". Thus a knowledge of the factors governing the formation and the stability of carbanions might be of help in the understanding of the mechanism of electrophilic substitution at saturated carbon.

An estimate of the standard enthalpy change for the formation of the methyl anion and a hydrogen cation from methane may be obtained by a calculation based on Hess's Law, as shown in Table 1.

TABLE 1

STANDARD ENTHALPY CHANGES (kcal.mole^{-1}) FOR THE FORMATION OF THE METHYL ANION FROM METHANE

	ΔH°_{298}
$CH_4(g) = CH_3\cdot(g) + H\cdot(g)$	104
$CH_3\cdot(g) = CH_3^-(g)$	-27
$H\cdot(g) = H^+(g)$	315
$CH_3^-(g) = CH_3^-(aq)$	-76
$H^+(g) = H^+(aq)$	-265
$CH_4(aq) = CH_4(g)$	3
$CH_4(aq) = CH_3^-(aq) + H^+(aq)$	54

In these calculations, the electron affinity of the methyl radical has been taken[1] as 27 kcal.mole^{-1}. The other enthalpy terms are all well-known quantities; the enthalpies of hydration of individual ions have been assigned as done by Valis'ev (see ref. 2) and the enthalpy of hydration of the gaseous methyl anion has been taken as that of the bromide ion. It can be seen from Table 1 that not only is the formation of the methyl anion energetically very unfavoured in the gas phase, but it is also endothermic to the extent of 54 kcal.mole^{-1} in aqueous solution. A check on this final result can be made by consideration of the standard entropy change for the reaction

$$CH_4(aq) = CH_3^-(aq) + H^+(aq) \tag{1}$$

A value of about -7 cal.deg^{-1}.mole^{-1} for the smaller carbon acids seems reasonable[3] (that for the dissociation of HCN in aqueous solution is -7.4), and combination of this value with the standard enthalpy change yields a value of 56 kcal.mole^{-1} for the standard free energy change at 298 °K for reaction (1), and hence an equilibrium constant of approximately 10^{-41}. This is in excellent agreement with the pK of methane of 40 on the MSAD scale (see Section 2., p. 7) and lends support to the calculations shown in Table 1.

Similar calculations to those in Table 1 can be made for the formation of the methyl anion from a number of organometallic compounds, and the final results are shown in Table 2.

The various carbon–metal bond energies needed have been taken from the compilation of Skinner[4], as have also the heats of vaporisation of the organometallic compounds. The heats of solution of liquid dimethylzinc and dimethylmercury have both been assumed to be $+2$ kcal.mole^{-1} (the heat of solution of dimethylzinc in water, to give dimethylzinc in aqueous solution, is, of course, a purely formal value).

TABLE 2

STANDARD ENTHALPY CHANGES (kcal.mole^{-1}) FOR THE FORMATION OF THE METHYL ANION FROM VARIOUS ORGANOMETALLIC COMPOUNDS

		ΔH°_{298}
MeH(g)	$= Me^-(g) + H^+(g)$	392
$\frac{1}{2}Me_2Hg(g)$	$= Me^-(g) + \frac{1}{2}Hg^{2+}(g)$	340
$\frac{1}{2}Me_2Zn(g)$	$= Me^-(g) + \frac{1}{2}Zn^{2+}(g)$	332
MeLi(g)	$= Me^-(g) + Li(g)$	158
MeH(aq)	$= Me^-(aq) + H^+(aq)$	54
$\frac{1}{2}Me_2Hg(aq)$	$= Me^-(aq) + \frac{1}{2}Hg^{2+}(aq)$	45
$\frac{1}{2}Me_2Zn(aq)$	$= Me^-(aq) + \frac{1}{2}Zn^{2+}(aq)$	10
MeLi(s)	$= Me^-(aq) + Li^+(aq)$	-17

The data in Table 2 lead to the expected conclusion that carbanion formation is thermodynamically more favoured when an electropositive group such as lithium is attached to the alkyl group. Furthermore, the carbanion formation is energetically more favoured in aqueous solution than in the gas phase, and no doubt in organic solvents of lower polarity than water the enthalpy change for carbanion formation will be more positive than in water itself. Hence for the alkyls of the less electropositive metals such as mercury and tin, solvents of high polarity will normally be required in order to induce carbanion formation. For the alkyls of the electro-positive metals such as lithium or magnesium, carbanion formation is thermodynamically quite favourable in polar solvents; unfortunately, such polar solvents are normally attacked with great violence by lithium and magnesium alkyls. On the other hand, use of non-polar solvents not only reduces the reactivity of the metal alkyl (with respect to carbanion formation) but leads, in addition, to considerable complications in the structure of the alkyls of the electropositive elements; the lithium alkyls are polymeric in solvents such as ethers and hydrocarbons, and the difficulties associated with the structure of Grignard reagents in ethereal solution are well known. The net result of all this is that kinetic measurements on the rate of carbanion formation from metal alkyls have, in the main, been restricted to alkyls of the less electropositive metals, using solvents of high polarity, and selecting various substituted alkyls in order to favour carbanion formation. The structural factors, in the alkyl group, favouring carbanion formation can be illustrated by a consideration of the acidity of hydrocarbons, and are discussed below.

2. Formation of carbanions from hydrocarbons

A carbanion can be regarded as the conjugate base of the corresponding carbon acid and thus the thermodynamic acidities of hydrocarbons and substituted hydrocarbons provide a very convenient quantitative measure of the extent of carbanion formation in the generalized reaction

$$R–H = R^- + H^+ \tag{2}$$

The acidities of carbon acids have been extensively discussed before, especially by Cram[5] who has constructed a scale of pK values based on the work of McEwan, Streitwieser, Applequist, and Dessy and known as the MSAD scale. Table 3 shows a selection of these pK values, and also values given by Reutov and co-workers[6] derived from polarographic measurements. It is clear from the values shown in Table 3 that substituents such as α-phenyl, α-benzyl, and α-carbethoxy increase the extent of carbanion formation from carbon acids very considerably,

TABLE 3

THERMODYNAMIC ACIDITIES OF SOME CARBON ACIDS ON THE MSAD SCALE[5] AND THE POLAROGRAPHIC SCALE[6]

Compound	pK		Compound	pK	
CH_3NO_2	10		$PhCH_3$	35	(35)[6]
Cyclopentadiene	15	(15)[6]	$CH_2 : CH_2$	36	(37)[6]
CH_3COPh	19		Benzene	37	(37)[6]
CH_3COMe	20		Me–H	40	
CH_3CO_2H	24		Et–H	42	(44)[6]
CH_3CO_2Et	24	(24)[6]	Pr^n–H		(49)[6]
CH_3CN	25		Bu^n–H		(50)[6]
Ph_3CH	33		Bu^i–H	44	
Ph_2CH_2	34		Pr^i–H	44	(60)[6]

and that such substituents could be used to advantage in the study of carbanion formation from the alkyls of the less reactive elements.

A further point of interest is in the assigned pK values to the unsubstituted alkanes. Although the MSAD scale and the polarographic scale do not agree very well in this region, the two scales give the same sequence of acidities, *i.e.*

$$Pr^i-H < Bu^i-H < Bu''-H < Pr''-H < Et-H < Me-H \qquad (3)$$

The order of sequence (3) is that of thermodynamic acidity, and not one of kinetic acidity. However, it appears[5] that the order of kinetic acidities of hydrocarbons rank as follows

$$Bu^t-H < (CH_2)_6 < Et-H < Me-H \qquad (4)$$

and hence that for these particular alkanes the rank of thermodynamic acidities is the same as that of kinetic acidities. Thus the ease of formation of carbanions from compounds of type R–L increases along the series

$$Bu^t-L < Pr^i-L < Bu^i-L < Bu''-L < Pr''-L < Et-L < Me-L \qquad (5)$$

This is the reverse sequence to that found for carbonium ion stabilities; sequence (5) is explainable in terms of the electron-releasing properties of the various alkyl groups, the greater the electron release (+I character) of the alkyl group the more destabilised is the negative charge on the carbanion.

3. Reactions and properties of carbanions

A carbanion is closely related to the corresponding derivative of nitrogen; the methyl anion is isoelectronic with ammonia itself. By analogy with the configuration of amines, carbanions should also possess a pyramidal structure (I) in which the pair of electrons of the carbanion is placed in an sp^3 orbital, rather than the alternative trigonal structure (II) with the pair of electrons in a p orbital.

sp³   sp²

(I)   (II)

Furthermore, just as the pyramidal structure in amines is inverting rapidly, so it would be expected that free carbanions will also invert their stereochemical configuration rapidly (III). Theoretical calculations have yielded[7] a value of about 11 kcal.mole^{-1} for the barrier to inversion of the methyl anion, and it may be expected that a carbanion formed free in solution will, on further reaction, yield products derived equally from the two possible configurations. Should the

(III)

carbanion be solvated asymmetrically, or should it exist as an ion pair, then it is possible for the carbanion to yield products in which one configuration predominates over the other[5]. For example, Davies and Roberts[8] have shown that the diethanolamine ester of optically active 1-phenylethylboronic acid is cleaved by OD$^-$ in boiling D$_2$O to yield (+)-1-deuterioethylbenzene with 54% net inversion of configuration. In this case the generated carbanion is shielded by the leaving group, and attack by the solvent then gives rise to the observed stereochemical result of inversion plus some racemisation, viz. (R = 1-phenylethyl)

$$
\text{DOD} \quad R-\overset{\overset{\displaystyle OR'}{|}}{\underset{\underset{\displaystyle OR'}{|}}{B}} \, OD^- \quad \rightleftharpoons \quad \text{DOD} \quad R-\overset{\overset{\displaystyle OR'}{|}}{\underset{\underset{\displaystyle OR'}{|}}{B}}-OD
$$

(6)

$$
DO^- \quad D-R \quad \overset{\overset{\displaystyle OR'}{|}}{\underset{\underset{\displaystyle OR'}{|}}{B}}-OD \quad \leftarrow \quad DOD \quad R^- \quad \overset{\overset{\displaystyle OR'}{|}}{\underset{\underset{\displaystyle OR'}{|}}{B}}-OD
$$

Carbanions may be considered as the conjugate bases of very weak acids, and hence if carbanions are generated in the presence of proton donors they will rapidly react to give the corresponding carbon acid

$$R^- + HX \rightarrow RH + X^- \tag{7}$$

Capture of carbanions by various electrophilic reagents can be predictably described, as in equations (8) and (9) for example,

$$R^- + HgCl_2 \rightarrow RHgCl + Cl^- \tag{8}$$

$$R^- + Br_2 \rightarrow RBr + Br^- \tag{9}$$

The numerous reactions of carbanions have been detailed by Cram[5] and by Ayres[9].

REFERENCES

1 H. O. PRITCHARD, *Handbook of Chemistry and Physics*, Chemical Rubber Co., Cleveland, 1967, 48th. edn., p. E-68.
2 M. F. C. LADD AND W. H. LEE, *Progr. Solid-State Chem.*, 1 (1964) 37; 2 (1965) 378.
3 R. H. BOYD AND C.-H. WANG, *J. Am. Chem. Soc.*, 87 (1965) 430.
4 H. A. SKINNER, *Advan. Organometal. Chem.*, 2 (1964) 49.

5 D. J. CRAM, *Fundamentals of Carbanion Chemistry*, Academic Press, New York, 1965.
6 K. P. BUTIN, I. P. BELETSKAYA, A. N. KASHIN AND O. A. REUTOV, *Dokl. Akad. Nauk SSSR*, 175 (1967) 1055; *Proc. Acad. Sci. USSR*, 175 (1967) 704.
7 G. W. KOEPPL, D. S. SAGATYS, G. S. KRISHNAMURTHY AND S. I. MILLER, *J. Am. Chem. Soc.*, 89 (1967) 3396.
8 A. G. DAVIES AND B. P. ROBERTS, *J. Chem. Soc. C*, (1968) 1474.
9 D. C. AYRES, *Carbanions in Synthesis*, Oldbourne Press, London, 1966.

Chapter 3

Mechanisms of Electrophilic Substitution at Saturated Carbon

In this Chapter are described the possible mechanisms of electrophilic substitution at saturated carbon, as a preliminary to the discussion of the kinetics of substitution. Additionally, there is a description of the nomenclature that has been used to date. There has been no general agreement on the nomenclature of the mechanisms of electrophilic substitution at saturated carbon, and the notation used in subsequent chapters in the present work can thus usefully be enumerated here. We deal first of all with the fundamental mechanisms, that is with mechanisms that do not involve rearrangement or nucleophilic (anionic) catalysis.

1. The fundamental mechanisms

1.1 UNIMOLECULAR ELECTROPHILIC SUBSTITUTION; S_E1

This mechanism was first envisaged by Hughes and Ingold[1] in 1935, although it was not observed with organometallic substrates[2] until 1963–4. A slow reversible ionisation to give a carbanion is followed by the rapid combination of the carbanion with an electrophile, E, viz.

$$R-MX_n \xrightarrow{\text{slow}} R^- + MX_n^+ \tag{1}$$

$$R^- + MX_n^+ \xrightarrow{\text{fast}} R-MX_n \tag{2}$$

$$R^- + E \xrightarrow{\text{fast}} R-E \tag{3}$$

A characteristic kinetic feature of this mechanism should be the observation of first-order kinetics (in the substrate RMX_n), at least for the case when the substrate is present at relatively low concentration and when the back reaction (2) is negligible in comparison with (3).

The nomenclature used to describe this mechanism is not a subject of controversy, and all workers[3–6] have used the symbol S_E1; that is substitution, electrophilic, unimolecular. The unimolecular process referred to is the elementary reaction (1).

1.2 BIMOLECULAR ELECTROPHILIC SUBSTITUTION *via* AN OPEN TRANSITION STATE; S_E2, S_E2(OPEN)

In this mechanism the incoming electrophile and the substrate react in a single bimolecular elementary reaction. There are two main possibilities for the stereochemical course of such a reaction, depending on whether the configuration of the carbon atom undergoing substitution is retained or inverted, *viz.*

$$R\!-\!MX_n \;+\; E \;\longrightarrow\; \left[\begin{array}{c} {}^{MX_n} \\ R \\ E \end{array} \right]^{\ddagger} \;\longrightarrow\; RE \;+\; MX_n \qquad (4)$$

(retention)

$$R\!-\!MX_n \;+\; E \;\longrightarrow\; \left[E\cdots R\cdots MX \right]^{\ddagger} \;\longrightarrow\; RE \;+\; MX_n \qquad (5)$$

(inversion)

Both retention of configuration and inversion of configuration have been observed in these electrophilic substitutions. The kinetic form of substitutions proceeding as in reactions (4) and (5) should be second-order overall, first-order in substrate and first-order in electrophile, when the two reactants are present in dilute solution at the same order of concentration.

Ingold[3] has used the term S_E2 to describe these bimolecular substitutions which proceed *via* open transition states (as shown in reaction (4)), but Reutov[4] uses the symbol S_E2 to describe all bimolecular electrophilic substitutions, including those in which cyclic transition states are formed as well as those in which the transition state is open. More recently, Abraham *et al.*[6] have suggested that bimolecular electrophilic substitutions in which an open transition state is formed could more explicitly be denoted by the term S_E2(open).

1.3 BIMOLECULAR ELECTROPHILIC SUBSTITUTION *via* A CYCLIC TRANSITION STATE; S_E2, S_Ei, S_F2, S_E2(CYCLIC)

In bimolecular electrophilic substitutions, the incoming electrophile may approach the saturated carbon atom undergoing substitution from the same side as that from which the leaving group departs. The possibility therefore arises that some nucleophilic centre in the electrophile could co-ordinate to the metal atom in the leaving group as substitution occurs. This gives rise to a cyclic, or synchronous, mechanism first described by Winstein *et al.*[7] who suggested the symbol S_Ei (substitution, electrophilic, internal). For the case in which the electrophile may be denoted as E–N, the mechanism is written as

$$R\!-\!MX_n \;+\; E\!-\!N \;\longrightarrow\; \left[\begin{array}{c} MX_n \\ R \qquad N \\ E \end{array} \right]^{\ddagger} \;\longrightarrow\; RE \;+\; NMX_n \qquad (5)$$

However, the mechanism is not limited to four-centred transition states, and cyclic six-centred transition states formed by synchronous electrophilic substitution and internal coordination have been postulated[7], *e.g.*

$$Bu-HgBu + CH_3CO_2H \longrightarrow \left[\begin{array}{c} Bu \\ | \\ Hg-O \\ Bu \diagdown \diagup \diagdown C-CH_3 \\ H-O \end{array} \right]^{\ddagger} \tag{6}$$

$$\downarrow$$

$$BuH + BuHgO \cdot COCH_3$$

As in reaction (4), reactions such as (5) and (6) should lead to retention of configuration at the centre of substitution and to second-order kinetics, first-order in substrate and first-order in electrophile, at least for the case when the two reactants are of the same order in concentration. It is thus not an easy matter to distinguish between bimolecular substitutions involving open transition states and those involving cyclic transition states if retention of configuration is observed.

The nomenclature used in describing bimolecular electrophilic substitutions involving cyclic transition states reflects, in part, the above-mentioned difficulty. Ingold[3] has adopted the nomenclature of Winstein *et al.*[7] and refers to such substitutions as S_Ei, but to the present author this is not a particularly appropriate choice since it does not indicate the bimolecular nature of the substitution. Dessy *et al.*[8] have used the term S_F2 to describe a mechanism, such as that in reaction (5), in which a four-centred transition state is formed, but not only is such a term too restricted, it also provides no indication that the mechanism is one of electrophilic substitution. The view of Reutov[4] is that the cyclic, synchronous mechanism is very close to the "open" mechanism and that both can be described as S_E2 mechanisms. Dessy and Paulik[9] used the term "nucleophilic assisted mechanisms" to describe these cyclic, synchronous mechanisms and Reutov[4, 10] has recently referred to them in terms* of "internal nucleophilic catalysis", "internal nucleophilic assistance", and "nucleophilic promotion". Abraham, *et al.*[6] have attempted to reconcile these various descriptions and have denoted such mechanisms as S_E2(cyclic).

1.4 ELECTROPHILIC SUBSTITUTION *via* CO-ORDINATION

This mechanism is the logical extension of the S_E2(open) and S_E2(cyclic) mechanisms, and occurs when the reagent first co-ordinates to the metal atom in the

---

* Both Dessy[9] and Reutov[4, 10] use these terms to describe not only the S_E2(cyclic) mechanism, but also the S_E2(co-ord) mechanism.

substrate to form a co-ordination complex, C. This complex then decomposes to the products and it is in this step that the actual electrophilic substitution takes place. Where the reagent can be denoted as E–N, the mechanism can be written as

$$R-MX_n + E-N \rightleftharpoons R-MX_n \rightarrow R \quad MX_n \tag{7}$$

$$\begin{array}{ccc} & \uparrow & | \quad | \\ & E-N & E \quad N \end{array}$$

although in other cases the electrophilic cleavage of the group R may occur by a 1,2 or 1,4 shift (etc.) rather than by the 1,3 shift illustrated in equation (7). In all of these cases, however, the mechanism may entail no less than three elementary reactions, as shown in (8), where the substrate and reagent are denoted simply by A and B, and the products by P, *viz.*

$$A + B \underset{k_1'}{\overset{k_2}{\rightleftharpoons}} C \overset{k_1}{\rightarrow} P \tag{8}$$

The set of elementary reactions shown in process (8) gives rise, in general, to a complex kinetic situation. There are two limiting cases in which a reduction to simpler kinetics takes place.

(*i*) The complex is present at any given time but in very small concentration. Application of the steady-state approximation, $dC/dt = 0$, then leads to the result that

$$-\frac{dA}{dt} = -\frac{dB}{dt} = \frac{dP}{dt} = \frac{k_1 k_2}{k_1 + k_1'}[A][B] \tag{9}$$

In this case, the kinetics will be those of a simple second-order reaction, in which the observed second-order rate coefficient, k_2^{obs}, is given by

$$k_2^{obs} = k_1 k_2 / (k_1 + k_1') \tag{10}$$

Even now, there are two further limiting cases, for if $k_1' \ll k_1$ then $k_2^{obs} = k_2$, whilst if $k_1' \gg k_1$ then $k_2^{obs} = k_1 k_2 / k_1' = k_1 K$, where K is the equilibrium constant for complex formation*. Experimentally it will be a matter of extreme difficulty to distinguish either of these possibilities from mechanism S_E2(open) or S_E2 (cyclic), since all of these mechanisms require the reaction to follow second-order kinetics. Indeed, Reutov[4] appears to include a situation such as (7), if the complex is present but in very low concentration, under the mechanistic title of S_E2. This is also the nomenclature used by Traylor and co-workers[11], but Abraham and Hill[5] refer to such a situation as S_EC (substitution, electrophilic, via co-ordination).

It seems to the author that assignment of a suitable notation to a substitution following the scheme of equations (7)–(10) is not at all easy; one based on

---

* The kinetic form $k_2^{obs} = k_1 K$ will also be found even if the complex is present in large amount, provided that $k_1' \gg k_1$ (*i.e.* provided that the complex is always in equilibrium with the reactants).

molecularity runs into the difficulty that the two forward elementary reactions are bimolecular and unimolecular respectively. Abraham et al.[6] have suggested that such a mechanism could be denoted as S_E2(co-ord), and this seems to be a reasonable compromise with respect to the various previous suggestions.

(ii) The complex is formed rapidly and irreversibly on allowing the reactants A and B to come into contact, and the complex then decomposes slowly to the products. The kinetic form of equation (8) then becomes

$$\frac{dP}{dt} = k_1[C] \tag{11}$$

Since the initial concentration of complex at $t = 0$ will be equal to the calculated initial concentration of either A or B (depending on whether A or B is in excess), then experimentally the observed kinetic form is

$$\frac{dP}{dt} = k_1[A] \qquad ([A]_{t=0} < [B]_{t=0}) \tag{12}$$

Assignment of a suitable notation in this case seems to the author to be impossible, and it is altogether more informative to describe such mechanisms in full than to attempt to use some misleading short-hand nomenclature.

1.5 SUMMARY OF THE NOMENCLATURE OF THE FUNDAMENTAL MECHANISMS

Before describing the more complicated mechanisms, it is appropriate to sum-

TABLE 1

NOMENCLATURE USED TO DESCRIBE MECHANISMS OF ELECTROPHILIC SUBSTITUTION AT SATURATED CARBON

Mechanism	Section	Previous notation	Recommended notation
Substitution, electrophilic, unimolecular	1.1	S_E1	S_E1
Substitution, electrophilic, bimolecular, via an open transition state	1.2	S_E2	S_E2(open)
Substitution, electrophilic, bimolecular, via a cyclic transition state	1.3	S_Ei, S_E2, S_F2	S_E2(cyclic)
Substitution, electrophilic, in which the substrate and electrophile react via a complex present only in low concentration	1.4	S_E2, S_EC	S_E2(co-ord)

marise in Table 1 the nomenclature used to date for the fundamental mechanisms and to outline the notation that will be used throughout the present work.

One of the most useful features of the recommended notation is that the unmodified term S_E2 may be used to describe an electrophilic substitution in which the kinetic form is second-order (first-order in substrate, first-order in electrophile) even though it may not be apparent which of the various possible S_E2 mechanisms is in force. The nomenclature of catalysed substitutions, and of allylic substitutions follows quite simply.

2. Substitution with rearrangement

This section deals, in general, with electrophilic substitution in allylic systems. For more complicated molecular rearrangements which may accompany electrophilic substitution, there is no simple method of description other than by a full statement of the mechanism. The recommended nomenclature is given in square brackets, [ ].

2.1 UNIMOLECULAR ELECTROPHILIC SUBSTITUTION WITH REARRANGEMENT; S_E1', $[S_E1']$

Should an allylic anion be formed as a result of an initial ionisation, attack of an electrophile on the anion could take place either at the (original) α carbon atom or at the (original) γ carbon atom, the latter leading to substitution with rearrangement. An illustration for the case of a crotyl compound follows, *viz.*

$$CH_3CH:CHCH_2^- \xrightarrow[S_E1]{+E} CH_3CH:CHCH_2\text{-}E \qquad (13)$$

$$CH_3CH:CHCH_2MX_n \Big\langle$$

$$CH_3\overset{-}{C}HCH:CH_2 \xrightarrow[S_E1']{+E} CH_3\underset{|}{C}HCH:CH_2 \qquad (14)$$
$$E$$

2.2 BIMOLECULAR ELECTROPHILIC SUBSTITUTION WITH REARRANGEMENT; S_E2', $[S_E2' \text{ (open)}]$

The α or γ carbon atom in an allylic organometallic compound could also be attacked by an electrophile in a single bimolecular encounter; the latter would give rise to mechanism S_E2'(open) if the transition state was of the open type and is again illustrated for a crotyl compound, *viz.*

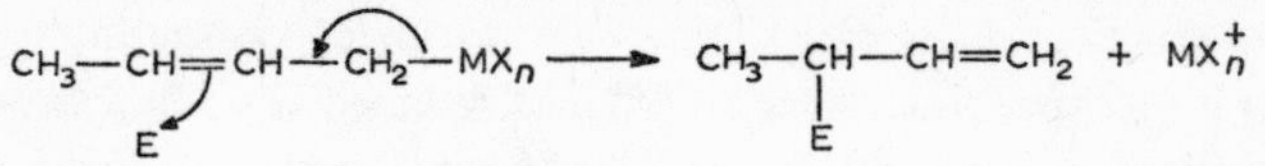

$$CH_3-CH\!=\!CH-CH_2-MX_n \longrightarrow CH_3-CH-CH\!=\!CH_2 \ + \ MX_n^+ \qquad (15)$$
$$\quad\quad\quad\quad\quad E \qquad\qquad\qquad\qquad\qquad E$$

Reaction (15) actually involves electrophilic attack at an unsaturated carbon atom and leads, as might be expected, to rates of substitution very different to those for the corresponding saturated compounds.

2.3 BIMOLECULAR ELECTROPHILIC SUBSTITUTION, *via* A CYCLIC TRANSITION STATE, WITH REARRANGEMENT; $S_Ei'[S_E2'$ (cyclic)]

For a crotyl compound undergoing such a substitution by an electrophile E–N, the mechanism is written

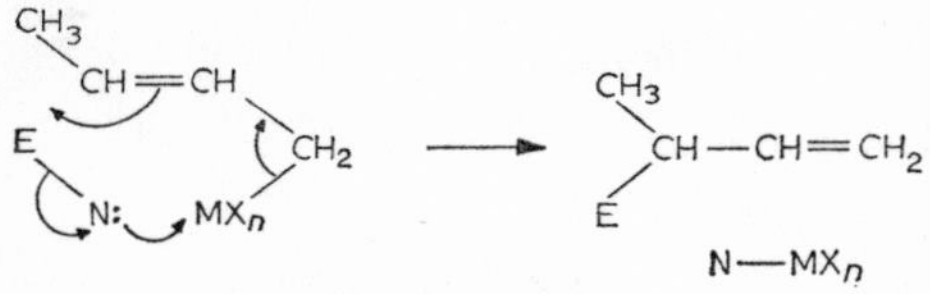

$$\qquad\qquad (16)$$

Again, the transition state may not always be six-centered (as in equation (16)) and it is obviously possible for seven- or eight-centred transition states to be formed if the structure of the electrophile so permitted.

2.4 ELECTROPHILIC SUBSTITUTION *via* CO-ORDINATION, WITH REARRANGEMENT; $[S_E2'$ (co-ord)]

The same substrate used in the example in Section 2.3 serves as a convenient illustration, *viz.*

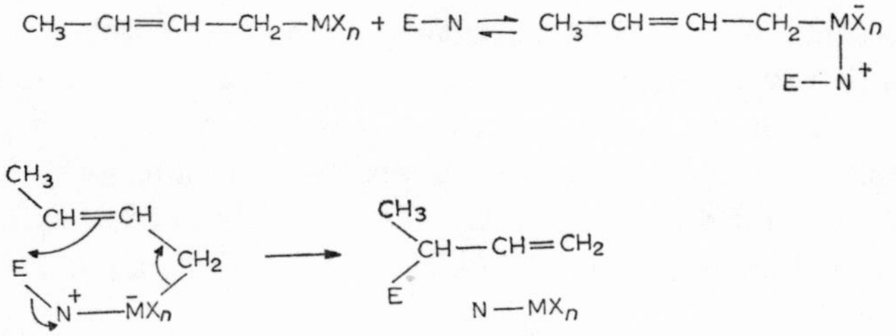

$$CH_3-CH\!=\!CH-CH_2-MX_n + E-N \rightleftharpoons CH_3-CH\!=\!CH-CH_2-\overset{-}{M}X_n$$
$$\qquad\qquad\qquad\qquad\qquad\qquad\qquad\qquad E-\overset{+}{N}$$

$$\qquad\qquad (17)$$

3. Mechanisms involving nucleophilic or anionic catalysis

Nearly all of the mechanisms that have been discussed in Sections 1 and 2 (pp.

11 and 16) may be aided, or accelerated, by a nucleophile co-ordinating to either the substrate or to the electrophile. Reutov[4, 10] has referred to such a process as "external nucleophilic catalysis", in contrast to the assistance afforded by the nucleophilic pole of the reagent (in mechanisms S_E2(cyclic) and S_E2(co-ord)) to which he refers as "internal nucleophilic catalysis" or "nucleophilic promotion" (see Section 1.3, p. 12). We shall use the expressions "nucleophilic catalysis" and "anion catalysis" in the sense of "external nucleophilic catalysis", the nucleophile or anion being some reagent other than that considered to be the electrophile.

3.1 THE SOLVENT AS A NUCLEOPHILIC CATALYST

Gielen et al.[12] have suggested that a molecule of a nucleophilic solvent might formally be incorporated into a transition state. They described the S_E2(open) mechanism of substitution of tetraalkyltins by bromine in solvent methanol as

$$R_4Sn + Br_2 \longrightarrow \left[\begin{array}{c} SnR_3 \leftarrow ---OHMe \\ R \\ Br---Br \end{array} \right]^{\ddagger} \longrightarrow RBr + R_3SnBr \qquad (18)$$

(I)

and opined that the role of the solvent was a major factor in governing the S_E2 (open) and S_E2(cyclic) mechanisms; when a bimolecular reaction takes place in a polar, nucleophilic solvent, then co-ordination of the solvent to the metal atom in the transition state [cf. equation (18)] would lead to mechanism S_E2(open), but if the solvent has but little nucleophilic character, a nucleophilic pole in the reagent could compete effectively with the solvent for the metal atom and thus bring about the S_E2(cyclic) mechanism. It would seem to the author that the nucleophilic power of the solvent may also be a critical feature of the S_E1 mechanism, since the leaving group cation is no doubt susceptible to nucleophilic attack.

The reaction order with respect to the solvent is, however, not determinable and thus there can be no direct kinetic evidence as to the number of solvent molecules in the transition state. In view of this, it is not reasonable to attempt to include nucleophilic catalysis by the solvent in any formal nomenclature of the mechanism, although it is clear that the role of the solvent is of great importance in studies of electrophilic substitutions.

The situation is quite different when the nucleophile acting as the catalyst is present in relatively small concentration (as compared to the solvent), since the kinetic order with respect to the nucleophile can, in principle, then be obtained. The mechanistic nomenclature can then be modified so as to include mention of the nucleophilic catalyst.

3.2 THE UNIMOLECULAR MECHANISM, S_E1, AND NUCLEOPHILIC CATALYSIS

Should a nucleophile (and this includes anions) complex reversibly with the substrate in a pre-rate-determining step, the set of reactions (19) and (20) may take place, where B denotes a nucleophile or Lewis base.

$$RMX_n + B \rightleftharpoons RMBX_n \tag{19}$$

$$RMBX_n \rightarrow R^- + MBX_n \tag{20}$$

Now the rate of carbanion formation will be proportional to $[RMX_n][B]$, and the mechanism denoted as S_E1-B. For instance Ingold[3] has described a unimolecular mechanism catalysed by two bromide ions as S_E1-2Br$^-$.

As well as the situation shown in (19) and (20), the observation of a rate proportional to $[RMX_n][B]$ could be interpreted as due to a single bimolecular encounter of the nucleophile with the substrate, viz.

$$R \overset{\frown}{-} MX_n \overset{\frown}{:}B \longrightarrow R^- + MBX_n \tag{21}$$

No doubt (21) would also be described as S_E1-B, although the significance of the "1" is somewhat obscure.

3.3 THE BIMOLECULAR MECHANISM, S_E2 (open), AND NUCLEOPHILIC CATALYSIS

Nucleophilic catalysis in the unimolecular mechanism is straightforward, since there is but one reactant that can bring the nucleophile into the transition state. If, however, a bimolecular substitution is found to be subject to nucleophilic catalysis and if it is concluded that, say, one molecule of the nucleophile, B, is involved in the transition state, then two main possibilities exist. The nucleophile may be brought into the transition state either by the organometallic substrate as in process (22) or by the electrophile as in process (23), viz.

$$R-MX_n + B \rightleftharpoons R-MBX_n$$
$$R-MBX_n + E-N \rightarrow RE + NMX_n + B \tag{22}$$

$$E-N + B \rightleftharpoons E(B)N$$
$$R-MX_n + E(B)N \rightarrow RE + NMX_n + B \tag{23}$$

Normally (22) will be followed*, since the entity $R-MBX_n$ should be more reactive

---

* This applies to catalysis involving one molecule of a nucleophile; should two such molecules be involved, then the situation is not amenable to generalisations.

(towards an electrophile) than R–MX$_n$, whereas the entity E(B)N is almost certain to be less electrophilic than the simple reagent E–N. Process (22) could be denoted as S$_E$2(open)-B when the transition state was of the open type, or more simply as an S$_E$2(open) reaction between the substrate RMBX$_n$ and an electrophile E–N.

3.4 THE BIMOLECULAR MECHANISM, S$_E$2(cyclic), AND NUCLEOPHILIC CATALYSIS

As for the S$_E$2 (open) mechanism, the two possible situations can be represented by (22) and (23), but if the transition states are of the cyclic type, a new situation obtains. The entity RMBX$_n$ will be less susceptible to internal nucleophilic attack by the reagent than will the entity RMX$_n$, whilst the reagent E(B)N is very likely to be more effective in completing a cyclic transition state than is the simple reagent E–N. Hence whilst (22) is preferred over (23) if both (22) and (23) proceed by mechanism S$_E$2(open), (23) is preferred to (22) if both (22) and (23) proceed by mechanism S$_E$2(cyclic)*.

The general process (23) in which mechanism S$_E$2(cyclic) operates may be described in two ways: either as an S$_E$2(cyclic) reaction with the substrate RMX$_n$ and the reagent E(B)N, or as an S$_E$2(cyclic) reaction with the substrate RMX$_n$ and the reagent EN catalysed by the nucleophile B. Ingold[3] has used the former description, and this is perhaps the simplest procedure.

3.5 SUBSTITUTION *via* CO-ORDINATION, S$_E$2(co-ord), AND NUCLEOPHILIC CATALYSIS

The arguments of Sections 3.3 and 3.4 apply here with even greater force, and lead to the conclusion that the sequence shown in (23) must apply; the full scheme is

$$E\text{–}N + B \rightleftharpoons E(B)N$$

$$RMX_n + E(B)N \rightleftharpoons Complex$$

$$Complex \rightarrow RE + NMX_n + B \tag{24}$$

---

* Again, if two molecules of the nucleophile are involved, no generalisation can be made.

3.6 MECHANISMS INVOLVING SUBSTITUTION WITH REARRANGEMENT AND NUCLEOPHILIC CATALYSIS

The various mechanisms outlined in Section 2 (p. 16) can, in principle, all be subject to nucleophilic catalysis. Without enumerating them again, the arguments and discussions of Section 3 can readily be applied to these substitutions with rearrangement.

REFERENCES

1 E. D. Hughes and C. K. Ingold, *J. Chem. Soc.*, (1935) 244.
2 O. A. Reutov, B. Praisner, I. P. Beletskaya and V. I. Sokolov, *Izv. Akad. Nauk SSSR, Otd. Khim. Nauk*, (1963) 970; *Bull. Acad. Sci. USSR, Div. Chem. Sci.*, (1963) 884; E. D. Hughes, C. K. Ingold and R. M. G. Roberts, *J. Chem. Soc.*, (1964) 743.
3 C. K. Ingold, *Helv. Chim. Acta*, 47 (1964) 1191; *Record Chem. Progr.*, 25 (1964) 145.
4 O. A. Reutov, *Usp. Khim.*, 36 (1967) 44; *Russ. Chem. Rev., Engl. Transl.*, 36 (1967) 163.
5 M. H. Abraham and J. A. Hill, *J. Organometal. Chem.*, 7 (1967) 11.
6 M. H. Abraham, M. Gielen and J. Nasielski, to be published.
7 S. Winstein, T. G. Traylor and C. S. Garner, *J. Am. Chem. Soc.*, 77 (1955) 3741; S. Winstein and T. G. Traylor, *J. Am. Chem. Soc.*, 77 (1955) 3747.
8 R. E. Dessy, W. L. Budde and C. Woodruff, *J. Am. Chem. Soc.*, 84 (1962) 1172; R. E. Dessy and F. Paulik, *Bull. Soc. Chim. France*, (1963) 1373.
9 R. E. Dessy and F. E. Paulik, *J. Chem. Educ.*, 40 (1963) 185.
10 O. A. Reutov and I. P. Beletskaya, *Izv. Akad. Nauk SSSR, Ser. Khim.*, (1966) 955; *Bull. Acad. Sci. USSR*, (1966) 914.
11 H. Minato, J. C. Ware and T. G. Traylor, *J. Am. Chem. Soc.*, 85 (1963) 3024.
12 M. Gielen, J. Nasielski, J. E. Dubois and P. Fresnet, *Bull. Soc. Chim. Belges*, 73 (1964) 293.

Chapter 4

The Unimolecular Mechanism, S_E1

1. Substitutions in organomercury compounds

1.1 THE FUNDAMENTAL MECHANISM, S_E1

In 1959–1960, Reutov et al.[1] reported that the electrophilic substitution of α-carbethoxybenzylmercuric bromide by mercuric bromide, labelled with ^{203}Hg, followed second-order kinetics, first-order in each reactant, in 70 % aqueous dioxan as solvent. The substitution may be represented as

$$PhCH(CO_2Et)HgBr + \overset{*}{Hg}Br_2 = PhCH(CO_2Et)\overset{*}{Hg}Br + HgBr_2 \qquad (1)$$

where $\overset{*}{Hg}$ denotes the radioactive isotope ^{203}Hg. They later retracted this report and claimed[2] that reaction (1) in 70 % aqueous dioxan followed first-order kinetics, first-order in the alkylmercuric bromide and zero-order in mercuric bromide. This is the kinetic form to be expected for a reaction proceeding via mechanism S_E1. Ingold and co-workers[3] then showed that reaction (1) did follow second-order kinetics, as originally stated[1], when 70 % aqueous dioxan was used and hence that mechanism S_E1 could not obtain in this particular solvent.

Meanwhile, Reutov et al.[4] had reported that reaction (1) followed first-order kinetics, first-order in alkylmercuric bromide and zero-order in mercuric bromide, when anhydrous dimethylsulphoxide (DMSO) was used as the solvent. This report was confirmed by Ingold and co-workers[3] and hence is the first authentic record* of a reaction following mechanism S_E1. Over a six-fold variation in the initial concentration of α-carbethoxybenzylmercuric bromide and a three-fold variation in that of mercuric bromide, only the first-order rate coefficient with respect to the alkylmercuric bromide remained constant[3]. The rate coefficients reported by the two sets of workers are given in Table 1, together with the reported Arrhenius equations.

Reutov et al.[4] also examined the effect of various p-substituents in the α-carbethoxybenzylmercuric bromide on the rate coefficient, and found the following relative rates in solvent DMSO at 30 °C: p-NO₂(4.04), p-I(1.31), p-H(1),

---

* That is amongst substrates in which the carbon atom undergoing substitution is attached to a metal atom. Almost ten years previously, Cram had observed mechanism S_E1 with other leaving groups (see ref. 5).

TABLE 1

FIRST-ORDER RATE COEFFICIENTS (sec^{-1}) FOR SUBSTITUTION OF α-CARBETHOXY-BENZYLMERCURIC BROMIDE BY MERCURIC BROMIDE IN SOLVENT DMSO

Temp. (°C)	$10^4 k_1$	Ref.
25	0.411	4
29.5	0.52	3
30	0.551	4
35	0.980	4
40	1.446	4
44.4	1.43	3
58.9	3.75	3

$k_1(\text{sec}^{-1}) = 2.5 \times 10^5 \exp. (-13,400/RT)$ (ref. 3)
$k_1(\text{sec}^{-1}) = 4.0 \times 10^{13} \exp. (-17,200/RT)$ (ref. 4)

and p-Bu^t(0.71). These rates are in accord with a mechanism which is promoted by groups stabilising a carbanion of the type p-X–C$_6$H$_4$CHCO$_2$Et. When bromide was replaced by chloride in each reactant, the rate coefficient was reduced[3] from 3.75×10^{-4} sec^{-1} to 2.05×10^{-4} sec^{-1} at 58.9 °C, again in agreement with a rate-determining ionisation of the alkylmercuric halide, with the more electronegative halide reducing the rate. Ingold and co-workers[3] were able to resolve the $(\pm)$-α-carbethoxybenzylmercuric bromide and to obtain the $(-)$-form. They then showed[3] that reaction (1), in DMSO, proceeded with racemisation of the $(-)$-α-carbethoxybenzylmercuric bromide and that the rate of racemisation was identical to the rate of Hg exchange; furthermore, the rate of racemisation was the same in the presence or absence of mercuric bromide[3].

All of this evidence supports the proposed[3, 4] S_E1 mechanism for reaction (1) in solvent DMSO, viz.

$$\text{PhCH(CO}_2\text{Et)HgBr} \underset{\text{fast}}{\overset{\text{slow}}{\rightleftharpoons}} \text{PhCH(CO}_2\text{Et)}^- + \text{HgBr}^+ \tag{2}$$

$$\text{PhCH(CO}_2\text{Et)}^- + \overset{*}{\text{HgBr}}_2 \xrightarrow{\text{fast}} \text{PhCH(CO}_2\text{Et)}\overset{*}{\text{HgBr}} + \text{Br}^- \tag{3}$$

$$\text{HgBr}^+ + \text{Br}^- \xrightarrow{\text{fast}} \text{HgBr}_2 \tag{4}$$

As suggested in Chapter 2, Section 3 (p. 8), the carbanion must invert its stereochemical configuration very rapidly.

Reutov et al.[6] have also studied the exchange reaction

$$p\text{-NO}_2\text{-C}_6\text{H}_4\text{CH}_2\text{HgBr} + \overset{*}{\text{HgBr}}_2 = p\text{-NO}_2\text{-C}_6\text{H}_4\text{CH}_2\overset{*}{\text{HgBr}} + \text{HgBr}_2 \tag{5}$$

using DMSO as solvent. They showed that the exchange followed first-order kinetics (first-order in alkylmercuric bromide and zero-order in mercuric bromide) over a wide range of initial concentrations of the reactants. Hence mechanism S_E1 is operative and an analogous sequence to reactions (2)–(4) may thus be written with the carbanion now having the form $p\text{-}NO_2\text{-}C_6H_4CH_2^-$. The rate coefficients for reaction (5) at various temperatures are given in Table 2, together with the reported[6] Arrhenius equation. The reproducibility of the rate coefficients is not very good, and Reutov et al.[6] give the activation energy as 18 ± 1 kcal.mole^{-1}.

TABLE 2

FIRST-ORDER RATE COEFFICIENTS (sec^{-1}) FOR SUBSTITUTION OF p-NITROBENZYL-MERCURIC BROMIDE BY MERCURIC BROMIDE IN DMSO[6]

Temp. (°C)	40	50	60	70
$10^6 k_1$	3.1	7.2	17.2	37.8

$k_1(sec^{-1}) = 1.2 \times 10^7 \exp. (-18,000/RT)$

The two substitutions dealt with above represent the only authentic cases fo organomercury compounds reacting *via* the fundamental mechanism S_E1. Jensen and Heyman[7] have shown that the supposed unimolecular ionisation[8] of di-*sec*-.butylmercury to a *sec.*-butyl anion, in solvent dimethylformamide, does not take place. The purported S_E1 protolysis of dibenzylmercury[9], in a number of solvents, has also been shown[10] to be an erroneous claim. Both Jensen and Heyman[7] and also Kitching and co-workers[10] demonstrated that the supposed unimolecular ionisations did not occur when reactions were run under a nitrogen atmosphere.

1.2 THE UNIMOLECULAR MECHANISM, S_E1, AND NUCLEOPHILIC CATALYSIS

The racemisation of $(-)$-α-carbethoxybenzylmercuric bromide in DMSO provides not only an example of mechanism S_E1, but also the most complete example of the anion-catalysed S_E1 mechanism. Ingold and co-workers had briefly investigated the effect of added salts on the rate of racemisation; their results are shown in Table 3. In contrast to the mild acceleration or retardation produced

TABLE 3

FIRST-ORDER RATE COEFFICIENTS (sec^{-1}) FOR THE RACEMISATION OF $(-)$-α-CARBETHOXYBENZYLMERCURIC BROMIDE IN DMSO AT 59.2 °C[3]

Initial concentration of substrate 0.03 M.

Added salt	None	KClO$_4$	LiClO$_4$	LiNO$_3$	Et$_4$NBr
Concentration of added salt (M)	None	0.03	0.03	0.03	0.0085
$10^4 k_1$	3.40	2.45	0.95	5.05	4.40

TABLE 4

FIRST-ORDER RATE COEFFICIENTS, k_1^{obs} (sec^{-1}), FOR THE RACEMISATION OF $(-)$-α-CARBETHOXYBENZYLMERCURIC BROMIDE IN DMSO AT 29.5 °C IN THE PRESENCE OF TETRAETHYLAMMONIUM BROMIDE[3]

Initial concentration of substrate 0.04 M.

$[Et_4NBr]$ (M)	0	0.0075	0.010	0.0125	0.015	0.016	0.018	0.020
$10^4 k_1^{obs}$	0.52	1.65	2.58	3.49	4.76	6.92	8.26	8.85
k_3		1.90	2.06	1.90	1.89	2.50	2.39	2.08

by the perchlorates or by lithium nitrate, addition of bromide ions leads to a marked increase in rate as shown in Table 4. If the increase in the observed first-order rate coefficient, k_1^{obs}, is due to some catalytic effect in addition to the un-catalysed reaction, then we may write

$$k_1^{obs} = k_1 + k_2[\text{Br}^-] + k_3[\text{Br}^-]^2 + \text{etc.} \tag{6}$$

An analysis[3] of the results shown in Table 4 revealed that only the term quadratic in $[\text{Br}^-]$ was important; the catalytic constant, k_3, was calculated[3] to have an average value of 2.1 l^2.mole^{-2}.sec^{-1}.

These elegant experiments suggest[3] that two bromide ions are incorporated into the transition state for the catalysed process, and the rate-determining ionisation can thus be written either in terms of the mechanism (7)–(9) or of the mechanism (10) and (11), where $(-)$-α-carbethoxybenzylmercuric bromide is denoted as RHgBr.

$$\text{RHgBr} + \text{Br}^- \rightleftharpoons \text{RHgBr}_2^- \tag{7}$$

$$\text{RHgBr}_2^- + \text{Br}^- \rightleftharpoons \text{RHgBr}_3^{2-} \tag{8}$$

$$\text{RHgBr}_3^{2-} \xrightarrow{\text{slow}} \text{R}^- + \text{HgBr}_3^- \tag{9}$$

$$\text{RHgBr} + \text{Br}^- \rightleftharpoons \text{RHgBr}_2^- \tag{10}$$

$$\text{RHgBr}_2^- + \text{Br}^- \xrightarrow{\text{slow}} \text{R}^- + \text{HgBr}_3^- \tag{11}$$

In both of the pre-rate-determining equilibria, (7) and (8), the position of equilibrium must lie well to the left. Ingold and co-workers[3] wrote this anion-catalysed reaction as S_E1-2Br$^-$, and, since there appears to be no simple method of experimentation to allow a choice between the two possible mechanisms, this nomenclature can conveniently be allowed to cover both.

Reutov and co-workers[11] have also observed the acceleration of the S_E1 exchange of p-nitrobenzylmercuric bromide, reaction (5), by added bromide ions in solvent DMSO. Their results are summarised in Table 5. The situation is rather different from that for the racemisation of $(-)$-α-carbethoxybenzylmercuric

TABLE 5

FIRST-ORDER RATE COEFFICIENTS, k_1^{obs} (sec^{-1}), FOR THE EXCHANGE BETWEEN p-NITROBENZYLMERCURIC BROMIDE AND ^{203}MERCURIC BROMIDE IN DMSO AT 18 °C IN THE PRESENCE OF POTASSIUM BROMIDE[11]

Initial concentration of substrate 0.03 M and of mercuric bromide 0.06 M.

$[KBr]$ (M)	0	0.01	0.02	0.03	0.04	0.05	0.06	0.07	0.09
$10^6 k_1^{obs}$	0.41	0.66	0.92	1.30	1.96	3.36	71.1	1900	20000

bromide, since now the added bromide ions will be preferentially removed by the mercuric bromide

$$HgBr_2 + Br^- \rightleftharpoons HgBr_3^- \tag{12}$$

and when the concentration of added potassium bromide is less than that of mercuric bromide (0.06 M) only a small fraction of the total added bromide ion is available for complex formation by the alkylmercuric bromide*. As soon as there is an excess of bromide ions over that demanded by equilibrium (12), the p-nitrobenzylmercuric bromide can then take part in equilibria of the type shown in (7) and (8). An analysis of the data shown in Table 5 suggests to the author that, again, the quadratic term in bromide ion is much more important than the linear term in equation (6). Hence this is another example of mechanism S_E1-2Br$^-$, with a sequence of elementary reactions such as (7)–(9) or a sequence such as (10) and (11), where R = p-nitrobenzyl.

The acidolysis of the 4-pyridiomethylmercuric chloride ion (I) in aqueous perchloric acid provides another example of the S_E1-anion mechanism[13].

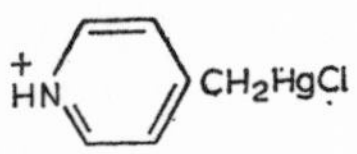

Although (I) is stable in acid solution, it decomposes on addition of chloride ion, *viz.*

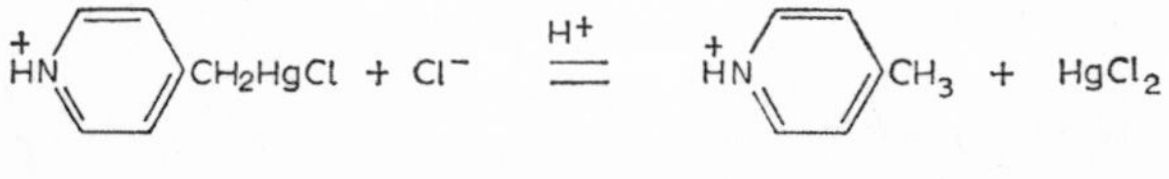

$$\tag{13}$$

Since the initial concentration of (I) was set at 10^{-4} to 10^{-5} M[13], the decomposition of (I) was first-order with respect to (I) in the presence of 0.01 M perchloric acid and 0.01 M chloride ion, as the latter two entities were in large excess. Variations in the acid concentration did not affect the rate of decomposition, and the

---

* The equilibrium constant for reaction (12) is likely to be much higher in solvent DMSO than in solvent water; even in aqueous solution, the equilibrium constant has a value[12] of about 200 l.mole^{-1}.

TABLE 6

FIRST-ORDER RATE COEFFICIENTS, $k_1^{obs}(sec^{-1})$, FOR THE ACIDOLYSIS OF THE 4-PYRIDIO-
METHYLMERCURIC CHLORIDE ION IN 0.01 M AQUEOUS PERCHLORIC ACID AND AT
AN IONIC STRENGTH OF 0.51 M, AT 65 °C[13]

Added chloride ion (M)	0	0.1	0.125	0.20	0.25	0.40	0.50
$10^6 k_1^{obs}$	0	5.37	7.29	16.5	23.9	56.3	90.2

effect of chloride ion is shown in Table 6. The increase in the value of k_1^{obs} with
increasing chloride ion concentration may be represented by equation (14), where
the value of k_1 is clearly zero (from Table 6), viz.

$$k_1^{obs} = k_1 + k_2[Cl^-] + k_3[Cl^-]^2 + \text{etc.} \qquad (14)$$

Coad and Johnson[13] calculated that $k_2 = 2 \times 10^{-5}$ l.mole^{-1}.sec^{-1} and that
$k_3 = 3.2 \times 10^{-4}$ l^2.mole^{-2}.sec^{-1} under the experimental conditions shown in
Table 6. Hence with $[Cl^-]$ ranging from 0.10 M to 0.50 M, the proportion of the
decomposition due to the quadratic term in equation (14) is from twice to eight
times that due to the linear term.

Two possible mechanisms of decomposition may be written, as shown for the
one-anion catalysis, viz.

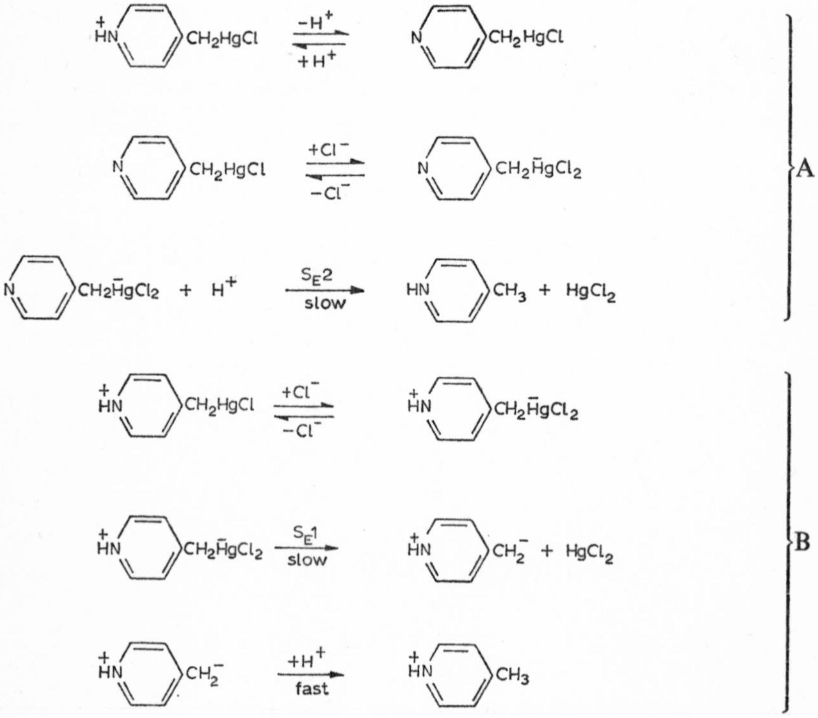

Coad and Johnson[13] deduced that mechanism **B** was operating, since (*i*) the ion

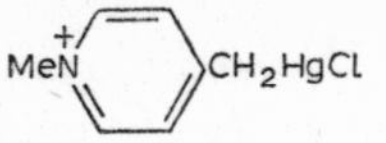

was almost as reactive as the ion (**I**), and (*ii*) 2-pyridiomethylmercuric chloride ion was about one-tenth as reactive as the 4-isomer, but the 3-isomer was completely unreactive. The latter observation can be accounted for on the basis of mechanism S_E1-Cl^- and S_E1-$2Cl^-$, since the transition state is stabilised by developing quinonoid character in the case of the 2- and the 4-isomers, but not for the 3-isomer, *viz.*

(II)

(III)

Dodd and Johnson[14] later pointed out that if mechanism **B** were correct, then mercury(II) and the acid medium would be expected to compete for the intermediate pyridiomethylide ion (**II**). Hence the observed first-order rate coefficient should be reduced on addition of mercury(II) and should be increased on addition

TABLE 7

FIRST-ORDER RATE COEFFICIENTS (sec^{-1}) FOR THE ACIDOLYSIS OF THE 4-PYRIDIO-METHYLMERCURIC CHLORIDE ION IN AQUEOUS PERCHLORIC ACID AND IN THE PRESENCE OF 0.5 M CHLORIDE ION AT AN IONIC STRENGTH OF 0.50 M AND AT 65 °C[14]

Additive (M)		
$HgCl_2$	$HClO_4$	$10^5 k_1$
0	0.01	9.26 ($k_1{}^0$)
0.002	0.01	7.62
0.005	0.01	5.99
0.010	0.01	4.54
0.015	0.01	3.55
0.025	0.01	2.49
0.01	0.005	2.93
0.01	0.010	4.54
0.01	0.020	6.21
0.01	0.040	7.62

of acid. The rate coefficients in Table 7 show how these expectations were borne out; these coefficients have been corrected[14] to an ionic strength of 0.500, since the actual ionic strength varied from 0.51 to 0.55 over the set of kinetic experiments. Dodd and Johnson deduced that the first-order rate coefficients should follow the equation

$$k_1^0/k_1 = 1 + \frac{\gamma[\text{Hg}^{\text{II}}]}{([\text{H}^+] + \varepsilon[\text{H}_2\text{O}])} \tag{15}$$

where γ represents the ratio of the average reactivity of mercury(II) species to the reactivity of acid (*i.e.* reactivity towards the pyridiomethylide ion), and where ε represents the ratio of the reactivity of water to that of acid. Under the conditions given in Table 7, the value of ε is 0 and the value of γ is 1.07. Further calculations enabled γ to be broken down into competition factors for the various mercury(II) species involved, *viz.* $\gamma(\text{HgCl}_2) = 1.4$, $\gamma(\text{HgCl}_3^-) = 5.3$, and $\gamma(\text{HgCl}_4^{2-}) = 0.3$. The scheme of mechanism B, above, must therefore be extended so as to include the capture of (II) by these mercury(II) species. The results of Dodd and Johnson[14] are particularly interesting in that they demonstrate the ability of a carbanion (albeit a resonance-stabilised carbanion) to discriminate between electrophilic reagents, even in the fast step of the acidolysis.

In the presence of mercuric chloride, the chloro complexes of 4-pyridiomethyl-mercuric chloride also react directly with the electrophiles HgCl_2, HgCl_3^-, and HgCl_4^{2-} in a series of S_E2 mercury-for-mercury exchanges. These exchanges, which do not result in any net chemical change, do not affect the rate of acidolysis but can be detected[15] through isotopic exchange using $^{203}\text{HgCl}_2$. Mechanism B can therefore be expanded so as to include the concurrent S_E2 path as well as the S_E1 process; this is illustrated for one particular set of complexes as

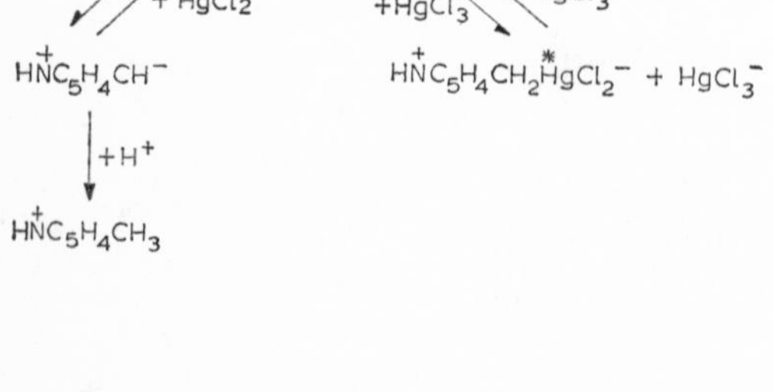

$$\tag{16}$$

2. Substitutions in other organometallic compounds

2.1 THE SUBSTITUTION OF 1-CYANO-1-CARBETHOXYPENTYL-(TRIPHENYL-PHOSPHINE)-GOLD(I) BY ALKYLMERCURIC SALTS

Kinetics of the substitution reaction

$$
\begin{array}{cc}
\text{CN} & \text{CN} \\
| & | \\
\text{Bu–C–AuPPh}_3 + \text{RHgX} = \text{Bu–C–HgR} + \text{XAuPPh}_3 \\
| & | \\
\text{CO}_2\text{Et} & \text{CO}_2\text{Et}
\end{array}
\qquad (17)
$$

have been followed[16] by a spectrophotometric method, using solvent dioxan at 24.7 °C. Under these conditions, reaction (17) (R = Me, X = Cl) was found to proceed by a second-order process, first-order in each reactant, and some type of S_E2 mechanism is therefore indicated[16].

In solvent DMSO, the rate of reaction (17) (R = Me, X = Cl) was too fast to follow, but using the mixture DMSO–dioxan (1 : 9 v/v) at 24.7 °C, rate coefficients for the substitution of the gold(I) complex by several alkylmercuric salts were obtained. It was found that in this mixed solvent, reaction (17) followed first-order kinetics, first-order in the gold(I) complex and zero-order in the alkylmercuric salt. Furthermore, the first-order rate coefficient had the same value no matter what alkylmercuric salt was used (methylmercuric acetate, methylmercuric chloride, methylmercuric bromide, and ethylmercuric chloride were the salts used). At 24.7 °C, the first-order rate coefficient has the value 0.0083 sec^{-1}, with a standard deviation of 0.0006 sec^{-1}.

It was suggested[16] that in solvent DMSO–dioxan (1 : 9 v/v), reaction (17) proceeds by a rate-determining heterolysis of the carbon–gold bond. We may therefore write for this S_E1 mechanism the equations

$$
\text{BuC(CN)(CO}_2\text{Et)AuPPh}_3 \xrightarrow{\text{slow}} \text{BuC(CN)CO}_2\text{Et}^- + \text{AuPPh}_3^+ \qquad (18)
$$

$$
\text{BuC(CN)CO}_2\text{Et}^- + \text{RHgX} \xrightarrow{\text{fast}} \text{BuC(CN)(CO}_2\text{Et)HgR} + \text{X}^- \qquad (19)
$$

2.2 THE HYDROLYSIS OF 4-PYRIDIOMETHYL-MANGANESE, -MOLYBDENUM, AND -TUNGSTEN COMPOUNDS

The hydrolysis of a number of 4-pyridiomethylmetal ions in aqueous acidic solution to give 4-methylpyridine has been studied by Johnson and Winterton[17]. The hydrolyses were run using a large excess of aqueous acid and were found to follow overall first-order kinetics (first-order in the 4-pyridiomethyl metal ion).

TABLE 8

FIRST-ORDER RATE COEFFICIENTS (sec^{-1}) FOR THE HYDROLYSIS OF 4-PYRIDIO-
METHYLMETAL COMPOUNDS IN 0.01 M AQUEOUS HYDROCHLORIC ACID AT 65 °C[17]

Compound	$10^5 k_1$
4-H$\overset{+}{N}$C$_5$H$_4$CH$_2$Mn(CO)$_5$	5.46[a]
4-H$\overset{+}{N}$C$_5$H$_4$CH$_2$Mo(π-C$_5$H$_5$)(CO)$_3$	1.17
4-H$\overset{+}{N}$C$_5$H$_4$CH$_2$W(π-C$_5$H$_5$)(CO)$_3$	0.044
4-MeN$\overset{+}{}$C$_5$H$_4$CH$_2$Mn(CO)$_5$	5.54
4-MeN$\overset{+}{}$C$_5$H$_4$CH$_2$Mo(π-C$_5$H$_5$)(CO)$_3$	2.13

[a] $E_a = 25.2 \pm 1.0$ kcal.mole^{-1}, $\Delta S^{\ddagger} = -4 \pm 4$ cal.deg.$^{-1}$ mole^{-1}.

Since the observed first-order rate coefficients were but little affected by change in either hydrogen ion concentration or chloride ion concentration, it was concluded[17] that the hydrolyses proceed by mechanism S_E1, following the reactions

$$H\overset{+}{N}C_5H_4CH_2M \xrightarrow{slow} H\overset{+}{N}C_5H_4\overset{-}{C}H_2 + M^+ \tag{20}$$

$$H\overset{+}{N}C_5H_4\overset{-}{C}H_2 + H^+ \xrightarrow{fast} H\overset{+}{N}C_5H_4CH_3 \tag{21}$$

Some of the reported rate coefficients are collected in Table 8.

2.3 ALKYL–METAL COMPOUNDS STUDIED BY PROTON MAGNETIC RESONANCE SPECTROSCOPY (PMR)

Quite recently, it has been observed that the PMR spectra of a number of alkyl-metal compounds is temperature-dependent. For several primary Grignard reagents, a detailed examination[18, 19] indicated that inversion of configuration at the –CH$_2$Mg centre was occurring rapidly at room temperature, although secondary Grignard reagents inverted so slowly that inversion rates could not be measured[19]. The inversion process for primary Grignard reagents in solvent ether appears to have a kinetic order greater than one[18, 19] and the mechanism of inversion remains obscure; the sequence of reactivity primary alkyl > secondary alkyl is compatible with both an S_E1 and an S_E2 process[20]. Activation energies for inversion of a number of Grignard reagents are given in Table 9.

Witanowski and Roberts[21] showed that the inversion rate of 3,3-dimethylbutyl (R) compounds increased along the series R$_2$Hg, R$_3$Al < R$_2$Zn < R$_2$Mg < RLi; the order of the inversion process was one for both the alkyllithium and the dialkyl-

TABLE 9

ACTIVATION ENERGIES (kcal.mole^{-1}) FOR THE INVERSION OF SOME PRIMARY ALKYL-METAL COMPOUNDS, STUDIED BY PMR

Compound	Solvent	Kinetic order	E_a^a	Ref.
MeCH$_2$CHMe·CH$_2$MgBr	Anisole		10.2	18
MeCH$_2$·CHMe·CH$_2$MgBr	Ether	2.5	12.6	18
MeCH$_2$·CHMe·CH$_2$MgBr	THF		17.2	18
Me$_3$C·CH$_2$·CH$_2$MgCl	Ether	>1	11	19
Me$_2$CH·CHPh·CH$_2$MgCl	Ether		12	19
Me$_2$CH·CHPh·CH$_2$MgCl	THF		18	19
(MeCH$_2$·CHMe·CH$_2$)$_2$Mg	Dioxan		5.5	18
(MeCH$_2$·CHMe·CH$_2$)$_2$Mg	THF		11.1	18
(MeCH$_2$·CHMe·CH$_2$)$_2$Mg	Ether	2.0	18.8	18
Me$_3$C·CH$_2$·CH$_2$·Li	Ether	1.0	15	21
(Me$_3$C·CH$_2$·CH$_2$)$_2$Mg	Ether	1.0	20	21
(Me$_3$C·CH$_2$·CH$_2$)$_2$Zn	Ether		26	21
(Me$_3$C·CH$_2$·CH$_2$)$_2$Hg	Ether } Inversion too slow			21
(Me$_3$C·CH$_2$·CH$_2$)$_3$Al	Ether } to measure			21
MeCH$_2$·CHMe·CH$_2$Li	n-Pentane	1.0	5.1	22
MeCH$_2$·CHMe·CH$_2$Li	Ether	1.0	8.6	22
MeCH$_2$·CHMe·CH$_2$Li	Toluene	1.0	11.1	22
Tri-isohexylaluminium	Toluene	1.0	28.6	22
Tri-isohexylaluminium	Ether	Inversion too slow to measure		22

^{a} Where $\Delta H^{\ddagger}$ was given, 0.6 kcal.mole^{-1} has been added to obtain E_a in kcal.mole^{-1}.

magnesium (solvent ether in each case) and it was suggested[21] that inversion of configuration at the –CH$_2$Mg centre took place through a unimolecular process of the S_E1 type, viz.

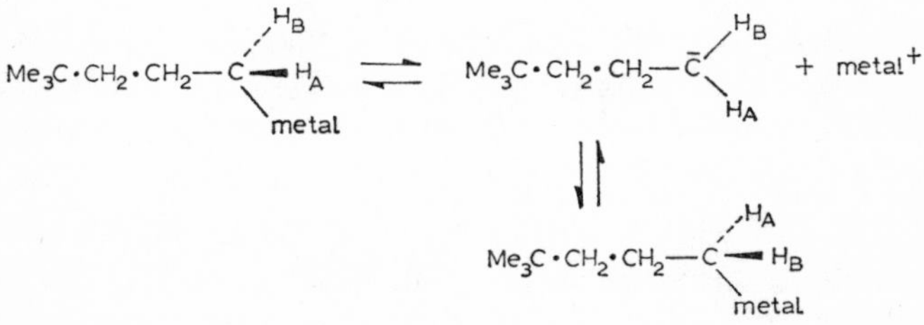

$$\text{(22)}$$

It is interesting that the observed sequence of reactivity (above, and Table 9) is strikingly similar to the calculated sequence for the ease of carbanion formation from various methyl derivatives, viz. Me$_2$Hg < Me$_2$Zn < MeLi (see Table 2, Chapter 2, p. 6).

Although Witanowski and Roberts[21] found the inversion of di(3,3-dimethyl-butyl)magnesium to be a first-order process, Fraenkel and Dix[18] have observed the inversion of di(2-methylbutyl)magnesium to be a second-order process. Since the same solvent, ether, was used in each study, and since the dialkylmagnesiums were both prepared by the same route — the action of magnesium on dialkyl-mercury — the cause of the contrasting results is not clear. However, the inversion of 2-methylbutyllithium was found to be a first-order process[22], and the general results of Fraenkel et al.[18, 22] are in agreement with those of Roberts and co-workers[19, 21], as may be seen from the various activation energies collected in Table 9.

2.4 THE CLEAVAGE OF ORGANOMETALLIC COMPOUNDS OF GROUP IVA BY ALKALI

Eaborn et al.[23-26] have studied the cleavage of a number of aralkyl groups from silicon, germanium, and tin under the influence of aqueous or aqueous–methanolic alkali. The cleavages proceed by formation of a carbanion, the latter then reacting with the solvent as in reactions (23) and (24), where OY^- may be OH^- or OMe^-.

$$YO^- + R_3M \cdot CH_2Ar = R_3MOY + \overset{-}{C}H_2Ar \tag{23}$$

$$\overset{-}{C}H_2Ar + HOY \quad = CH_3Ar + OY^- \tag{24}$$

The alkali is regenerated in reaction (23) and thus, as expected, the cleavage of the organometallic substrate follows simple first-order kinetics (first-order in substrate). The first-order rate coefficients were converted into second-order rate coefficients (k_2) by division by the concentration of alkali in a given run; $k_2 = k_1/[OY^-]$.

The first set of substrates studied were compounds of the type $Me_3SiCH_2C_6H_4X$; Eaborn and Parker[23] obtained a value of 0.48×10^{-6} l.mole^{-1}.sec^{-1} for k_2 for the parent compound (X = H) at 49.7 °C in the solvent 39 wt.% water–methanol. The Arrhenius parameters were $E_a = 29.5$ kcal.mole^{-1} and $A = 4.5 \times 10^{13}$ l.mole^{-1}.sec^{-1}. A number of substrates with substituents in the benzyl group were examined, and it was shown that electron-withdrawing substituents markedly aided reaction[23].

In later work[26] the effect of the metal atom was investigated and rate coefficients (k_2) calculated for silicon, germanium, and tin compounds. These are given in Table 10.

The order of reactivity with respect to the metal atom is thus Sn > Si > Ge; in addition, the uncleaved alkyl groups are seen to exert an effect in the sense $Me_3M > Et_3M$, perhaps steric in origin. Again, electron-withdrawing groups aid reaction as shown[26] by the relative rate coefficients for cleavage of Me_3SnCH_2 C_6H_4X under the conditions given in Table 10, viz. m-Cl(32.8), p-Br(12.8), p-Cl(79.), H(1.0), m-Me(0.73), p-F(0.55), and p-Me(0.21). This sequence sug-

TABLE 10

SECOND-ORDER RATE COEFFICIENTS ($l.mole^{-1}.sec^{-1}$) FOR THE ALKALI CLEAVAGE OF
SOME ORGANOMETALLICS IN SOLVENT 80 % v/v AQUEOUS METHANOL AT 50 °C[26]

Substrate	$10^6 k_2$	Substrate	$10^6 k_2$
Me_3SiCH_2Ph	0.336	Me_3SiCPh_3	630
Me_3SnCH_2Ph	10.1	Me_3GeCPh_3	29.8
$Me_3SiCHPh_2$	480	Et_3GeCPh_3	0.16
$Me_3GeCHPh_2$	3.7		
$Et_3SiCHPh_2$	1.1		
$Et_3GeCHPh_2$	0.05		

gests[26] that there must be considerable negative charge on the leaving benzyl group in the transition state, although the timing of the cleavage cannot be determined exactly. A concerted bond-making and bond-breaking process could occur as in (25), illustrated for the substrate Me_3SnCH_2Ph, *viz.*

$$(25)$$

Another possibility[26] is that the species (IV) is an intermediate and that the rate-determining transition state lies between (IV) and the products $YO-SnMe_3$ and CH_2Ph. If this is so, then the structure of this transition state would still be expected to be close to that of (IV)[26].

The mechanism shown in (25) can be described as an S_N2 substitution at tin, or as an S_E1 substitution at carbon catalysed by the ion OY^-, *i.e.* as mechanism S_E1-OY^-. A similar mechanism to that of (25) has also been postulated[27] for the base-catalysed cleavage of 3-phenallyl derivatives of silicon, germanium, and tin; these cleavages are more fully discussed in Chapter 10, Section 1 (p. 195).

Eaborn and co-workers[28] have reported product isotope ratios for the cleavage of benzyl–silicon, benzyl–tin, and aryl–tin bonds using an equimolar mixture of MeOH and MeOD as the solvent. A free carbanion would not be expected to discriminate between MeOH and MeOD in the fast step (24), and hence[28] the product isotope ratio k_H/k_D should be unity*. The values of k_H/k_D in Table 11 indicate that the carbanions are not entirely free, but that some degree of electrophilic attack by the solvent at the benzyl carbon atom takes place, as in (V).

---

* Dodd and Johnson[14] have shown, however, that the carbanion $\overset{+}{H}NC_5H_4\overset{-}{C}H_2$ does discriminate significantly between various electrophiles (see p. 30).

TABLE 11

PRODUCT ISOTOPE RATIOS, $k_H/k_D{}^a$, FOR THE ALKALI CLEAVAGE OF SOME BENZYL AND
ARYL COMPOUNDS IN MeOH : MeOD AT 50 °C[28]

X	$XC_6H_4CH_2SiMe_3$	$XC_6H_4CH_2SnMe_3$	$XC_6H_4SnMe_3$
H	1.4	2.8	4.4
p-Me	1.5	2.5	3.4
m-Cl	1.6	2.4	3.8
m-CF$_3$	1.6	2.0	4.6

^{a} k_H/k_D, ±0.3.

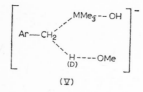

($\underline{V}$)

REFERENCES

1 O. A. REUTOV, *Acta Chim. Acad. Sci. Hung.*, 18 (1959) 439;
   O. A. REUTOV, I. P. BELETSKAYA AND YANG-TS'E WU (Y.-C. WU), *Kinetica i Kataliz, Akad. Nauk SSSR, Sb. Statci*, (1960) 43.
2 O. A. REUTOV, V. I. SOKOLOV AND I. P. BELETSKAYA, *Dokl. Akad. Nauk SSSR*, 136 (1961) 631; *Proc. Acad. Sci. USSR*, 136 (1961) 115; *Izv. Akad. Nauk SSSR, Otd. Khim. Nauk*, (1961) 1217, 1427; *Bull. Acad. Sci. USSR, Div. Chem. Sci.*, (1961) 1127, 1328.
3 E. D. HUGHES, C. K. INGOLD AND R. M. G. ROBERTS, *J. Chem. Soc.*, (1964) 3900.
4 O. A. REUTOV, B. PRAISNER, I. P. BELETSKAYA AND V. I. SOKOLOV, *Izv. Akad. Nauk SSSR, Otd. Khim. Nauk*, (1963) 970; *Bull. Acad. Sci. USSR, Div. Chem. Sci.*, (1963) 884.
5 D. J. CRAM, *Fundamentals of Carbanion Chemistry*, Academic Press, New York, 1965.
6 O. A. REUTOV, V. A. KALYAVIN AND T. A. SMOLINA, *Dokl. Akad. Nauk SSSR*, 156 (1964) 95; *Proc. Acad. Sci. USSR*, 156 (1964) 460.
7 F. R. JENSEN AND D. HEYMAN, *J. Am. Chem. Soc.*, 88 (1966) 3438.
8 C. R. HART AND C. K. INGOLD, *J. Chem. Soc.*, (1964) 4372.
9 O. A. REUTOV, I. P. BELETSKAYA AND L. A. FEDOROV, *Dokl. Akad. Nauk SSSR*, 163 (1965) 1381; *Proc. Acad. Sci. USSR*, 163 (1965) 794.
10 B. F. HEGARTY, W. KITCHING AND P. R. WELLS, *J. Am. Chem. Soc.*, 89 (1967) 4816.
11 V. A. KALYAVIN, T. A. SMOLINA AND O. A. REUTOV, *Dokl. Akad. Nauk SSSR*, 157 (1964) 919; *Proc. Acad. Sci. USSR*, 157 (1964) 762.
12 G. B. DEACON, *Rev. Pure Appl. Chem.*, 13 (1963) 189.
13 J. R. COAD AND M. D. JOHNSON, *J. Chem. Soc. B*, (1967) 633.
14 D. DODD AND M. D. JOHNSON, *J. Chem. Soc. B*, (1969) 1071.
15 D. DODD, C. K. INGOLD AND M. D. JOHNSON, *J. Chem. Soc. B*, (1969) 1076.
16 B. J. GREGORY AND C. K. INGOLD, *J. Chem. Soc. B*, (1969) 276.
17 M. D. JOHNSON AND N. WINTERTON, *J. Chem. Soc. A*, (1970) 511.
18 G. FRAENKEL AND D. T. DIX, *J. Am. Chem. Soc.*, 88 (1966) 979; G. FRAENKEL, D. T. DIX and D. G. ADAMS, *Tetrahedron Letters*, (1964) 3155.
19 G. M. WHITESIDES, M. WITANOWSKI AND J. D. ROBERTS, *J. Am. Chem. Soc.*, 87 (1965) 2854; G. M. WHITESIDES AND J. D. ROBERTS, *J. Am. Chem. Soc.*, 87 (1965) 4878.
20 M. H. ABRAHAM AND J. A. HILL, *J. Organometal. Chem.*, 7 (1967) 11.
21 M. WITANOWSKI AND J. D. ROBERTS, *J. Am. Chem. Soc.*, 88 (1966) 737.

22 G. Fraenkel, D. T. Dix and M. Carlson, *Tetrahedron Letters*, (1968) 579.
23 C. Eaborn and S. H. Parker, *J. Chem. Soc.*, (1955) 126.
24 C. Eaborn and J. C. Jeffrey, *J. Chem. Soc.*, (1957) 137.
25 C. Eaborn and S. H. Parker, *J. Chem. Soc.*, (1957) 955.
26 R. W. Bott, C. Eaborn and T. W. Swaddle, *J. Chem. Soc.*, (1963) 2342.
27 R. M. G. Roberts and F. El Kaissi, *J. Organometal. Chem.*, 12 (1968) 79.
28 R. Alexander, C. Eaborn and T. G. Traylor, *J. Organometal. Chem.*, 21 (1970) P 65.

Chapter 5

Mercury-for-Mercury Exchanges

Much of the fundamental kinetic and mechanistic work on electrophilic substitution at saturated carbon has involved the study of reactions in which an organomercury substrate undergoes substitution by an electrophilic mercuric compound. Ingold and co-workers[1] have concluded that these "mercury-for-mercury" exchanges occur only through the one-alkyl (1), the two-alkyl (2), and the three-alkyl (3) mercury exchange, *viz.*

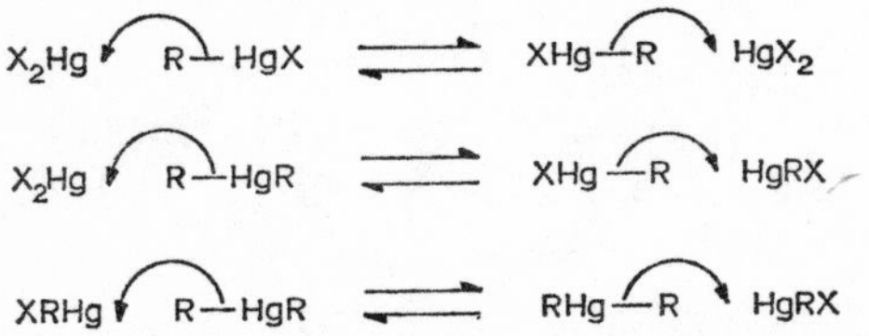

$$X_2Hg \quad R\text{—}HgX \quad \rightleftharpoons \quad XHg\text{—}R \quad HgX_2 \tag{1}$$

$$X_2Hg \quad R\text{—}HgR \quad \rightleftharpoons \quad XHg\text{—}R \quad HgRX \tag{2}$$

$$XRHg \quad R\text{—}HgR \quad \rightleftharpoons \quad RHg\text{—}R \quad HgRX \tag{3}$$

The two-alkyl exchange (2) in the direction left to right is sometimes referred to as a *syn*-proportionation, and in the direction right to left is called a symmetrisation; the position of the equilibrium in reaction (2) is normally well to the right. The work of Ingold and his school on the one-alkyl, two-alkyl (*syn*-proportionation), and three-alkyl exchange has been summarised by Ingold and by Thorpe[2]. Reutov[3] has reviewed the extensive work of his own school on the symmetrisation reaction, and on the other alkyl exchanges.

A number of one-alkyl mercury-for-mercury exchanges, proceeding by mechanism S_E1, have been considered in Chapter 4, Sections 1.1 (p. 23) and 1.2 (p. 25).

1. The one-alkyl mercury-for-mercury exchange

In this exchange the products are chemically identical to the reactants and the exchange has normally been followed by the transfer of radioactive mercury (^{203}Hg, denoted as $\overset{*}{Hg}$) from the mercuric salt to the alkylmercuric salt, *viz.*

$$RHg X + \overset{*}{Hg}X_2 \rightleftharpoons R\overset{*}{Hg}X + HgX_2 \tag{4}$$

1.1 STEREOCHEMICAL STUDIES

The exchange between *cis-* (and also *trans-*)-2-methoxycyclohexylmercuric chloride and mercuric chloride in solvents dioxan, acetone, and isobutyl alcohol proceeded with retention of configuration at the carbon atom undergoing substitution[4], a result fully confirmed by Ingold and co-workers using the simple optical isomers (−)-*sec.*-butylmercuric acetate (exchanging with mercuric acetate in ethanol[5]) and (−)-α-carbethoxybenzylmercuric bromide (exchanging with mercuric bromide in 70 % aqueous dioxan)[6]. Ingold and co-workers also showed that the anion-catalysed exchange (4) (R = (+)-*sec.*-butyl, X = Br or I) in acetone as solvent proceeded with full retention of configuration. Since the above exchanges were shown[5-7] to be bimolecular, it follows that the stereochemical course of reaction (4) is that of retention of configuration at the site of substitution for both mechanism S_E2(open)[5] and S_E2(cyclic)[7].

1.2 EXCHANGE BETWEEN SIMPLE ALKYLMERCURIC SALTS AND MERCURIC SALTS

The kinetics of such exchanges were first investigated by Nefedov and Sintova[8-11] who showed that the exchange (4) (R = Me, Et, Pr^n, X = Br) in solvent ethanol followed second-order kinetics. They did not establish the order with respect to each of the reactants, and wrote the exchange mechanism as two successive two-alkyl substitutions, *viz.*

$$2 \text{ RHgBr} \quad \rightarrow R_2Hg + HgBr_2 \tag{5}$$

$$\overset{*}{R_2Hg} + \overset{*}{HgBr_2} \rightarrow 2 \text{ RHgBr} \tag{6}$$

The rate coefficients and activation energies recorded by Nefedov and Sintova are collected in Table 1.

In a re-investigation of reaction (4) (R = Me, X = Br) using solvent ethanol at 100 °C, Ingold and co-workers[5] observed that the kinetic order was second overall, first-order in each reactant. This latter observation excludes the Nefedov–

TABLE 1

SECOND-ORDER RATE COEFFICIENTS (l.mole^{-1}.sec^{-1}) AND ACTIVATION ENERGIES FOR REACTION (4); SOLVENT ETHANOL

Reactants	E_a (kcal.mole^{-1})	$10^5 k_2$ (100 °C)	Ref.
MeHg$\overset{*}{Br}$+HgBr$_2$	18.3	35	9
EtHg$\overset{*}{Br}$+HgBr$_2$	11.5		8
Pr^{n}Hg$\overset{*}{Br}$+HgBr$_2$	20.0	9.7	11

Sintova mechanism as shown in equations (5) and (6) and it was suggested[5] that a single bimolecular elementary reaction took place, *viz.*

$$RHg\overset{*}{X} + HgX_2 \xrightarrow{\text{ethanol}} R\overset{*}{Hg}X + HgX_2 \qquad (7)$$

The rate of reaction (7) (R = Me, X = Br) was increased on addition of lithium nitrate or on addition of water to the solvent, and mechanism $S_E2(\text{open})$ was therefore indicated[5]. The rate coefficients for reaction (7) (R = Me) increased in the sequence $X = Br < I \ll OAc \ll NO_3$, again suggesting mechanism $S_E2(\text{open})$ rather than mechanism $S_E2(\text{cyclic})$. In Table 2 are collected the various rate coefficients reported[5, 12]; for the substitution of methylmercuric bromide by mercuric bromide in ethanol the Arrhenius parameters[5] were $A = 5 \times 10^7$ l.mole^{-1}.sec^{-1} and $E_a = 19.8$ kcal.mole^{-1}, which I have calculated correspond to $\Delta H^{\ddagger} = 19.2$ kcal.mole^{-1} and $\Delta S^{\ddagger} = -25$ cal.deg^{-1}.mole^{-1} at 333 °K.

TABLE 2

SECOND-ORDER RATE COEFFICIENTS (l.mole^{-1}.sec^{-1}) FOR REACTION (7); SOLVENT ETHANOL[5, 12]

		$10^5 k_2$			
R	X	100.2 °C	59.8 °C	0.0 °C	Relative rate
Me	Br	12.8	0.50	0.0007	1
Me	I	101			7.9
Me	OAc		500		1000
Me	NO$_3$			169	240,000
Me	Br	12.8			1
Et	Br	5.4			0.42
Pe^{neo}	Br	4.2			0.33
Me	OAc		500		1
Bu^s	OAc		30.9		0.06

Hughes and Volger[12] explained the reactivity sequence (Table 2), Me (1) > Et(0.42) > Pe^{neo}(0.33) > Bu^s(0.06), as due to steric non-bonded compressions in the transition state. For example, in the ethyl transition state (I), there will be extra interactions between the α-methyl group and the two mercury atoms, as compared with the methyl transition state.

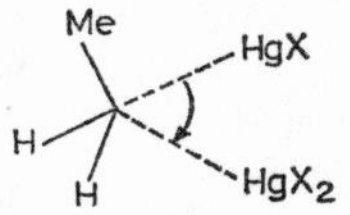

(I)

Added lithium nitrate ([LiNO$_3$] = 0.05 M) produces a 50 % increase in the rate coefficient for reaction (7) (R = Me, X = Br)[5]. The effect of added lithium

TABLE 3

SECOND-ORDER RATE COEFFICIENTS (l.mole^{-1}.sec^{-1}) FOR REACTION (7) (R = Me, X = Br) AT 59.8 °C IN SOLVENT ETHANOL WITH ADDED LITHIUM BROMIDE[7]

Initial concentration of methylmercuric bromide and of mercuric bromide 0.095 M.

Added LiBr(M)	0	0.035	0.065	0.105	0.161	0.203	0.271	0.311
$10^5 k_2$	0.49	12.8	21.5	39	69	89	121	146

bromide is much more pronounced; details are given in Table 3.

If the initial concentration of mercuric bromide is denoted as a, and the concentration of added lithium bromide as c, then a plot of k_2 versus c is found[7] to be a straight line up to the point at which $c = a$. After this point, in the region $c > a$, the linear dependence of k_2 on c is still maintained but with an increased slope. This behaviour was interpreted as follows[2,7]. When $c < a$, the bromide ion is nearly all taken up to form the species $HgBr_3^-$; this species is a very reactive electrophile and leads to a transition state (II) of the S_E2(cyclic) type*, viz.

$$\overset{*}{H}gBr_2 + Br^- \rightleftharpoons \overset{*}{H}gBr_3^- \tag{8}$$

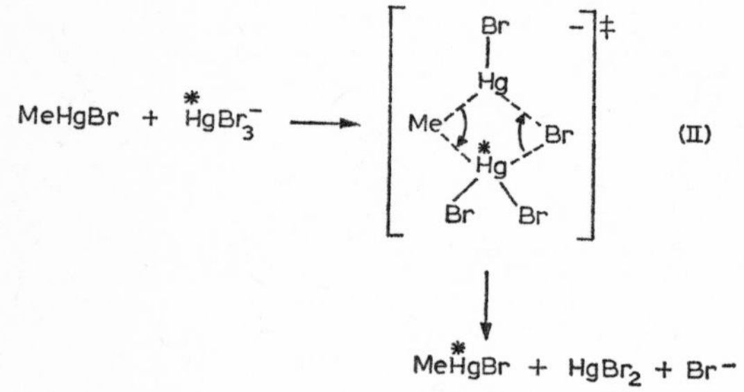

$$\text{(9)}$$

That is, although the bromide ion is strongly held in the initial state, it is even more strongly held in the transition state where it can "bridge" the two mercury atoms.

In the region where $c > a$, then the bromide ion not taken up in the equilibrium

---

* There is no reason (cf. refs. 35 and 53) why two bromide ions should not bridge the mercury atoms in the transition state to give the symmetrical arrangement shown in transition state (IIa)

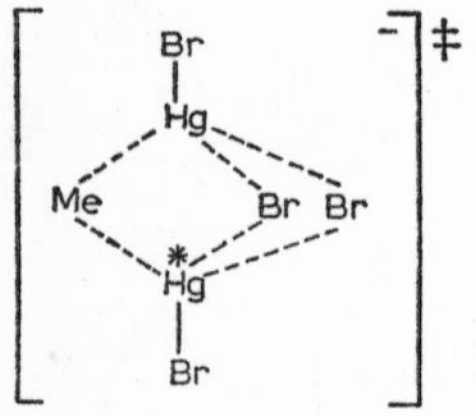

(IIa)

(8) is effectively $c - a$. This residual bromide ion is available to take part in the equilibria (10) and (11) below. It was suggested that it is equilibrium (11) in conjunction with the equilibrium (8) which leads to another S_E2(cyclic) transition state (III), now incorporating two bromide ions.

$$\overset{*}{Hg}Br_3^- \; + \; Br^- \; \rightleftharpoons \; \overset{*}{Hg}Br_4^{2-} \tag{10}$$

$$MeHgBr \; + \; Br^- \; \rightleftharpoons \; MeHgBr_2^- \tag{11}$$

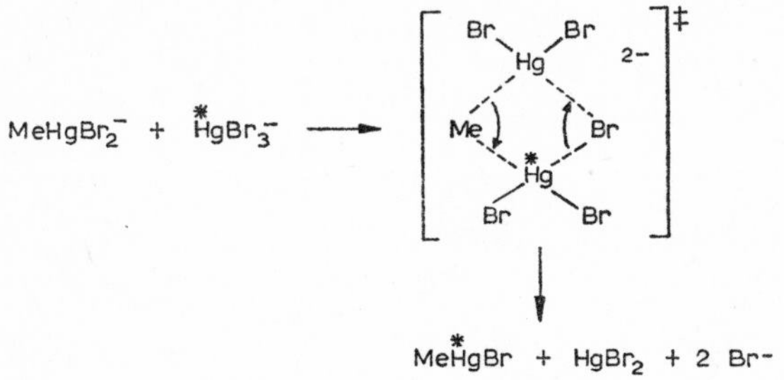

$$MeHgBr_2^- \; + \; \overset{*}{Hg}Br_3^- \; \longrightarrow \; \left[\begin{array}{c} Br \quad Br \\ \diagdown Hg \diagup \\ Me \bigg) \quad \bigg(Br \\ \overset{*}{Hg} \\ \diagup \quad \diagdown \\ Br \quad Br \end{array} \right]^{\ddagger\,2-}$$

$$\downarrow$$

$$\overset{*}{Me}HgBr \; + \; HgBr_2 \; + \; 2\; Br^- \tag{12}$$

Catalysis was also found with other anions[7], the catalytic power being in the order $NO_3^- < OAc^- < Cl^- < Br^- < I^-$, i.e. that of increased co-ordinating power of the anions to mercuric salts and to alkylmercuric salts. The one-alkyl exchange of sec.-butylmercuric salts[7] and neopentylmercuric salts[12] was also subject to one-anion and two-anion catalysis, and such catalysis was observed[7] when acetone was used as a solvent as well as with ethanol solvent.

Jensen and Rickborn[35] and Matteson[53] have argued that the uncatalysed one-alkyl exchange (7) cannot proceed through an open transition state such as (I); they considered that the principle of microscopic reversibility was contravened in that the overall reaction path was not symmetrical with respect to the forward and back reactions. This problem has been considered[54] in some detail and it has been established that there are several possible reaction paths for the exchange (7) that are overall symmetrical with respect to the forward and back reaction. One such possible reaction path is shown in Fig. 1; the principle of microscopic reversibility is obeyed and yet, contrary to the views of Jensen and Rickborn and of Matteson, reaction proceeds through an open transition state of type (I). The transition state for the forward reaction is denoted in the figure as (I$_f$) and that for the back reaction as (I$_b$). The cyclic symmetrical species (I$_i$) is a high-energy intermediate, the presence of which is necessary to retain overall symmetry of the reaction path.

A complicated reaction path is also necessary to retain overall symmetry in the one-anion-catalysed exchange if the transition state is represented as (II). However, if the transition state is (IIa), the pathway is automatically symmetrical. Similarly transition state (III) leads to a symmetrical pathway.

Symmetrical transition states in symmetrical exchanges therefore lead to symmetrical pathways that are in accord with the principle of microscopic reversibility.

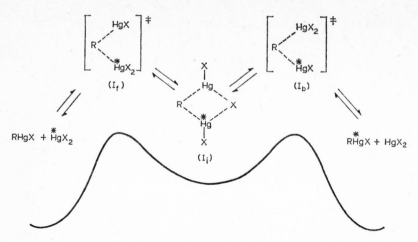

Fig. 1. A possible reaction path and energy profile for the uncatalysed one-alkyl mercury-for-mercury exchange reaction.

The principle, however, can never be used to exclude an unsymmetrical transition state provided that a pathway of overall symmetry with respect to forward and back reactions can be constructed (*e.g.* the pathway in Fig. 1).

1.3 EXCHANGE BETWEEN BENZYLMERCURIC BROMIDE AND MERCURIC BROMIDE

Although the above exchange takes place too slowly to follow in solvents such as toluene, alcohol, acetone, etc., it proceeds at a reasonable rate in solvent quinoline[13, 14]. The exchange is second-order overall, first-order in each reactant, and the second-order rate coefficient remained constant over a four-fold range of initial concentration in each reactant. Second-order rate coefficients were reported[13, 14] to be 0.183×10^{-4} (60 °C), 0.411×10^{-4} (70 °C), 0.897×10^{-4} (80 °C) and 1.78×10^{-4} (88 °C) in l.mole^{-1}.sec^{-1}, and the Arrhenius parameters were given[13, 14] as $A = 3.33 \times 10^7$ l.mole^{-1}.sec^{-1}, and $E_a = 18.8 \pm 0.9$ kcal.mole^{-1}; from these parameters I have calculated that $\Delta H^{\ddagger} = 18.2$ kcal.mole^{-1} and $\Delta S^{\ddagger} = -26$ cal.deg^{-1}.mole^{-1} at 298 °K.

The observation of the above kinetic form rules out the Nefedov and Sintova mechanism (equations (5) and (6)), and thus Reutov *et al.*[13, 14], concluded that the exchange proceeded via a single bimolecular reaction. They also suggested[13, 14] that a complex of mercuric bromide with quinoline, Q, was the actual electrophile, but that the possibility of benzylmercuric bromide reacting via a quinoline complex could not be excluded[13]. The transition states (IV)[13] and (V)[14] were proposed, the mechanism thus being regàrded as S_E2(cyclic), but since the order with respect to the solvent quinoline cannot be determined, it is clearly not possible to specify completely the number of solvent molecules in the transition state.

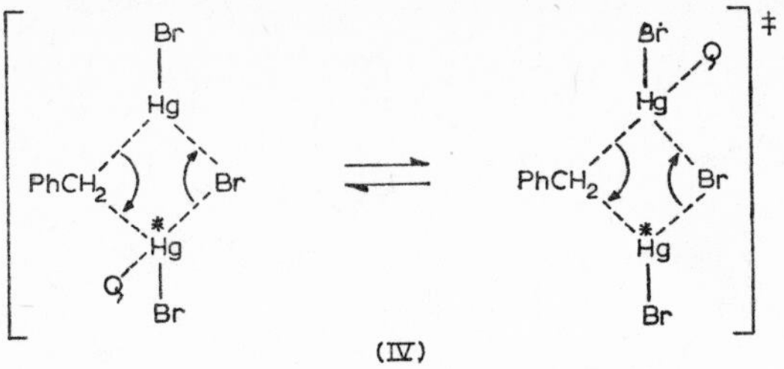

(IV)

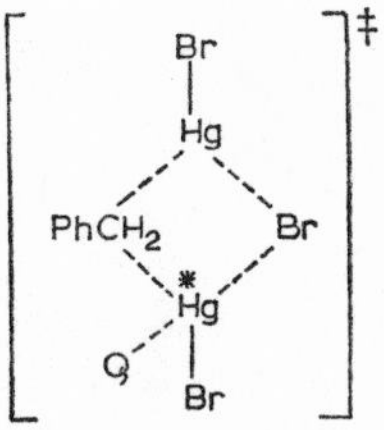

(V)

The significance of the formulation of the two isomeric entities within the transition state (IV) is not clear to the present author, but the transition state (V) is quite straightforward.

In later work[15], the exchange in solvent DMSO was investigated; the rate is somewhat higher than in quinoline ($k_2(70\,°C) = 1.03 \times 10^{-4}$ l.mole^{-1}.sec^{-1} in DMSO and $k_2(70\,°C) = 0.41 \times 10^{-4}$ l.mole^{-1}.sec^{-1} in quinoline), but the kinetic order remained first-order in benzylmercuric bromide and first-order in mercuric bromide[15]. The influence of p-substituents on the exchange

$$p\text{-X–C}_6\text{H}_4\text{CH}_2\overset{*}{\text{H}}\text{gBr} + \text{HgBr}_2 \xrightarrow{\text{Br}^-} p\text{-X–C}_6\text{H}_4\text{CH}_2\text{HgBr} + \overset{*}{\text{H}}\text{gBr}_2 \qquad (13)$$

TABLE 4

SECOND-ORDER RATE COEFFICIENTS, 10^4k_2 (l.mole^{-1}.sec^{-1}), FOR THE EXCHANGE (13) AT 70 °C[13,15]

Initial concentrations of benzylmercuric bromides and of mercuric bromide 0.06 M.

Solvent	Added KBr (M)	p-X				
		F	Cl	H	Me	Pr^i
Quinoline	0	0.34	0.36	0.41	0.57	0.66
DMSO	0	0.83	0.83	1.03	1.22	1.22
DMSO	0.09	27	40	21	21	19
DMSO	0.12	55	95	40	40	30

was also studied[13, 15] and details are given in Table 4. For the runs in solvent DMSO, the rate coefficients given in Table 4 cannot be accurate to better than 5 % since I have calculated these coefficients from exchange half-lives recorded by Reutov *et al.* to the nearest hour or nearest minute.

The substituent effect on the uncatalysed reaction (13) is the same in both solvents (Table 4; added KBr = 0), and indicates that electron-donating substituents aid reaction by increasing the charge density at the carbon atom undergoing substitution and thereby facilitating attack by an electrophile.

Reaction (13) is catalysed strongly by bromide ions in solvent DMSO; with each reactant (X = H) at 0.06 M, rate coefficients at 70 °C are

Added KBr (M)	0	0.06	0.09	0.12	0.18
$10^4 k_2$ (*l.mole*$^{-1}$.*sec*$^{-1}$)	1.03	4.1	21	40	60

Interpretation of these results[15] follows the scheme of Ingold for the one-anion and two-anion-catalysed exchange between methylmercuric bromide and mercuric bromide. At low concentrations of added bromide ion the $HgBr_3^-$ complex is formed, and attacks the benzylmercuric bromide to give an S_E2(cyclic) transition state (VI); at higher concentrations of added bromide ion the reactants are actually $HgBr_3^-$ and p-X–$C_6H_4CH_2HgBr_2^-$ yielding the S_E2(cyclic) transition state (VII).

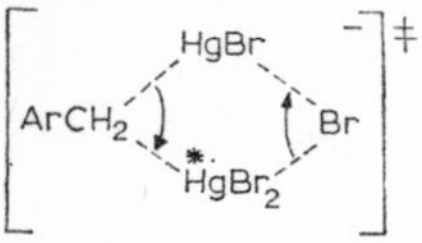

(VI) 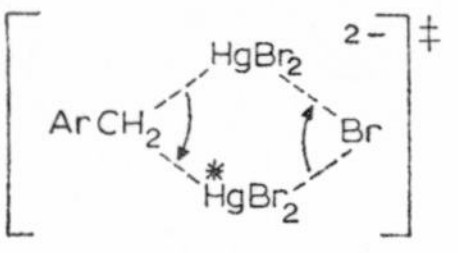   (VII)

The reversal in the substituent effects (Table 4; added KBr = 0.09 M and 0.12 M) for the catalysed exchange may then be due[15] to electron-donating substituents now retarding the formation of the complex $ArCH_2HgBr_2^-$; these results thus support the idea[15] that the second anion (in the two-anion catalysis) is brought into the transition state as $RHgX_2^-$, rather than as HgX_4^{2-}.

The exchange between p-nitrobenzylmercuric bromide and mercuric bromide, proceeding by mechanism S_E1, is dealt with in Chapter 4, Section 1.1 (p. 23), and the same exchange proceeding by mechanism S_E1-anion catalysed is dealt with in Chapter 4, Section 1.2 (p. 25).

1.4 EXCHANGE BETWEEN α-CARBETHOXYBENZYLMERCURIC BROMIDE AND MERCURIC BROMIDE

The exchange

$$\text{Ph}\overset{*}{\text{C}}\text{H(CO}_2\text{Et)HgBr} + \text{HgBr}_2 = \text{PhCH(CO}_2\text{Et)}\overset{*}{\text{Hg}}\text{Br} + \text{HgBr}_2 \qquad (14)$$

was reported by Reutov et al.[16], in 1959–60, to follow second-order kinetics, first-order in each reactant, in solvent 70 % aqueous dioxan. In 1961, Reutov et al.[17, 18], re-investigated this exchange and found that first-order kinetics were followed (first-order in the alkylmercuric bromide and zero-order in mercuric bromide), although there were a number of experimental difficulties with regard to impurities in the dioxan. This work[17, 18] dealt with substituent effects in (14), and it was shown that electron-donating p-substituents in the phenyl group retarded reaction and electron-attracting p-substituents accelerated reaction in accord with the (presumed) S_E1 mechanism of substitution, when 70 % aqueous dioxan was the solvent. Ingold and co-workers[6] later claimed that reaction (14) in 70 % aqueous dioxan as solvent followed second-order kinetics, as Reutov et al.[16] originally reported, and hence that the mechanism was S_E2 and not S_E1. Evidence based on kinetic and stereochemical experiments was put forward[6] in support of the contention that reaction (14) proceeded by mechanism S_E2 and it seems to the author that the conclusions of Ingold and co-workers must be accepted. In particular, the fact that reaction (14) proceeds in 70 % aqueous dioxan with retention of configuration, using (−)-α-carbethoxybenzylmercuric bromide[6], is very strong evidence against the presumed S_E1 mechanism.

Reutov et al. also studied reaction (14) using pyridine[17, 19], dimethylformamide (DMF)[20], and 80 % aqueous ethanol[21] as solvents. In all of these solvents the reaction followed second-order kinetics, first-order in each reactant, although good kin-

TABLE 5

SECOND-ORDER RATE COEFFICIENTS AT 60 °C (l.mole^{-1}.sec^{-1}) AND ACTIVATION PARAMETERS FOR REACTION (14)

Solvent	k_2	log A	E_a	$\Delta H_{298}^{\ddagger}$	$\Delta S_{298}^{\ddagger a*}$	Refs.
70 % aq. dioxan	2.8×10^{-3}					6
Pyridine	6.6×10^{-2}	9.5	16.3	15.7	−17	17, 19
DMF[b]	2.35×10^{-4}	6.9	16.1	15.5	−29	20
80 % aq. ethanol[c]	1.03×10^{-3}	13.4	25.1	24.5	+1	21

[a] Recalculated by the author.
[b] Reutov records log $A = 8.7$ and $\Delta S^{\ddagger} = -19$ based on rate coefficients in l.mole^{-1}.min^{-1}.
[c] Reutov records log $A = 15.2$ and $\Delta S^{\ddagger} = +11$ based on rate coefficients in l.mole^{-1}.min^{-1}.

* A in l.mole^{-1}.sec^{-1}, E_a and $\Delta H^{\ddagger}$ in kcal.mole^{-1}, and $\Delta S^{\ddagger}$ in cal.deg^{-1}.mole^{-1}.

etic plots were obtained in solvent DMF only by allowing the mercuric bromide to stand in DMF solution for five hours at 70 °C prior to reaction. It may also be mentioned that, whereas Reutov et al.[20] studied reaction (14) using solvent DMF over the temperature range 60–80 °C, Ingold and co-workers[6] reported that they had failed to detect any reaction at 60 °C using the same solvent. The various rate coefficients and activation parameters are collected in Table 5.

Reaction (14), in the solvents listed in Table 5, no doubt proceeds by mechanism S_E2; Reutov et al.[17] write a transition state corresponding to mechanism S_E2(cyclic) for the reaction in pyridine and point out that the electrophile is probably a pyridine complex of mercuric bromide rather than the unsolvated bromide itself.

The exchange between α-carbethoxybenzylmercuric bromide and mercuric bromide, proceeding by mechanism S_E1, is dealt with in Chapter 4, Section 1.1 (p. 23), and the racemisation of (−)-α-carbethoxybenzylmercuric bromide proceeding by mechanism S_E1-anion catalysed is dealt with in Chapter 4, Section 1.2 (p. 25).

2. The two-alkyl (*syn*-proportionation) mercury-for-mercury exchange

This exchange may be represented by equation (15), where R is some organic radical and X an inorganic group.

$$R_2Hg + HgX_2 = 2 RHgX \tag{15}$$

2.1 STEREOCHEMICAL STUDIES

Diastereoisomers in the cyclohexane series are cleaved by mercuric acetate in methanol[22] and by mercuric chloride in ether[23] with retention of configuration at the site of substitution. The diastereoisomer $[PhCH(CO_2(-)\text{-menthyl})]_2Hg$ reacts with mercuric bromide in acetone at room temperature, again with retention of configuration[24]; previously it had been claimed[25] that this diastereoisomer reacted with mercuric bromide in acetone at 65 °C with inversion of configuration, but this was later shown[24] to arise from a secondary racemisation that occurred at the higher temperature. Using simple optical isomers, it has been unambiguously demonstrated that reaction (15) proceeds with complete retention of configuration at the site of substitution for the cases (R, HgX_2, solvent): $(-)$-Bu^s, $HgBr_2$, ethanol[26, 27]; $(-)$-1,4-dimethylpentyl, $HgBr_2$, $-$[28]; $(-)$-Bu^s, $Hg(OAc)_2$, ethanol[26]; $(-)$-Bu^s, $Hg(NO_3)_2$, ethanol[26]; $(-)$-Bu^s, $Hg(NO_3)_2$, 1 : 1 water : ethanol[26].

2.2 *Syn*-PROPORTIONATION OF DIALKYLMERCURYS AND MERCURIC SALTS

Reaction (15) is the longest-known of the mercury redistributions, but was studied kinetically only as recently as 1959 by Ingold *et al.*[26]. These workers showed that the substitution of di-*sec.*-butylmercury by mercuric bromide in solvent acetone was first-order in each reactant, and that the second-order rate coefficient remained constant over a range of initial concentrations of the reactants. Substitutions by other mercuric salts also followed the second-order rate equation, and the various rate coefficients obtained are collected in Table 6. The

TABLE 6

SECOND-ORDER RATE COEFFICIENTS ($l.mole^{-1}.sec^{-1}$) FOR THE SUBSTITUTION OF DI-*sec*-BUTYLMERCURY BY MERCURIC SALTS[26]

Salt	Solvent	Temp. (°C)	k_2	Relative rate[a]
$HgBr_2$	Acetone	25	2.4	
$HgBr_2$	Ethanol	25	0.39	1
$Hg(OAc)_2$	Ethanol	0	5.3	75
$Hg(NO_3)_2$	Ethanol	−46.6	7.6	4000

[a] Calculated by the author assuming values of 11 $kcal.mole^{-1}$ and 9 $kcal.mole^{-1}$ for E_a in the case of acetate and nitrate respectively.

substitution of di-*sec.*-butylmercury by mercuric bromide was strongly retarded by added lithium bromide, thus indicating that the species $HgBr_3^-$ is unreactive as an electrophile, and giving the sequence of reactivity (see Table 6) as $HgBr_3^- < HgBr_2 < Hg(OAc)_2 < Hg(NO_3)_2$. This is the order of increasing positive charge on the mercury atom, and is the expected order of increasing electrophilic power of the mercury salts. The above substitutions of di-*sec.*-butylmercury were postulated[26] to proceed *via* mechanism S_E2(open), for which a transition state (VIII) may be written, since if the mechanism S_E2(cyclic) obtained, the reactivity sequence would be expected to be the reverse of the sequence observed, with the more ionic mercuric salts being less able to co-ordinate with the leaving group and so complete the cyclic transition state (IX)*.

substitution of di-*sec.*-butylmercury by mercuric bromide was strongly retarded

---

* For a criticism of this argument see pp. 51–53.

Ingold and co-workers[26] did not write a reaction profile for these substitutions. As for the uncatalysed one-alkyl exchanges, several possibilities exist; one such possible pathway is shown in Fig. 2. It should be noted that the mechanism given

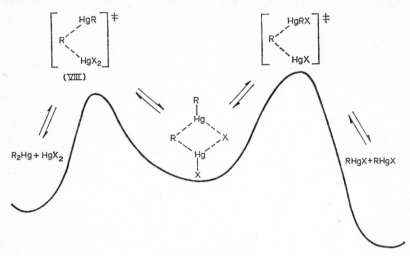

Fig. 2. A possible reaction path and energy profile for the two-alkyl mercury-for-mercury exchange reaction.

in Fig. 2 for the reverse process, the uncatalysed symmetrisation of RHgX, has no bearing on the mechanism of the catalysed symmetrisations discussed in Section 3.1 because the catalysed and uncatalysed symmetrisations must proceed through different transition states (*e.g.* the transition state for the catalysed symmetrisation includes two molecules of ammonia).

More complex transition states, (X) and (XI) below, for electrophilic substitution were suggested by Dessy *et al.*[29], as a result of their observation that the cleavage of ethylphenylmercury by radiomercuric chloride led to two equally radioactive products, *viz.*

$$\overset{*}{PhHgEt} + \overset{*/2}{HgCl_2} = \overset{*/2}{PhHgCl} + EtHgCl \tag{16}$$

Later work[30, 31], however, showed that all of the radioactive mercury appeared in the phenylmercuric chloride. Hence (16) is actually an electrophilic substitution at an aromatic carbon, and the transition states (X) and (XI) are no longer necessary to account for the experimental observations*.

---

* Nesmeyanov and Reutov[30] also showed that in the analogous cleavage of *n*-butylphenylmercury by radiomercuric bromide, the resulting phenylmercuric bromide carried all of the original activity, thus casting doubt on the observation[32] that all of the original activity appeared in the *n*-butylmercuric bromide.

TABLE 7

SECOND-ORDER RATE COEFFICIENTS (l.mole^{-1}.sec^{-1}) AT 25 °C AND ACTIVATION
PARAMETERS FOR THE SUBSTITUTION OF DIALKYLMERCURYS BY MERCURIC IODIDE

R_2Hg	Solvent	k_2	$\Delta H^{\ddagger}$ (kcal.mole^{-1})	$\Delta S^{\ddagger}$ (cal.deg^{-1}.mole^{-1})	Ref.
Me$_2$Hg	Dioxan	Too slow to measure			
Et$_2$Hg	Dioxan	0.0163	12.3	−28	33
Pr$^n_2$Hg	Dioxan	0.0186	12.2	−28	33
Pr$^i_2$Hg	Dioxan	0.0160	12.0	−29	33
cyclo-Pr$_2$Hg^a	Dioxan	0.0767	12.8	−22	33
Ph$_2$Hg	Dioxan	1.97	12.8	−16	33
Pr$^n_2$Hg	Benzene	0.112	11.0	−28	29
Ph$_2$Hg	Benzene	29.2	7.6	−28	29

^{a} Note that values of the rate coefficient are recorded in ref. 33 as 0.14×10^{-2} (35 °C) and
0.24×10^{-2} (45 °C), but these would seem to be misprints for 14×10^{-2} and 24×10^{-2}.

(X)   (XI)

The two-alkyl exchange (15) has also been studied by Dessy *et al.*[29, 33]; the
reported rate coefficients at 25 °C and the activation parameters are collected in
Table 7. Dessy and Lee[33] suggested that the dialkylmercurys were attacked by
mercuric iodide in dioxan to give a four-centre transition state (XII) (of the
S_E2(cyclic) type) or a transition state (XIII) derived from an ion-pair attack, *viz.*

(XII)   (XIII)

Rate coefficients for the two-alkyl exchange (15) (R = Me, X = Cl, Br, I) have
been reported by Rausch and Van Wazer[34], and are collected in Table 8.

Jensen and Rickborn[35] have criticised the use of relative reactivities of mercuric
salts in reactions such as (15) as a basis for the deduction of reaction mechanism.
They point out that, whereas cyclic transition states involving mercuric halides
as electrophiles must, of necessity, be four-centred (*e.g.* (XIV)), the electrophiles
mercuric acetate and mercuric nitrate could give rise to six-centred transition states
(*e.g.* (XV)) that might be energetically more favoured than the four-centred.

TABLE 8

SECOND-ORDER RATE COEFFICIENTS ($l.mole^{-1}.sec^{-1}$) AT 36 °C FOR THE SUBSTITUTION OF DIMETHYLMERCURY BY MERCURIC HALIDES[34]

HgX_2	Solvent	
	Dioxan	Methanol
$HgCl_2$	0.00259	1.05
$HgBr_2$	0.00354	0.30
HgI_2	0.00154	0.05

Hence the observation that, for example, mercuric acetate reacts with a given substrate in a given solvent faster than does mercuric bromide can be interpreted in at least two ways; (i) the mechanism of reaction is S_E2(open) and mercuric acetate is a more powerful electrophile than is mercuric bromide, and (ii) the mechanism of reaction is S_E2(cyclic) and mercuric acetate is better able to act as a bridging group in a six-centred transition state than is mercuric bromide in a four-centred transition state. The possibility that the two salts might be reacting by different mechanisms must also be considered.

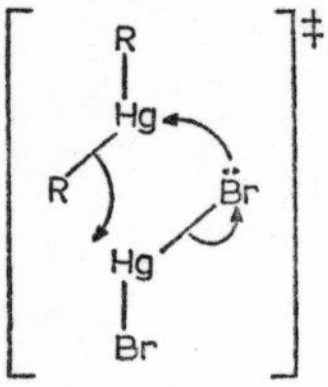

(XIV) 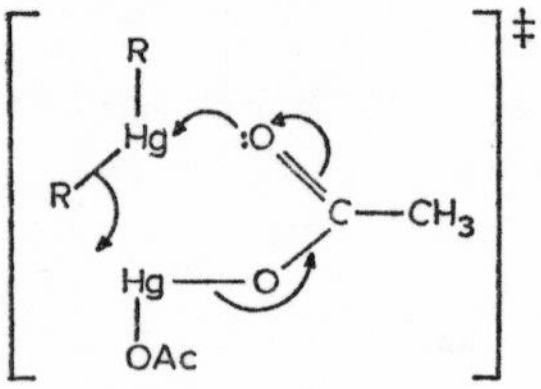   (XV)

In addition to the above argument, a further difficulty arises in that it is not possible to assign an unambiguous order of electrophilic power to the mercuric salts. The stepwise stability constants[36a]

$$K(HgI_2, Br^-) = 76, \quad K(HgBr_2, Br^-) = 190$$

indicate that, towards the bromide ion, mercuric bromide is a slightly stronger Lewis acid than is mercuric iodide (in solvent 0.5 M aqueous sodium perchlorate at 25 °C) whereas the stepwise stability constants[36a]

$$K(HgI_2, I^-) = 4700, \quad K(HgBr_2, I^-) = 71$$

demonstrate that, towards the iodide ion, mercuric iodide is a much stronger Lewis acid than is mercuric bromide (again in solvent 0.5 M aqueous sodium perchlorate at 25 °C). Furthermore, the relative catalytic powers of the mercuric halides in the racemisation of α-methylbenzylchloride (first-order in organic halide and first-order in mercuric halide) have been shown by Satchell[36b] to be markedly solvent-dependent, viz.

$$HgCl_2 \simeq HgI_2 < HgBr_2 \quad \text{(solvent acetone)}$$

$$HgI_2 \ll HgCl_2 < HgBr_2 \quad \text{(solvent nitrobenzene)}$$

It seems that use of relative reactivities of mercuric salts in order to distinguish between mechanisms S_E2(open) and S_E2(cyclic) is not a generally valid procedure.

3. The two-alkyl (symmetrisation) mercury-for-mercury exchange

3.1 SYMMETRISATION OF α-CARBALKOXYBENZYLMERCURIC BROMIDES BY AMMONIA IN CHLOROFORM

These reactions have been extensively investigated by Reutov et al.[37-39]. They showed[37], using the diastereoisomeric (−)-menthyl esters of α-carboxybenzylmercuric bromide, that symmetrisation proceeded with retention of configuration at the site of substitution and also[38] that the overall stoichiometry was that of equation (17) (R = ethyl- and (−)-menthyl ester of the α-carboxybenzyl group).

$$2\,RHgBr + 2\,NH_3 \xrightarrow{\text{CHCl}_3} R_2Hg + HgBr_2 \cdot 2\,NH_3\downarrow \tag{17}$$

Kinetics of symmetrisation were followed using a nephelometric method, normally with the ammonia present in a large excess; under these conditions, reaction (17) was shown to follow second-order kinetics with respect to the alkylmercuric bromide[38]. Variation of the concentration of the excess ammonia showed[38] that reaction (17) was also second-order in ammonia, and hence[38] that the reaction is of the fourth-order overall, viz.

$$\text{rate} = k_4[RHgBr]^2[NH_3]^2 \tag{18}$$

when R = ethyl and menthyl ester of the α-carboxybenzylgroup. Reutov[3, 39] has advanced two possible mechanisms for the symmetrisation reaction, A and B, and has stated[39] that he considered mechanism A to be the more probable, although mechanism B was compatible with the kinetic evidence.

$$2\,RHgBr \rightleftharpoons R_2Hg + HgBr_2 \tag{19}$$
$$HgBr_2 + 2\,NH_3 \rightleftharpoons HgBr_2 \cdot 2\,NH_3 \tag{20}$$

A

$$RHgBr + NH_3 \rightleftharpoons RHgBr \cdot NH_3 \tag{21}$$
$$2\,RHgBr \cdot NH_3 \rightleftharpoons R_2Hg + HgBr_2 \cdot 2\,NH_3 \tag{22}$$

B

The transition state for the symmetrisation reaction (17) was written[3, 39] as the cyclic activated complex

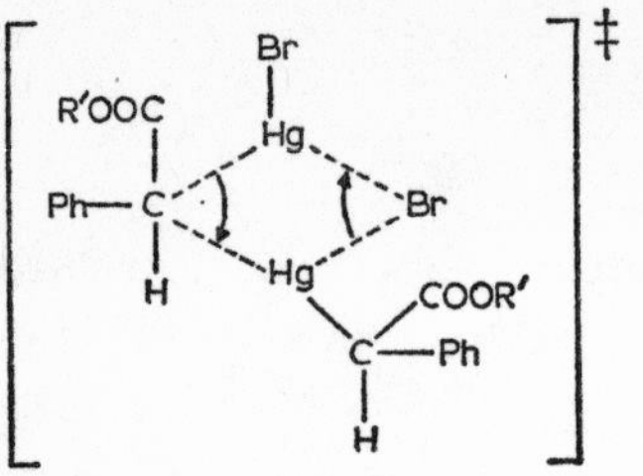

(XVI)

More recently, Jensen and Rickborn[40] severely criticised the assignment of mechanism A. They pointed out that if the forward reaction (19) is rate-controlling, then the overall kinetic form should be

$$\text{rate} = k_{(19)}[\text{RHgBr}]^2 \tag{23a}$$

and if the forward reaction (20) is rate-controlling, then the overall kinetic form would be

$$\text{rate} = \frac{k_{(19)}k_{(20)}}{k_{(-19)}} \cdot \frac{[\text{RHgBr}]^2[\text{NH}_3]^2}{[\text{R}_2\text{Hg}]} \tag{23b}$$

Neither (23a) nor (23b) is in accord* with the observed kinetic expression (18), and hence mechanism A must be incorrect[40]. There now seems to be general agreement[40-43] that reaction (17) (R = α-carbalkoxybenzyl) proceeds either by mechanism B (normally written as shown in B') or by mechanism C.

$$\text{RHgBr} + \text{NH}_3 \overset{\text{fast}}{\rightleftharpoons} \text{RHgBr·NH}_3 \qquad\qquad \text{B}' \left\{ \begin{array}{l} \quad(24) \\ \\ \quad(25) \end{array}\right.$$

$$2\,\text{RHgBr·NH}_3 \overset{\text{slow}}{\rightleftharpoons} \text{R}_2\text{Hg} + \text{HgBr}_2\text{·2 NH}_3$$

$$\text{RHgBr} + 2\,\text{NH}_3 \overset{\text{fast}}{\rightleftharpoons} \text{RHgBr·2 NH}_3 \qquad\qquad \text{C} \left\{ \begin{array}{l} \quad(26) \\ \\ \quad(27) \end{array}\right.$$

$$\text{RHgBr·2 NH}_3 + \text{RHgBr} \overset{\text{slow}}{\rightleftharpoons} \text{R}_2\text{Hg} + \text{HgBr}_2$$

Jensen et al.[43] have confirmed both the stoichiometry of equation (17), and the fourth-order rate expression (18) for the symmetrisation of the tert.-butyl ester of α-carboxybenzylmercuric bromide in a study by nuclear magnetic resonance. The fourth-order rate coefficient remained constant at about $15 \times 10^{-4}\ \text{l}^3.\text{mole}^{-3}.$ sec^{-1} at 31.4 °C over a wide range of initial concentrations of both the mercuric compound and of ammonia. Of mechanisms B' and C, these workers preferred C.

The effects of various groups X and R' on the rate of symmetrisation of compounds $\text{X–C}_6\text{H}_4\text{CH(CO}_2\text{R')HgBr}$ have been studied by Reutov et al.[44-46];

---

* Reutov[41] has reported that addition of R_2Hg has no effect on the rate of symmetrisation and that an earlier statement[38] to the contrary was in error.

usually the ammonia concentration was fixed as 1 M and under these conditions of a large excess of ammonia the rate equation (18) reduces to the second-order expression

$$\text{rate} = k_2^{obs}[\text{RHgBr}]^2 \qquad (28)$$

where $k_4[\text{NH}_3]^2 = k_2^{obs}$. For the particular case of $[\text{NH}_3] = 1\ M$, then k_4 is numerically equal to k_2^{obs} when the rate coefficients are expressed in units of

TABLE 9

SUBSTITUENT EFFECTS IN THE SYMMETRISATION OF $X\text{–}C_6H_4CH(CO_2R')HgBr$ BY
AMMONIA (1 M) IN CHLOROFORM AT 20 °C

X	R'	$10^3 k_2^{obs}$ $(l.mole^{-1}.sec^{-1})$	Ref.
H	Me	181	44
H	Et	110	44
H	Pr^i	25.9	44
H	Bu^t	0^a	44
H	Nonyl	10.2	44
H	Menthyl	6.9	44
p-NO$_2$	Et	17,730^b	45
m-Br	Et	1445	45
p-I	Et	676	45
p-Br	Et	540	45
p-Cl	Et	470	45
o-Br	Et	426	45
p-F	Et	148	45
H	Et	110	45
m-Me	Et	71	45
p-Pr^i	Et	41	45
p-Et	Et	40	45
p-Me	Et	34	45
o-Me	Et	32	45
p-Bu^t	Et	28	45

^{a} A value of about 0.7 l.mole^{-1}.sec^{-1} at 20 °C would be expected from the value of $k_4 = 15 \times 10^{-4}$ l^3.mole^{-3}.sec^{-1} obtained[43] at 31.4 °C.
^{b} This value is one obtained[45] by extrapolation of the other results assuming Hammett's equation with $\rho = 2.85$.

l^3.mole^{-3}.sec^{-1} and l.mole^{-1}.sec^{-1} respectively*. Details of substituent effects are given in Table 9; where the work has been repeated, only the most recent data are given. The effects of the substituents in the aromatic ring, X, have been

* Note that Reutov invariably expresses concentrations in mole/ml and the second-order rate coefficients in equation (28) as ml.mole^{-1}.sec^{-1} throughout his work on symmetrisations. In addition, later papers do not state the temperature at which kinetic runs were performed, although from the conditions given in earlier papers[38] it appears that kinetics were normally followed at 20 °C.

discussed by Reutov several times; Reutov and Beletskaya[44] suggest that since electron-acceptor substituents (*e.g.* NO_2, Halogen) accelerate reaction and since electron-donator substituents (*e.g.* alkyl) retard reaction, then in the transition state (*e.g.* XVI) cleavage of the old C–Hg bond is more important than formation of the new C–Hg bond. In other words, the S_E2 reaction is in a boundary region near the S_E1 reaction[46].

However, Jensen *et al.*[40, 43], have correctly pointed out that the fourth-order coefficient (k_4 in equation (18)) or the second-order coefficient (k_2^{obs} in equation (28)) are actually complex coefficients and include K, the equilibrium constant for reaction (24) or (26). In terms of mechanism B′, $k_4 = K_{(24)} \times k_{(25)}$ and in terms of mechanism C, $k_4 = K_{(26)} \times k_{(27)}$. Thus substituent effects may well refer to the equilibrium (24) or (26) rather than to the actual electrophilic substitution, reaction (25) or (27). In this connection it is worth recalling that in the bimolecular one-alkyl exchange (13) the sequence of *p*-substituents in the benzyl group is *p*-Cl < H < *p*-Pr^i, but in the anion-catalysed exchange (13), which takes place *via* a pre-rate-determining equilibrium, the sequence is *p*-Cl > H > *p*-Pr^i (see Table 4, p. 45); it seems to the author that the substituent effects shown in Table 9 may also be explained as effects on the equilibrium constants $K_{(24)}$ or $K_{(26)}$.

It may be noted that Reutov normally writes the transition states of these symmetrisations without incorporation of any ammonia molecules. Although he specifically states[46] that ammonia molecules can be incorporated, Reutov regards such incorporation as not affecting the fundamental cyclic four-centred aspect of the transition state.

The full transition states corresponding to reactions (25) and (27), in mechanisms B′ and C, may be written as follows, where R is an ester of an α-carboxybenzyl group.

(XVII)   (XVIII)

Thus (XVII) represents attack by the electrophile $RHgBr \cdot NH_3$ on the substrate $RHgBr \cdot NH_3$, and (XVIII) represents attack by the electrophile $RHgBr$ on the substrate $RHgBr \cdot 2 NH_3$.

3.2 CO-SYMMETRISATION OF α-CARBETHOXYBENZYLMERCURIC BROMIDES BY AMMONIA IN CHLOROFORM

Reutov *et al.*[47] have shown that the symmetrisation of, for example, a mixture of a *p*-methyl substituted ester with a *p*-bromo substituted ester proceeds more readily than expected from the separate reactivities of the two esters. Furthermore, when the *p*-bromo ester was labelled with a radioactive mercury isotope, most of the activity was found in the resulting $HgBr_2 \cdot 2\,NH_3$ precipitated in the co-symmetrisation. The co-symmetrisation was thus written[47] as

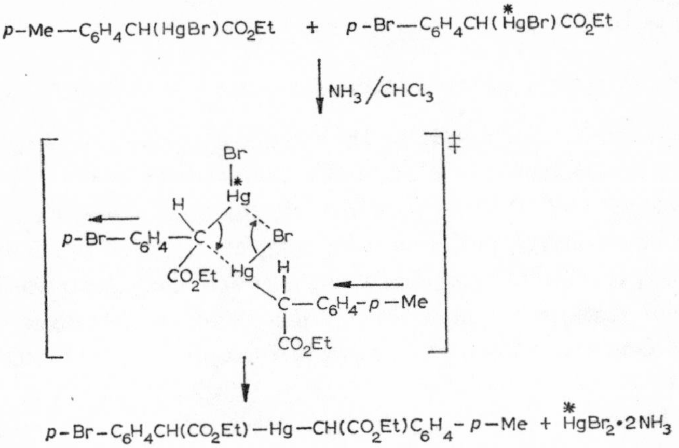

$$\begin{aligned} p\text{-Me}-C_6H_4CH(HgBr)CO_2Et \quad + \quad p\text{-Br}-C_6H_4CH(\overset{*}{Hg}Br)CO_2Et \end{aligned} \tag{29}$$

$$\downarrow NH_3/CHCl_3$$

$$p\text{-Br}-C_6H_4CH(CO_2Et)-Hg-CH(CO_2Et)C_6H_4-p\text{-Me} + \overset{*}{Hg}Br_2 \cdot 2NH_3$$

The breaking of the $C-\overset{*}{Hg}$ bond and the breaking of the Hg–Br bond are aided[47] by the *p*-bromo and *p*-methyl substituents respectively.

Again, Jensen et al.[43] have suggested that such interpretations may not be entirely correct, since the influence of the various substituents on the pre-rate-determining equilibrium (24) or (26) has also to be taken into account.

3.3 SYMMETRISATION OF α-CARBETHOXYBENZYLMERCURIC BROMIDE BY DIPHENYLMERCURY

No kinetic studies of this reaction have been reported, but product analyses[48] and radiomercuric labelling studies[49] indicate that the reaction proceeds in two stages, of which the first is very rapid and the second is slow, steps (30) and (31)

$$RHgBr + Ph_2Hg \overset{fast}{\rightleftharpoons} RHgPh + PhHgBr\downarrow \tag{30}$$

$$RHgPh + RHgBr \overset{slow}{\rightleftharpoons} R_2Hg + PhHgBr\downarrow \tag{31}$$

where RHgBr is α-carbethoxybenzylmercuric bromide.

4. The three-alkyl mercury-for-mercury exchange

The technique of double-labelling has been used both by Reutov *et al.*[50, 51] and by Ingold and co-workers[52] to show that the exchange (3) is an independent reaction (32) ($\overset{\circ}{R}$ = optically active alkyl group, $\overset{*}{Hg}$ = radioactive mercury).

$$\overset{\circ}{R}-\overset{*}{Hg}R + RHgBr \rightleftharpoons R-\overset{*}{Hg}R + BrHg\overset{\circ}{R} \qquad (32)$$

Reaction (32) may be described as electrophilic substitution of the substrate R–HgR by the electrophilic reagent $R\overset{*}{Hg}Br$. The kinetics of (32) were followed both by measurement of the optical activity, and of the radioactivity, of the recovered alkylmercuric bromide and showed that the electrophilic substitution (32) (R = Bu^s or 1,5-dimethylhexyl) proceeded with complete retention of configuration, each elementary act of substitution involving retention of configuration[51, 52]. Rate coefficients for reaction (32) and also for the corresponding substitutions by alkylmercuric acetate and nitrate are collected in Table 10. The rate coefficients for reaction (32) (R = Bu^s) and reaction (32) (R = 1,5-dimethylhexyl) at 35 °C in solvent ethanol agree remarkably well, considering that both of the alkyl groups are secondary ones and hence that polar and steric effects should be of the same order of magnitude (as observed). Activation parameters for reaction (32) (R = 1,5-dimethylhexyl) are[51] E_a = 15.3 kcal.mole^{-1} and $\Delta S^{\ddagger}$ = −31.6 cal.deg^{-1}.mole^{-1}.

TABLE 10

SECOND-ORDER RATE COEFFICIENTS (l.mole^{-1}.sec^{-1}) FOR THE SUBSTITUTION OF DI-*sec*-BUTYLMERCURY BY *sec*-BUTYLMERCURIC SALTS IN SOLVENT ETHANOL

Bu^{s}HgX	Added salt	Temp. (°C)	10^5k_2	Ref.
Bu^{s}HgBr	None	35	4.7	52
Bu^{s}HgBr	0.1 M LiOAc	35	5.0	52
Bu^{s}HgBr	0.1 M LiNO$_3$	35	6.0	52
Bu^{s}HgBr	0.15 M LiNO$_3$	35	7.5	52
Bu^{s}HgBr	0.1 M LiBr	35	10.0	52
Bu^{s}HgBr	0.1 M LiClO$_4$	35	12.0	52
Bu^{s}HgOAc	None	35	27	52
Bu^{s}HgNO$_3$	None	0	3400	52
Bu^{s}HgNO$_3$	0.15 M LiNO$_3$	0	4900	52
RHgBr^a	None	65	51.9	51
RHgBr^a	None	60	37.8	51
RHgBr^a	None	55	25	51
RHgBr$^{a, b}$	None	35	5.0	51

^{a} R = 1,5-dimethylhexyl, equation (32).
^{b} Rate coefficient calculated by the author by a graphical extrapolation.

It can be seen from Table 10 that the order of reactivity of the electrophilic reagent is $Bu^sHgBr < Bu^sHgOAc < Bu^sHgNO_3$, and that the addition of various electrolytes accelerates both substitution by *sec.*-butylmercuric bromide and substitution by *sec.*-butylmercuric nitrate. All of this is evidence[52] for mechanism S_E2(open), and we may therefore write (XIX) as a general transition state for all the above substitutions, in solvent ethanol.

$$\left[\begin{array}{c} \overset{\delta(+)}{HgR} \\ R \diagdown \\ \overset{\delta(-)}{Hg}-X \\ R \end{array} \right]^{\ddagger}$$

(XIX)

In order that the reaction paths for the forward reaction (32) and the back reaction (32) should be similar (see Jensen and Rickborn[35]), it is necessary to postulate a rather more complicated mechanism than just the formation of the transition state (XIX). A possible mechanism that still retains the S_E2(open) transition state is shown in Fig. 3, together with the corresponding energy profile. In Fig. 3, (XIXf) is the transition state for the forward reaction, (XIXb) is the transition state for the back reaction, and (XIXi) is a high-energy cyclic intermediate. Just as in the uncatalysed one-alkyl exchange (Fig. 1, p. 44), this cyclic intermediate is required in order that a halogen atom may be transferred from one mercury atom to the other, whilst still keeping a symmetrical mechanism with respect to the forward and back reactions.

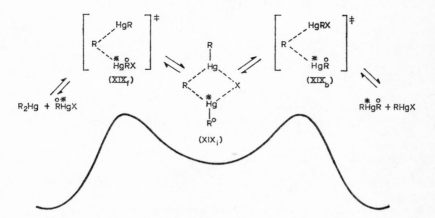

Fig. 3. A possible reaction path and energy profile for the three-alkyl mercury-for-mercury exchange reaction.

REFERENCES

1 H. B. CHARMAN, E. D. HUGHES AND C. K. INGOLD, *J. Chem. Soc.*, (1959) 2523.
2 C. K. INGOLD, *Helv. Chim. Acta*, 47 (1964) 1191; *Record Chem. Progr.*, 25 (1964) 145; F. G. THORPE, in *Studies on Chemical Structure and Reactivity*, J. H. RIDD (Ed.), Methuen, London, 1966, p. 247.
3 O. A. REUTOV, *Record Chem. Progr.*, 22 (1961) 1.
4 O. A. REUTOV, P. KNOLL' AND U. IAN-TSEI (Y.-C. WU), *Dokl. Akad. Nauk SSSR*, 120 (1958) 1052; *Proc. Acad. Sci. USSR*, 120 (1958) 477.
5 E. D. HUGHES, C. K. INGOLD, F. G. THORPE AND H. C. VOLGER, *J. Chem. Soc.*, (1961) 1133.
6 E. D. HUGHES, C. K. INGOLD AND R. M. G. ROBERTS, *J. Chem. Soc.*, (1964) 3900.
7 H. B. CHARMAN, E. D. HUGHES, C. K. INGOLD AND H. C. VOLGER, *J. Chem. Soc.*, (1961) 1142.
8 V. D. NEFEDOV AND E. N. SINTOVA, *Collected Works on Radiochemistry*, Leningrad University Press, 1955, pp. 110–113.
9 V. D. NEFEDOV, E. N. SINTOVA AND N. YA. FROLOV, *Zh. Fiz. Khim.*, 30 (1956) 2356.
10 V. D. NEFEDOV AND E. N. SINTOVA, *Zh. Neorg. Khim.*, 2 (1957) 1162.
11 E. N. SINTOVA, *Zh. Neorg. Khim.*, 2 (1957) 1205.
12 E. D. HUGHES AND H. C. VOLGER, *J. Chem. Soc.*, (1961) 2359.
13 O. A. REUTOV, T. A. SMOLINA AND V. A. KALYAVIN, *Dokl. Akad. Nauk SSSR*, 139 (1961) 389; *Proc. Acad. Sci. USSR*, 139 (1961) 697.
14 O. A. REUTOV, T. A. SMOLINA AND V. A. KALYAVIN, *Zh. Fiz. Khim.*, 36 (1962) 119; *Russ. J. Phys. Chem. Engl. Transl.*, 36 (1962) 59.
15 O. A. REUTOV, T. A. SMOLINA AND V. A. KALYAVIN, *Dokl. Akad. Nauk SSSR*, 155 (1964) 596; *Proc. Acad. Sci. USSR*, 155 (1964) 273.
16 O. A. REUTOV, *Acta Chim. Acad. Sci. Hung.*, 18 (1959) 439; O. A. REUTOV, I. P. BELETSKAYA and YANG-TS'E WU (Y.-C. WU), *Kinetica i Kataliz, Akad. Nauk SSSR, Sb. Statei*, (1960) 43.
17 O. A. REUTOV, V. I. SOKOLOV AND I. P. BELETSKAYA, *Dokl. Akad. Nauk SSSR*, 136 (1961) 631; *Proc. Acad. Sci. USSR*, 136 (1961) 115.
18 O. A. REUTOV, V. I. SOKOLOV AND I. P. BELETSKAYA, *Izv. Akad. Nauk SSSR, Otd. Khim. Nauk*, (1961) 1217, 1427; *Bull. Acad. Sci. USSR, Div. Chem. Sci.*, (1961) 1127, 1328.
19 O. A. REUTOV, V. I. SOKOLOV AND I. P. BELETSKAYA, *Izv. Akad. Nauk SSSR, Otd. Khim. Nauk*, (1961) 1213; *Bull. Acad. Sci. USSR, Div. Chem. Sci.*, (1961) 1123.
20 O. A. REUTOV, V. I. SOKOLOV AND I. P. BELETSKAYA, *Izv. Akad. Nauk SSSR, Otd. Khim. Nauk*, (1961) 1561; *Bull. Acad. Sci. USSR, Div. Chem. Sci.*, (1961) 1458.
21 O. A. REUTOV, V. I. SOKOLOV, I. P. BELETSKAYA AND YU. S. RYABOKOBYLKO, *Izv. Akad. Nauk SSSR, Otd. Khim. Nauk*, (1963) 965; *Bull. Acad. Sci. USSR, Div. Chem. Sci.*, (1963) 879.
22 G. F. WRIGHT, *Can. J. Chem.*, 30 (1952) 268.
23 S. WINSTEIN, T. G. TRAYLOR AND C. S. GARNER, *J. Am. Chem.Soc.*, 77 (1955) 3741.
24 A. N. NESMEYANOV, O. A. REUTOV, WU YANG-CH'IEH AND LU CHING-CHU, *Izv. Akad. Nauk SSSR, Otd. Khim. Nauk*, (1958) 1327; *Bull. Acad. Sci. USSR, Div. Chem. Sci.*, (1958) 1280.
25 A. N. NESMEYANOV, O. A. REUTOV AND S. S. PODDUBRAYA, *Dokl. Akad. Nauk SSSR*, 88 (1953) 479; *Izv. Akad. Nauk SSSR, Otd. Khim. Nauk* (1953) 850; *Bull. Acad. Sci. USSR, Div. Chem. Sci.*, (1953) 753.
26 H. B. CHARMAN, E. D. HUGHES AND C. K. INGOLD, *J. Chem. Soc.*, (1959) 2530.
27 F. R. JENSEN, *J. Am. Chem. Soc.*, 82 (1960) 2469.
28 O. A. REUTOV AND E. V. UGLOVA, *Izv. Akad. Nauk SSSR, Otd. Khim. Nauk*, (1959) 1691; *Bull. Acad. Sci. USSR, Div. Chem. Sci.*, (1959) 1628.
29 R. E. DESSY, Y. K. LEE AND J.-Y. KIM, *J. Am. Chem. Soc.*, 83 (1961) 1163.
30 N. A. NESMEYANOV AND O. A. REUTOV, *Dokl. Akad. Nauk SSSR*, 144 (1962) 126; *Proc. Acad. Sci. USSR*, 144 (1962) 405; *Tetrahedron*, 20 (1964) 2803.
31 R. E. DESSY, W. KITCHING, T. PSARRAS, R. SALINGER, A. CHEN AND T. CHIVERS, *J. Am. Chem. Soc.*, 88 (1966) 460.
32 K. BRODERSEN AND U. SCHLENKER, *Chem. Ber.*, 94 (1961) 3304.
33 R. E. DESSY AND Y. K. LEE, *J. Am. Chem. Soc.*, 82 (1960) 689.
34 M. D. RAUSCH AND J. R. VAN WAZER, *Inorg. Chem.*, 3 (1964) 761.

35 F. R. JENSEN AND B. RICKBORN, *Electrophilic Substitution of Organomercurials*, McGraw-Hill, New York, 1968.

36aY. MARCUS, *Acta Chem. Scand.*, 11 (1957) 599, 811.

36bR. S. SATCHELL, *J. Chem. Soc.*, (1963) 5963; (1964) 5464.

37 A. N. NESMEYANOV, O. A. REUTOV AND S. S. PODDUBNAYA, *Dokl. Akad. Nauk SSSR*, 88 (1953) 479; *Izv. Akad. Nauk SSSR, Otd. Khim. Nauk*, (1953) 850; *Bull. Acad. Sci. USSR, Div. Chem. Sci.*, (1953) 753.

38 O. A. REUTOV, I. P. BELETSKAYA AND R. E. MARDALEISHVILI, *Dokl. Akad. Nauk SSSR*, 116 (1957) 617; *Proc. Acad. Sci. USSR*, 116 (1957) 901; *Zh. Fiz. Khim.*, 33 (1959) 152, 1962; *Russ. J. Phys. Chem., Engl. Transl.*, 33 (1959) 4 (Abstract), 240.

39 O. A. REUTOV, *Angew. Chem.*, 72 (1960) 198.

40 F. R. JENSEN AND B. RICKBORN, *J. Am. Chem. Soc.*, 86 (1964) 3784.

41 O. A. REUTOV, *Dokl. Akad. Nauk SSSR*, 163 (1965) 909; *Proc. Acad. Sci. USSR*, 163 (1965) 744.

42 O. A. REUTOV, I. P. BELETSKAYA AND G. A. ARTAMKINA, *Zh. Obshch. Khim.*, 34 (1964) 2817; *Russ. J. Gen. Chem.*, 34 (1964) 2850.

43 F. R. JENSEN, B. RICKBORN AND J. J. MILLER, *J. Am. Chem. Soc.*, 88 (1966) 340.

44 O. A. REUTOV AND I. P. BELETSKAYA, *Dokl. Akad. Nauk SSSR*, 131 (1960) 853; *Proc. Acad. Sci. USSR*, 131 (1960) 333.

45 O. A. REUTOV, I. P. BELETSKAYA AND G. A. ARTAMKINA, *Zh. Fiz. Khim.*, 36 (1962) 2582; *Russ. J. Phys. Chem., Engl. Transl.*, 36 (1962) 1407; *Izv. Akad. Nauk SSSR, Ser. Khim.*, (1964) 1737; *Bull. Acad. Sci. USSR, Div. Chem. Sci.*, (1964) 1651.

46 O. A. REUTOV, *Dokl. Akad. Nauk SSSR*, 163 (1965) 909; *Proc. Acad. Sci. USSR*, 163 (1965) 744.

47 O. A. REUTOV, I. P. BELETSKAYA AND G. A. ARTAMKINA, *Dokl. Akad. Nauk SSSR*, 149 (1963) 90; *Proc. Acad. Sci. USSR*, 149 (1963) 181. See also *Izv. Akad. Nauk SSSR, Otd. Khim. Nauk*, (1963) 765; *Bull. Acad. Sci. USSR, Div. Chem. Sci.*, (1963) 691.

48 F. R. JENSEN AND J. MILLER, *J. Am. Chem. Soc.*, 86 (1964) 4735.

49 O. A. REUTOV, I. P. BELETSKAYA AND G. A. ARTAMKINA, *Dokl. Akad. Nauk SSSR*, 166 (1966) 1347; *Proc. Acad. Sci. USSR*, 166 (1966) 242.

50 O. A. REUTOV, T. P. KARPOV, É. V. UGLOVA AND V. A. MALYANOV, *Tetrahedron Letters*, (1960) 6; *Dokl. Akad. Nauk SSSR*, 134 (1960) 360; *Proc. Acad. Sci. USSR*, 134 (1960) 1017; *Izv. Akad. Nauk SSSR, Otd. Khim. Nauk*, (1960) 1311; *Bull. Acad. Sci. USSR, Div. Chem. Sci.*, (1960) 1223.

51 O. A. REUTOV, T. P. KARPOV, É. V. UGLOVA AND V. A. MALYANOV, *Izv. Akad. Nauk SSSR, Ser. Khim.*, (1964) 1580; *Bull. Acad. Sci. USSR, Div. Chem. Sci.*, (1964) 1492.

52 H. B. CHARMAN, E. D. HUGHES, C. K. INGOLD AND F. G. THORPE, *J. Chem. Soc.*, (1961) 1121.

53. D. S. MATTESON, *Organometal. Chem. Rev.* 4A (1969) 263.

54 M. H. ABRAHAM, D. DODD, M. D. JOHNSON, E. S. LEWIS AND R. A. MORE O'FERRALL, *J. Chem. Soc.*, B, (1971) 762.

Chapter 6

Other Metal-for-Metal Exchanges

1. Substitution of alkyl–metal compounds by metal salts

The substitution reaction (1) (R = alkyl, M and M′ = metal, X = halogen or electronegative group)

$$R-M + M'-X = R-M' + M-X \tag{1}$$

is of great practical importance in the preparation of organometallic compounds. Until recently, only few kinetic studies of the reaction had been reported, but since about 1964 numerous detailed investigations have been published.

A number of substitutions of allyl derivatives of Group IVA metals by mercuric salts are described in Chapter 10, Section 2.1 (p. 197).

1.1 SUBSTITUTION OF METHYLMAGNESIUM HALIDES BY TRIALKYLSILICON HALIDES

In a pioneering study, Reid and Wilkins[1] reported that reactions of the type

$$MeMgX + R_3SiX \overset{\text{ether}}{=\!=\!=} MeSiR_3 + MgX_2 \tag{2}$$

followed second-order kinetics. They did not record any rate coefficients, but gave the activation parameters assembled in Table 1. On the assumption* that the methylmagnesium halides were present as the monomeric species, MeMgX, Reid and Wilkins[1] suggested that the exchange (2) proceeded through a cyclic transition state (I).

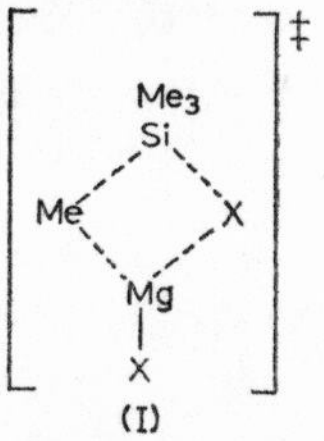

(I)

---

* The initial concentrations of MeMgX in the kinetic experiments ranged from 0.2 to 0.4 M. At such concentrations, the association factor, i, for $(MeMgI)_i$ is about 1.5–2.0 and the i factors for $(MeMgBr)_i$ and $(MeMgCl)_i$ are probably even greater[2].

TABLE 1

ACTIVATION PARAMETERS AT 273 °K FOR THE SUBSTITUTION OF METHYL-MAGNESIUM HALIDES BY TRIALKYLSILICON HALIDES IN SOLVENT ETHER, REACTION (2)[1]

$\Delta G^{\ddagger}$ in kcal.mole^{-1} (± 0.05); $\Delta H^{\ddagger}$ in kcal.mole^{-1} (± 0.6); $\Delta S^{\ddagger}$ in cal.deg^{-1}.mole^{-1} (± 2.5)

	MeMgCl			MeMgBr			MeMgI		
	$\Delta G^{\ddagger}$	$\Delta H^{\ddagger}$	$\Delta S^{\ddagger}$	$\Delta G^{\ddagger}$	$\Delta H^{\ddagger}$	$\Delta S^{\ddagger}$	$\Delta G^{\ddagger}$	$\Delta H^{\ddagger}$	$\Delta S^{\ddagger}$
Me$_3$SiF				20.86	8.8	−44	21.04	13.5	−27
Me$_3$SiCl	21.20	8.0	−48	20.47	8.1	−45	20.34	10.6	−34
Me$_3$SiBr				20.21	10.6	−35	19.32	11.3	−29
Me$_3$SiI							18.46	9.2	−34
ClCH$_2$SiMe$_2$Cl	19.40	8.3	−42	19.72	11.5	−30	20.52	9.8	−39
PhSiMe$_2$Cl							20.12	11.2	−33
p-TolylSiMe$_2$Cl							20.19	10.1	−37

The reactivity sequence, in the trimethylsilicon halides, X = F < Cl < Br < I suggests that the silicon-halogen group is involved in a bridged transition state, since the reverse order would be expected for a straightforward S_E2(open) attack by Me$_3$SiX on the methyl group of the Grignard reagent.

1.2 SUBSTITUTION OF DIALKYLZINCS BY PHENYLMERCURIC CHLORIDE

Abraham and Rolfe[3] have shown that the electrophile phenylmercuric chloride reacts rapidly with dialkylzincs in ether or tetrahydrofuran (THF) as solvent. The substitutions are second-order, first-order in each component, and this kinetic form together with the known stoichiometry[3] leads to

$$R_2Zn + PhHgCl \rightarrow RHgPh + RZnCl \tag{3}$$

as the simple reaction sequence. Second-order rate coefficients for reaction (3) (R = Et) were found to be 0.64 l.mole^{-1}.sec^{-1} at 0 °C in solvent THF and 6.0 l.mole^{-1}.sec^{-1} at 0 °C in solvent ether. Relative rate coefficients for reaction (3) in solvent ether at 35 °C were reported[3] to be: Me$_2$Zn(100), Et$_2$Zn(450), Pr$^n_2$Zn(1700), and Pr$^i_2$Zn(2200). This sequence was suggested[3] to arise as a result of mechanism S_E2(cyclic) in which coordination of the reagent to the zinc atom was more important than electrophilic attack at the alkyl group. Transition state (II) was postulated to obtain.

(II)

It was thought[3] that the similar rates of reaction (3) (R = Me, Et, Pr^n, Pr^i) in solvent THF at 25 °C might be due to an exact balance between steric and polar effects in mechanism S_E2(cyclic), or to the displacement of THF (coordinated to the zinc atom in the dialkylzincs) by phenylmercuric chloride being involved in the rate-determining step (3).

1.3 SUBSTITUTION OF ALKYLBORONIC ACIDS BY MERCURIC CHLORIDE

Exo-5-norbornene-2-boronic acid (*exo*-III) has been shown[4] to react with mercuric chloride much more rapidly than does the isomer *endo*-III; in each case the product was nortricyclylmercuric chloride (IV). Boron isotope effects,

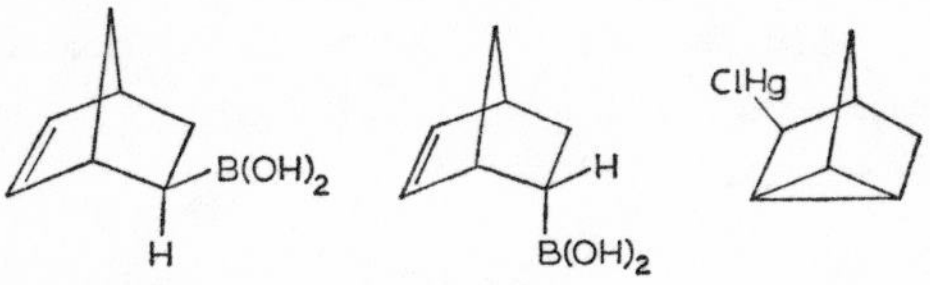

(*exo* –III)   (*endo* - III)   (IV)

k_{10_B}/k_{11_B}, of 1.033 and 1.027 were found[5] for the substitution of (*exo*-III) and (*endo*-III) respectively. Since the theoretical maximum for k_{10_B}/k_{11_B} is about 1.05–1.06, these results suggest[5] that the boron atom in (III) is involved in the rate-determining step. A kinetic study[6] showed that the substitution of (III) by mercuric chloride in solvent 75 % aqueous acetone, buffered with phthalate, and in the presence of an excess of chloride ion, obeyed the rate law

$$-d[(\text{III})]/dt = k_2[(\text{III})][\text{HgCl}_2][\text{phthalate}^-]/[\text{Cl}^-] \qquad (4)$$

In Table 2 are given the second-order rate coefficients reported by Matteson and

TABLE 2

SECOND-ORDER COEFFICIENTS (l.mole^{-1}.sec^{-1}) AND ACTIVATION PARAMETERS FOR THE SUBSTITUTION OF (*exo*-III) AND (*endo*-III) BY MERCURIC CHLORIDE

$\Delta G^{\ddagger}$ and $\Delta H^{\ddagger}$ in kcal.mole^{-1}, $\Delta S^{\ddagger}$ in cal. deg^{-1}.mole^{-1}, all at 298 °K.

	$k_2(25\,°C)$	$k_2(45.2\,°C)$	$\Delta G^{\ddagger}$	$\Delta H^{\ddagger}$	$\Delta S^{\ddagger}$
(*Exo*-III)	2.08×10^{-3}	8.1×10^{-3}	21.1	12.1 ± 0.7	-30 ± 2
(*Endo*-III)	$4.9 \ \times 10^{-6}$	3.0×10^{-5}	24.7	16.3 ± 0.7	-28 ± 2

Talbot[6], together with the activation parameters calculated from these coefficients by the author.

For the *exo*-isomer, the reaction mechanism was written as

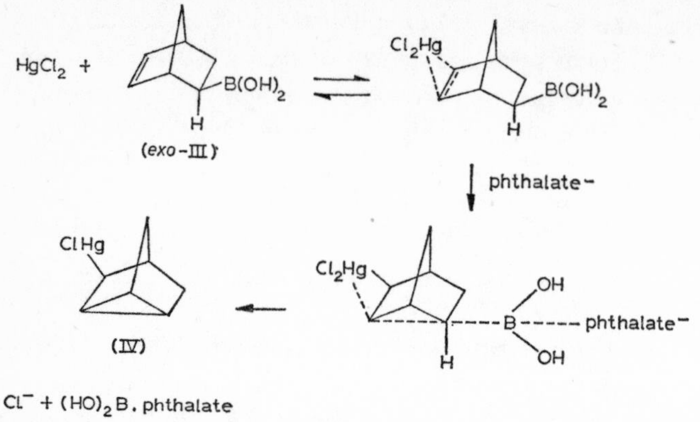

$$(5)$$

An alternative mechanism, involving attack by mercuric chloride at the $C-B(OH)_2$ site to yield an intermediate (V) which rearranges rapidly to the final product (IV) has been shown[7] to be unlikely since compound (VI), which is analogous to (V), does not rearrange to a tricyclic compound[7].

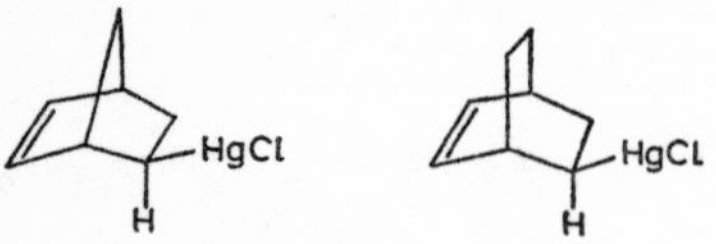

Matteson and Talbot[7] pointed out that mechanism (5) involves inversion of configuration at the carbon atom undergoing electrophilic substitution. The normal stereochemical course of the metal-for-metal exchange (1) is retention of configuration, and it is significant that the substitution of the butyl ester of 1-phenylethylboronic acid by mercuric chloride[8] and the substitution of norbornylmagnesium chloride by mercuric bromide[9] both proceed with retention of configuration; in these two metal-for-metal exchanges, no molecular rearrangement takes place.

In further studies, Matteson et al.[10, 11] reported relative rates of substitution of various boronic esters by mercuric chloride; details are given in Table 3. The solvent used was a mixture of ethanol (88%), water (8%), and glycerol (4%), buffered with sodium acetate and acetic acid, and reactions were run in the presence of added sodium chloride. Under these conditions, the kinetics of reaction of benzylboronic ester with mercuric chloride were found to obey the rate law

$$-d[ester]/dt = k[ester][HgCl_2][HO^-]$$

although certain discrepancies were noticed[11]. In the above rate expression, the term $[HO^-]$ is, of course, proportional to the term $[H_2O][AcO^-]/[HOAc]$.

Matteson and Kramer[11] suggested that a glycerol ester of the boronic acid* is converted into a hydroxide complex, as in equation (6), and this complex is then attacked by mercuric chloride in the rate-determining step (7).

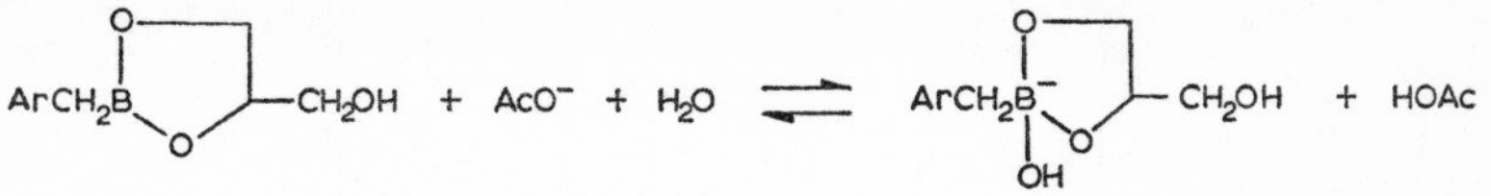

$$ArCH_2B \quad + \quad AcO^- + H_2O \quad \rightleftharpoons \quad ArCH_2B \quad \quad CH_2OH \quad + \quad HOAc \qquad (6)$$

$$ArCH_2B^- \quad CH_2OH \quad + \quad HgCl_2 \quad \longrightarrow \quad ArCH_2HgCl \quad + \quad HO-B \quad CH_2OH \quad + \quad Cl^- \qquad (7)$$

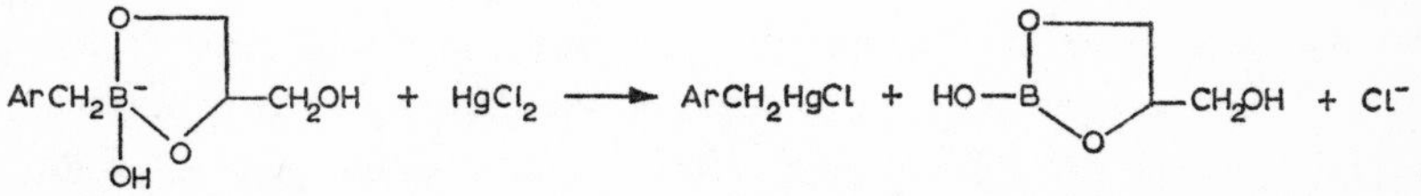

Relative rate-coefficients for the substitution of a number of benzylboronic esters are given in Table 3(b); a Hammett plot yields a value of $+0.93 \pm 0.08$ for ρ. If

TABLE 3

RELATIVE RATE COEFFICIENTS FOR THE SUBSTITUTION OF ESTERS OF BORONIC ACIDS, $RB(OH)_2$, BY MERCURIC CHLORIDE[10, 11]

R in $RB(OH)_2$	Relative rate	Boronic ester pK_a
$PhCH_2-$	1^a	
$PhCHMe-$	0.50	
$p\text{-}ClC_6H_4CHMe-$	0.04	
$p\text{-}CF_3C_6H_4CHMe-$	0.0015	
Bu^s-	0	
$p\text{-}MeC_6H_4CH_2-$	0.74^b	9.54
$PhCH_2-$	1	9.24
$p\text{-}ClC_6H_4CH_2-$	2.11	8.97
$m\text{-}CF_3C_6H_4CH_2-$	2.49	8.64

^{a} At 40 °C; ^{b} At 30 °C.

it is assumed[11] that the actual reactive species is the boronic ester/hydroxide ion complex, then the pK_a values of the esters (determined under the same conditions as those used for the kinetic experiments) may be used to calculate a value of ρ for the rate-determining step (7). The pK_a values themselves yield $\rho = +1.47 \pm 0.05$ and hence for the substitution (7), $\rho = -0.55 \pm 0.17$. Thus electron-donating substituents accelerate the displacement (7), and electron-withdrawing substituents retard the displacement. The following transition states were proposed for the electrophilic substitution (7)

---

* The boronic acids appear to have been introduced into the solvent medium as their di-*n*-butyl esters; presumably these butyl esters are rapidly converted into the glycerol esters.

Matteson and Kramer[11] refer to the mechanism as a "concerted front-side displacement", and make the interesting suggestion that some direct bonding interaction occurs between the boron atom and the mercury atom in the transition state. Their justification for this suggestion is that it seems sterically difficult to bring the two metal atoms close to the front-side of the carbon atom undergoing substitution without forcing some overlap of the orbitals on the boron and mercury atoms[11].

It is of some interest, therefore, to calculate the possible boron-to-mercury distance in the above transition states for reaction (7). Gielen and Nasielski[12] have shown that in substitutions at saturated carbon, the angle subtended by the entering and leaving groups at the carbon atom undergoing substitution has an optimum value of 77°. Thus in the triangle defined by the three atoms C, B, and Hg, we may take the angle B–C–Hg as 77°, the C–––B distance as 1.80 A (a little longer than the normal C–B bond length of 1.6 A), and the C–––Hg distance as 2.30 A (again, a little longer than the normal C–Hg bond length of 2.1 A). The boron-to-mercury distance may then be calculated to be 2.58 Å, that is about 0.3 A greater than the boron-to-mercury Van der Waals distance of 2.25 A. On this basis there can be but little direct boron-to-mercury interaction. Even if the B–C–Hg angle were reduced to 65°, the boron-to-mercury distance in the transition state would still be only 2.25 A; at this (Van der Waals) distance there would be an attraction between the boron and mercury atoms of about a few hundred calories per mole. By adjusting the three parameters (two lengths and one angle) for the CBHg triangle, it is possible to arrive at a whole range of values for the boron-to-mercury distance, but it is evident that a reasonable choice of values for the three parameters results in a boron-to-mercury distance that allows the boron and mercury atoms to be brought to the front side of the carbon atom without forcing some metal–metal interaction (other than weak Van der Waals attraction). There seems little justification on steric grounds for the suggestion of Matteson and Kramer, although such calculations as those presented above can neither prove nor disprove such a suggestion.

Second-order rate coefficients for the reaction of some boronic esters with mercuric chloride in methanol were reported by Matteson and Allies[13]; details are in Table 4. The solution was buffered with sodium acetate/acetic acid, and the observed rate equation was found to be

$$-d[HgCl_2]/dt = k[ester][HgCl_2][NaOAc]/[HOAc]$$

TABLE 4

SECOND-ORDER RATE COEFFICIENTS (l.mole^{-1}.sec^{-1}) AND ACTIVATION PARAMETERS
FOR THE SUBSTITUTION OF BORONIC ESTERS BY MERCURIC CHLORIDE IN SOLVENT
METHANOL[13]

	Temperature (°C)				$\Delta H^{\ddagger}$ (kcal. mole^{-1})	$\Delta S^{\ddagger}$ (cal. deg^{-1}. mole^{-1})
Ester	0.03	20	30	40		
CH$_3$[B(OMe)$_2$]		2.85×10^{-4}	6.7×10^{-4}	1.64×10^{-3}	15.4	-22
CH$_2$[B(OMe)$_2$]$_2$	0.032	0.088	0.137	0.278	8.8 ± 1.5	-33 ± 5
CH[B(OMe)$_2$]$_3$			$\simeq 0.27$			
C[B(OMe)$_2$]$_4$			$\simeq 0.16$			
ClHgCH$_2$B(OMe)$_2$			2×10^{-3}			
PhCH$_2$B(OMe)$_2$			2.4×10^{-4}		$\simeq 16.2$	$\simeq -21$

where the term [NaOAc]/[HOAc] is proportional to [OMe$^-$]. The high reactivity
of the ester CH$_2$[B(OMe)$_2$]$_2$ was attributed to a neighbouring group effect in the
transition state

Evidence in favour of the cyclic transition state was the observation that reaction
of CH$_2$[B(OMe)$_2$]$_2$ with HgCl$_2$ in methanol was not accelerated on addition of
10 % water or of 0.04 M NaNO$_3$.

1.4 SUBSTITUTION OF ALKYLTHALLIUM COMPOUNDS BY ALKYLMERCURIC COMPOUNDS

A number of electrophilic substitutions involving transfer of an alkyl group
from mercury to mercury, thallium to mercury, mercury to thallium, and thallium
to thallium have been reported by Hart and Ingold[14] to proceed through mechanism S$_E$1 in solvent dimethylformamide. It has since been shown by Jensen
and Heyman[15] that this report is not correct.

1.5 SUBSTITUTION OF TETRAALKYLTINS BY MERCURIC SALTS

Abraham and Spalding[16] have shown that tetraethyltin reacts with mercuric iodide in solvent 96 % methanol –4 % water by the rate-determining bimolecular reaction (8), followed by the rapid, reversible reaction (9), viz.

$$Et_4Sn + HgI_2 \rightarrow EtHgI + Et_3SnI \tag{8}$$

$$Et_3SnI + HgI_2 \rightleftharpoons Et_3Sn^+ + HgI_3^- \tag{9}$$

The kinetic form for the set of reactions (8) and (9) is complex, and is described by

$$dX/dt = k_2(A-X)(B-X-Y) \tag{10}$$

where the quantity Y is given by

$$Y = \frac{KB - [K^2B^2 - 4KX(K-1)(B-X)]^{\frac{1}{2}}}{2(K-1)} \tag{11}$$

In equations (10) and (11), the initial concentrations of tetraethyltin and mercuric iodide are denoted by A and B respectively, the concentration of ethylmercuric iodide at time t is denoted by X, and the concentration of Et_3Sn^+ (and also of HgI_3^-) at any time t is denoted by Y. The equilibrium constant for reaction (9) is K, and k_2 is the second-order rate coefficient for the electrophilic substitution (8). Equations (10) and (11) can be solved by the method of numerical analysis[17] and values of k_2 were thus obtained (values of K were determined[16] by direct experiments). It was shown that k_2 remained constant over a ten-fold range of initial concentration of tetraethyltin and of mercuric iodide. Values of the second-order rate coefficient and of the associated activation parameters are given in Table 5. Reaction (8) is thus characterised by a very negative activation entropy

TABLE 5

SECOND-ORDER RATE COEFFICIENTS (l.mole^{-1}.sec^{-1}) AND ACTIVATION PARAMETERS FOR THE SUBSTITUTION OF TETRAALKYLTINS BY MERCURIC IODIDE IN SOLVENT 96 % METHANOL–4 % WATER AT 298 °K[16,18]

$\Delta G^{\ddagger} \pm 20$, $\Delta H \pm 400$, and $T\Delta S \pm 400$ all in cal.mole^{-1} at 298 °K; $\Delta S^{\ddagger} \pm 1$ cal.deg^{-1}.mole^{-1}.

R_4Sn	$10^3 k_2$	$\Delta G^{\ddagger}$	$\Delta H^{\ddagger}$	$-T\Delta S^{\ddagger}$	$\Delta S^{\ddagger}$
Me$_4$Sn	710	17,700	9,200	8,500	−28.4
Et$_4$Sn	4.75	20,600	11,650	8,950	−30.1
Pr$^n_4$Sn	0.720	21,750	12,350	9,400	−31.5
Pr$^i_4$Sn^a	$<4 \times 10^{-5}$	$>28,000$	$>19,000$	9,000	−30
Bu$^n_4$Sn	0.758	21,700	12,500	9,200	−30.9
Bu$^i_4$Sn	5.03×10^{-2}	23,300	14,350	8,950	−30.1
MeSnBu$^n_3$	213	18,400	11,400	7,000	−23.4

^{a} The values for Pr$^i_4$Sn have been calculated from the observed rate coefficient, $10^3 k_2 < 2 \times 10^{-4}$ l.mole^{-1}.sec^{-1} at 313 °K, on the assumption that $\Delta S^{\ddagger} = -30$ cal.deg^{-1}.mole^{-1}.

$(-30 \text{ cal.deg}^{-1}.\text{mole}^{-1})$. Abraham and Spalding[16] considered that such a value was compatible with either mechanism S_E2(open) or S_E2(cyclic).

In subsequent work, the substitution of a number of tetraalkyltins by mercuric iodide was investigated[18]. All of the substitutions proceeded by the rate-determining bimolecular reaction

$$R_4Sn + HgI_2 \rightarrow RHgI + R_3SnI \tag{12}$$

followed again by equilibria analogous to (9). Values of the second-order coefficients for reaction (12), together with the activation parameters, are collected in Table 5. The relative rate coefficients for reaction (12), at 25 °C, in solvent 96 % methanol –4 % water, yield the very pronounced "steric" sequence

$$Me_4Sn(100), Et_4Sn(0.67), Pr''_4Sn(0.10),$$

$$Bu''_4Sn(0.11), Bu^i_4Sn(7 \times 10^{-3}), Pr^i_4Sn(<6 \times 10^{-6}) \tag{13}$$

The substitution of the unsymmetrical tetraalkyl, $MeSnBu''_3$, by mercuric iodide was shown to involve only the cleavage of the methyl–tin bond, and hence the overall rate coefficient for the substitution is also that for the cleavage of the methyl–tin bond. Rate coefficients and activation parameters (given in Table 5) are compared with those for the symmetrical tetraalkyls, in Table 6.

Abraham and Spalding[18] suggested that the sequences of rate coefficients given in Tables 5 and 6 (see also equation (13)) arise from various non-bonded interactions in transition states such as (VII)*:

TABLE 6

RELATIVE SECOND-ORDER RATE COEFFICIENTS AND ACTIVATION PARAMETERS FOR THE CLEAVAGE OF ALKYL–TIN BONDS BY MERCURIC IODIDE IN SOLVENT 96 % METHANOL–4 % WATER AT 298 °K[18]

$\Delta G^\ddagger$, $\Delta H^\ddagger$, and $-T\Delta S^\ddagger$ in cal.mole^{-1}; $\Delta S^\ddagger$ in cal.deg^{-1}.mole^{-1}.

	k_2(rel.)	$\delta\Delta G^\ddagger$	$\delta\Delta H^\ddagger$	$\delta(-T\Delta S^\ddagger)$	$\delta\Delta S^\ddagger$
Me–SnMe$_3$[a]	100	0	0	0	0
Me–SnBu''$_3$[a]	120	−100	2200	−2300	7.7
Bu''–SnBu''$_3$[a]	0.11	4000	3300	700	−4.5
Me–SnMe$_3$[b]	100	0	0	0	0
Me–SnBu''$_3$[b]	30	700	2200	−1500	5.0
Bu''–SnBu''$_3$[b]	0.11	4000	3300	700	−2.5

[a] A statistical factor of ¼ has been applied to the observed rate coefficients for the substitution of Me_4Sn and Bu''_4Sn.
[b] No statistical factor has been applied.

---

* Transition state (VII) corresponds to mechanism S_E2(open) for reaction (12). Evidence from salt effects strongly suggests[19] that reaction (12) proceeds *via* this mechanism, rather than by mechanism S_E2(cyclic), in solvent 96 % methanol −4 % water.

$$\left[\begin{array}{c} {\overset{\delta+}{\underset{R}{\diagup}} \overset{SnR_3}{\diagdown} } \\ {\diagdown \overset{\delta-}{HgI_2}} \end{array} \right]^{\ddagger}$$

(VII)

They considered the following possible interactions.

(a) Interactions between the alkyl group undergoing substitution and the incoming electrophile, HgI_2.

(b) Interactions between the alkyl group undergoing substitution and the leaving set of atoms SnC_3^{α} (this set of atoms remains constant no matter what alkyl groups make up the total leaving group, SnR_3).

(c) Interactions between the alkyl group undergoing substitution and the β or γ carbon and hydrogen atoms in the leaving group.

(d) Interactions between the incoming electrophile, HgI_2, and the β or γ carbon and hydrogen atoms in the leaving group (the interaction between HgI_2 and the set of atoms SnC_3^{α} remains constant through all of the substitutions and can thus be ignored).

In addition to interactions (a)–(d), initial-state contributions and interactions involving solvent molecules could also play a (minor) part.

It is clear from the data given in Table 6 that when the leaving group is increased in size from $SnMe_3$ to $SnBu^n{}_3$, the rate of cleavage of a methyl group from tin is not decreased. On the other hand, the relative rates of cleavage Me–$SnBu^n{}_3$ (1000) to Bu^n–$SnBu^n{}_3$(1) demonstrate that interactions involving the alkyl group undergoing substitution must be considerable. Hence interactions (c) and (d) must be very small, and the major non-bonded interactions are (a) and (b).

It is interesting to note (Table 6) that the relative rate coefficients for the cleavages Me–$SnMe_3$(100) and Me–$SnBu^n{}_3$(120) leads to the conclusion that the $SnBu^n{}_3$ leaving group slightly accelerates the Me–Sn cleavage compared with the effect of the $SnMe_3$ leaving group. Perhaps this could be due to a small inductive (+I) effect, in the sense Bu^n > Me, helping to stabilise the potential cation $^+SnR_3$ as it separates as the leaving group. Abraham and Spalding[18] pointed out, however, that the above rate coefficients are calculated on the basis that a full statistical factor of $\frac{1}{4}$ can be applied to the rate coefficient for substitution of tetramethyltin in order to obtain the rate coefficient for cleavage of a single Me–Sn bond. If the four sites of attack in the compound Me_4Sn are close enough to be considered not independent, then the full statistical factor may not be applicable, and the sequence Me–$SnMe_3$(100) to Me–$SnBu^n{}_3$(120) may to some extent be artificial.

Finally, it may be noted that the relatively high reactivity of $MeSnBu^n{}_3$ is due largely to a more favourable entropy of activation (Table 6). The activation enthalpies given in Tables 5 and 6 would, by themselves, lead to a sequence of rates at 25 °C of $Me_4Sn(100)$, $MeSnBu^n_3(2.9)$, and $Bu^n_4Sn(0.37)$ rather than to the sequences given in Table 6. The reason for this more favourable activation entropy is not clear, although Abraham and Spalding[18] suggested that the unsymmetrical (and perhaps very slightly polar) tetraalkyltin might enhance the solvent structure in the vicinity of the tetraalkyl. This would have the effect of creating solvent–solvent hydrogen bonds and thus decreasing both the enthalpy and entropy of the initial state. The activation enthalpy and activation entropy are hence both increased (*i.e.* the latter is made less negative).

A study[19] of the effect of added lithium perchlorate on the second-order rate coefficients for reaction (12) ($R = Et$, Pr^n, Bu^n) showed that all three substitutions, in solvent 96 % methanol–4 % water, were subject to marked positive kinetic salt effects. The effects were analysed in terms of the Brönsted–Bjerrum equation

$$k/k_0 = \gamma^{R_4Sn} \cdot \gamma^{HgI_2}/\gamma^{\ddagger} \qquad (14)$$

where k and k_0 are the rate coefficients in the presence, and in the absence, of lithium perchlorate respectively, γ^{R_4Sn} and γ^{HgI_2} are the molar activity coefficients of the tetraalkyltin and mercuric iodide, and $\gamma^{\ddagger}$ is the molar activity coefficient of the transition state. In the absence of lithium perchlorate all of these activity coefficients are set equal to unity.

Abraham and Spalding[19] separately determined values of γ^{Et_4Sn}, γ^{Pr_4Sn}, γ^{Bu_4Sn}, and γ^{HgI_2} at various concentrations of added lithium perchlorate, and combined these values with those of k/k_0 to yield values of $\gamma^{\ddagger}$ for the three transition states. The magnitude of the observed kinetic salt effects may be seen from Fig. 1 in

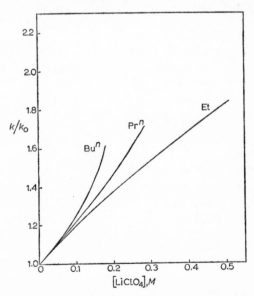

Fig. 1. Kinetic salt effects in the bimolecular substitution of tetraalkyltins (R_4Sn) by mercuric iodide in solvent 96 % methanol–4 % water. Et ≡ Et_4Sn, Pr^n ≡ Pr_4^nSn, Bu^n ≡ Bu_4^nSn.

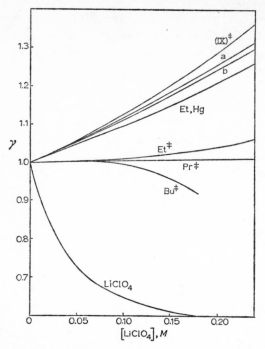

Fig. 2. The effect of lithium perchlorate on the molar activity coefficients of reactants and transition states in the bimolecular substitution of tetraalkyltins by mercuric iodide in solvent 96% methanol–4% water. $LiClO_4 \equiv LiClO_4$ (estimated); $Bu^\ddagger \equiv Bu^n{}_4Sn/HgI_2$ transition state; $Pr^\ddagger \equiv Pr^n{}_4Sn/HgI_2$ transition state; $Et^\ddagger \equiv Et_4Sn/HgI_2$ transition state; $Et \equiv Et_4Sn$; $Hg \equiv HgI_2$; $b \equiv Pr^n{}_4Sn$; $a \equiv Bu^n{}_4Sn$; $(IX)^\ddagger \equiv$ transition state (IX), see text.

which values of k/k_0 have been plotted against the concentration of lithium perchlorate. Fig. 2 shows the variation of the various activity coefficients with the concentration of added lithium perchlorate.

It is clear that the positive kinetic salt effect is due to the reactants being destabilised by the added lithium perchlorate (γ^{R_4Sn} and γ^{HgI_2} becoming greater than unity) and is not due to the stabilisation of the transition state, except to a very small extent in the $Bu^n{}_4Sn/HgI_2$ transition state. The variation of $\gamma^\ddagger$ with $[LiClO_4]$ is intermediate between that observed for non-polar non-electrolytes, such as the tetraalkyltins, and that observed for very polar solutes such as 1 : 1 electrolytes. It was suggested[19] that transition states (VIII) (R = Et, Pr^n, Bu^n) are reasonable models, but that (IX) (R = Et, Pr^n, Bu^n) is not a reasonable model.

TABLE 7

SECOND-ORDER RATE COEFFICIENTS $(l.mole^{-1}.sec^{-1})$ FOR THE SUBSTITUTION OF TETRAETHYLTIN BY MERCURIC CHLORIDE[a] IN SOLVENT 96 % METHANOL–4 % WATER AT 313 °K[20]

Added salt (M)		$10^2 k_2$
None		1.92 ± 0.02
LiClO$_4$,	0.020	2.00
LiClO$_4$,	0.038	2.07
LiClO$_4$,	0.058	2.20
LiClO$_4$,	0.077	2.33
LiClO$_4$,	0.175	2.71
LiClO$_4$,	0.222	2.85
LiClO$_4$,	0.305	3.23
Bu$^n_4$NClO$_4$,	0.147	2.73
Bu$^n_4$NClO$_4$,	0.196	2.93
Bu$^n_4$NClO$_4$,	0.246	3.10
NaCl,	0.005	1.93
NaCl,	0.010	1.93
NaCl,	0.020	2.03
NaCl,	0.050	2.00

[a] Initial concentrations were: $[Et_4Sn] = 0.0125 \ M$, $[HgCl_2] = 0.0100 \ M$.

Support for this conclusion was provided by Abraham and Spalding[19], who estimated the expected activity coefficients for a transition state such as (IX). These coefficients are also plotted in Fig. 2, and clearly do not agree at all with the observed transition state activity coefficients*. It was thus concluded that the substitution of tetraalkyltins by mercuric iodide in solvent 96 % methanol–4 % water proceeds by a rate-determining step, reaction (12), which follows mechanism $S_E2(open)$ through a transition state of type (VIII).

The related metal-for-metal exchange (15) (R = Et)

$$R_4Sn + HgCl_2 = RHgCl + R_3SnCl \tag{15}$$

was later studied by Abraham and Johnston[20]. They showed, using the solvent 96 % methanol–4 % water, that the substitution (15) (R = Et) followed straightforward second-order kinetics, first-order in tetraethyltin and first-order in mercuric chloride. The second-order rate coefficient was invarient over a four-fold range of initial concentration of tetraethyltin and a ten-fold range of initial concentration of mercuric chloride. Salt effects were also studied, and some of the results obtained are given in Table 7. Sodium chloride accelerates reaction (15) (R = Et) about as

---

* The activity coefficients shown in Fig. 2 are for (IX; R = Pr^n). Values for (IX; R = Et, Bu^n) do not differ significantly from those shown. For example, at 0.2 M LiClO$_4$, activity coefficients were calculated[19] to be 1.26 (IX; R = Et), 1.29 (IX; R = Pr^n), and 1.32 (IX; R = Bu^n).

TABLE 8

SECOND-ORDER RATE COEFFICIENTS $(l.mole^{-1}.sec^{-1})$ AND ACTIVATION PARAMETERS[a] FOR THE SUBSTITUTION OF TETRAAETHYLTIN BY MERCURIC CHLORIDE IN METHANOL–WATER MIXTURES[20]

MeOH (%)	Mole fraction MeOH in solvent	10^2k_2 (25 °C)	10^2k_2 (40 °C)	$\Delta G^{\ddagger}$	$\Delta H^{\ddagger}$	$\Delta S^{\ddagger}$
100	0.999	0.333	1.115	20,832	14,350	−21.8
98	0.956	0.455	1.477	20,648	13,950	−22.4
96	0.914	0.630	1.915	20,455	13,150	−24.5
94	0.874	0.838	2.458	20,286	12,700	−25.4
90	0.800	1.388	3.955	19,987	12,350	−25.6
85	0.716	2.442	7.022	19,653	12,450	−24.1
80	0.640	3.950	11.20	19,368	12,300	−23.7
70	0.510	9.317	26.30	18,859	12,250	−22.2

[a] $\Delta G^{\ddagger}(\pm 6)$ and $\Delta H^{\ddagger}(\pm 170)$ in cal.mole^{-1}, $\Delta S^{\ddagger}(\pm 0.6)$ in cal.deg^{-1}.mole^{-1}, all at 298 °K.

much as expected for a normal salt effect. Although little can be said about the possibility of reaction *via* $HgCl_3{}^-$, it is clear that the species $HgCl^+$ cannot contribute significantly to the overall rate*. Both lithium perchlorate and tetra-*n*-butylammonium perchlorate considerably accelerate the rate of reaction (15) (R = Et). It was calculated (using an assumption as to values of γ^{HgCl_2}) that values of $\gamma^{\ddagger}$, the activity coefficient of the transition state in reaction (15) (R = Et), varied with [LiClO$_4$] in much the same way as did those for the activity coefficient of the transition state in reaction (12) (R = Et); see Fig. 2, p. 74. On the basis of these salt effects, it was concluded[20] that reaction (15) (R = Et) proceeds by mechanism S_E2(open) and not by mechanism S_E2(cyclic).

Abraham and Johnston[20] also showed that reaction (15) (R = Et) was considerably accelerated as the solvent was changed from methanol to aqueous methanol; rate coefficients and activation parameters are given in Table 8. The relative standard deviation of the rate coefficients was given as not greater than 1 %. From the variation of $\Delta G^{\ddagger}_{298}$ with the dielectric constant of the solvent, it was deduced that in the transition state for reaction (15) (R = Et) a charge separation of as much as 0.7 units had occurred, *viz.*

$$\left[\begin{array}{c} \overset{+0.7}{SnEt_3} \\ Et \\ \underset{-0.7}{HgCl_2} \end{array} \right]^{\ddagger}$$

* The stepwise formation constant for formation of $HgCl_3{}^-$ is not known in solvent 96 % methanol −4 % water, but is likely to be quite small, perhaps of the order of 10 l.mole^{-1}. At the kinetic concentration used ([HgCl$_2$] = 0.01 *M*), such a formation constant would result in not more than 10 % of the original mercuric chloride present being converted into $HgCl_3{}^-$.

TABLE 9

SECOND-ORDER RATE COEFFICIENTS $(l.mole^{-1}.sec^{-1})$ AND ACTIVATION PARAMETERS[a] FOR THE SUBSTITUTION OF TETRAALKYLTINS BY MERCURIC CHLORIDE IN METHANOL–WATER MIXTURES AT 298 °K[21]

| R_4Sn | Mole fraction MeOH in solvent | | | |
| | 0.999 | | | |
	10^2k_2	$\Delta G^{\ddagger}$	$\Delta H^{\ddagger}$	$\Delta S^{\ddagger}$
Me$_4$Sn	155	17,193	10,600[b]	−22.1[b]
Et$_4$Sn	0.333	20,832	14,350	−21.7
Pr$^n_4$Sn	0.0628	21,821	15,100	−22.6
Pr$^i_4$Sn	$<10^{-8}$	$>28,000$	$>21,000$[d]	−25[d]
Bu$^n_4$Sn	0.0615	21,834	14,700	−23.9
Bu$^i_4$Sn	0.00800	23,042	16,150	−23.1

| R_4Sn | 0.914 | | | |
	10^2k_2	$\Delta G^{\ddagger}$	$\Delta H^{\ddagger}$	$\Delta S^{\ddagger}$
Me$_4$Sn	259	16,890	9,850	−23.7
Et$_4$Sn	0.630	20,455	13,150	−24.5
Pr$^n_4$Sn	0.113	21,472	13,850	−25.6
Pr$^i_4$Sn	$<10^{-8}$	$>28,000$	$>21,000$[d]	−25[d]
Bu$^n_4$Sn	0.104	21,521	14,150	−24.8
Bu$^i_4$Sn	0.0138	22,718	16,200	−21.9

| R_4Sn | 0.716 | | | |
	10^2k_2	$\Delta G^{\ddagger}$	ΔH	$\Delta S^{\ddagger}$
Me$_4$Sn	730	16,273	9,300[c]	−23.4[c]
Et$_4$Sn	2.44	19,653	12,450	−24.1
Pr$^n_4$Sn	0.392	20,737	12,750	−26.7
Pr$^i_4$Sn	$<10^{-8}$	$>28,000$	$>21,000$[d]	−25[d]
Bu$^n_4$Sn	0.323	20,850	13,480	−24.7
Bu$^i_4$Sn	0.0393	22,099	15,350	−22.6

[a] $\Delta G^{\ddagger}(\pm 6)$ and $\Delta H^{\ddagger}(\pm 200)$ in $cal.mole^{-1}$, $\Delta S^{\ddagger}(\pm 0.7)$ in $cal.deg^{-1}.mole^{-1}$.
[b] $\Delta H^{\ddagger}\pm 450$, $\Delta S^{\ddagger}\pm 1.5$.
[c] $\Delta H^{\ddagger}\pm 800$, $\Delta S^{\ddagger}\pm 3$.
[d] A value of -25 $cal.deg^{-1}.mole^{-1}$ for $\Delta S^{\ddagger}$ has been assumed.

Both salt effects and solvent effects thus lead to the conclusion[20] that reaction (15) (R = Et) proceeds by mechanism S_E2(open) in solvent methanol–water.

Rate coefficients and activation parameters have also been determined[21] for the substitution of a number of other tetraalkyltins by mercuric chloride in various methanol–water mixtures; details are given in Table 9. Most of the given rate coefficients are accurate to about $\pm 1 \%$, and the standard deviations in the activation parameters are hence reasonably low. The variation in the values of $\Delta G^{\ddagger}$ with solvent dielectric constant is about the same for all the tetraalkyltins studied. Abraham and Johnston[21] suggested that the S_E2(open) transition state

(X)

was in force for all the substitutions studied.

Relative rates of substitution of tetraalkyltins by mercuric halides are in Table 10. The similarity of the four sets of relative rates led Abraham and Johnston[21] to conclude that all four series of substitutions proceed by the same mechanism, S_E2(open). The pronounced "steric" sequences of relative rates (Table 10) may be considered in terms of the analysis of Abraham and Spalding (see p. 71); there is a very slight extra steric effect when mercuric chloride is the electrophile, but there is no obvious reason why this should be so.

The variation of $\Delta H^{\ddagger}$ for reaction (15) (R = Et) with methanol–water composition has been studied by Abraham et al.[22]. They dissected the solvent effect on $\Delta H^{\ddagger}$ into initial-state and transition-state contributions through the equation

$$\Delta H_t^0(\mathrm{Tr}) = \Delta H_t^0(\mathrm{Et_4Sn}) + \Delta H_t^0(\mathrm{HgCl_2}) + \Delta H_2^{\ddagger} - \Delta H_1^{\ddagger} \qquad (16)$$

where $\Delta H_t^0(\mathrm{Y})$ is the standard enthalpy of transfer of species Y from solvent 1 (methanol) to solvent 2 (any other solvent). The tetraethyltin–mercuric chloride transition state is denoted as Tr. Values of $\Delta H_t^0(\mathrm{Et_4Sn})$ and $\Delta H_t^0(\mathrm{HgCl_2})$ were found through calorimetric determinations of the heats of solution of tetraethyltin and mercuric chloride, and combined with the known activation enthalpies to yield $\Delta H_t^0(\mathrm{Tr})$. Details are in Table 11; it can be seen that the large reduction

TABLE 10

RELATIVE SECOND-ORDER RATE COEFFICIENTS FOR THE SUBSTITUTION OF TETRAAL-
KYLTINS BY MERCURIC HALIDES IN METHANOL–WATER MIXTURES AT 298 °K[18,21]

Mercuric halide	Mole fraction MeOH in solvent	Me_4Sn	Et_4Sn	Pr^n_4Sn	Bu^n_4Sn	Bu^i_4Sn	Pr^i_4Sn
				R_4Sn			
$HgCl_2$	1.0	100	0.215	0.0405	0.0396	0.00516	$<10^{-6}$
$HgCl_2$	0.91	100	0.243	0.0437	0.0403	0.00534	$<10^{-7}$
$HgCl_2$	0.72	100	0.333	0.0534	0.0441	0.00536	$<10^{-7}$
HgI_2	0.91	100	0.669	0.101	0.107	0.00709	$<10^{-5}$

TABLE 11

ENTHALPIES OF TRANSFER FROM METHANOL TO AQUEOUS METHANOL OF TETRA-
ETHYLTIN, MERCURIC CHLORIDE, AND THE TETRAETHYLTIN–MERCURIC CHLORIDE
TRANSITION STATE (cal.mole^{-1}) AT 298 °K[22]

Mole fraction MeOH in solvent	$\Delta H^{\ddagger}$	ΔH_t^{0}			
		Et_4Sn	$HgCl_2$	Tr	KBr^{a}
0.999	14,350	0	0	0	0
0.956	13,950	190	410	200	300
0.914	13,150	350	850	0	650
0.874	12,700	630	1170	150	900
0.800	12,350	1010	1770	780	1350
0.716	12,450	1410	2430	1940	1850
0.640	12,300	1770	2810	2530	2300
0.510	12,250	2200	3440	3540	2950

^{a} From ref. 23.

in $\Delta H^{\ddagger}$ observed as methanol is replaced by aqueous methanol is due to an increase in the enthalpy of the reactants. It is not possible at the moment to use the determined values of $\Delta H_t^{0}(\text{Tr})$ to make any deductions about the nature of the transition state because it appears as though species as widely different as non-polar non-electrolytes (e.g. Et_4Sn) and 1 : 1 electrolytes (e.g. KBr) give rise to large positive values of ΔH_t^{0} on transfer from methanol to aqueous methanol.

An analysis of the effect of methanol–water solvents on values of $\Delta G^{\ddagger}$ for substitution of tetraalkyltins by mercuric chloride has been carried out by Abraham[24]. The relevant equation is

$$\Delta G_t^{0}(\text{Tr}) = \Delta G_t^{0}(R_4Sn) + \Delta G_t^{0}(HgCl_2) + \Delta G_2^{\ddagger} - \Delta G_1^{\ddagger} \tag{17}$$

where ΔG_t^{0} refers to the standard free energy of transfer from solvent 1 (methanol) to solvent 2 (any other solvent). The concentration scale used was that of molar concentration (standard state 1 mole/l of solution, and unit activity). Values of $\Delta G_t^{0}(HgCl_2)$ had already been determined[25] through solubility measurements, and Abraham then determined the necessary $\Delta G_t^{0}(R_4Sn)$ values through measurement of Henry's law constants by a gas–liquid chromatographic method. Then

$$\Delta G_t^{0} = RT\ln(H_2^{0}/H_1^{0}) \tag{18}$$

where H_2^{0} and H_1^{0} are the Henry's law constants of a given solute in solvent 2 and solvent 1 respectively. A complete calculation of $\Delta G_t^{0}(\text{Tr})$ for the tetraethyltin–mercuric chloride reaction is shown in Table 12, where

$$\delta \Delta G^{\ddagger} = \Delta G_2^{\ddagger} - \Delta G_1^{\ddagger} \tag{19}$$

The enormous reduction in $\Delta G^{\ddagger}$ as methanol is gradually replaced by water can be seen to be due entirely to an increase in the free energy of the reactants; the

TABLE 12

FREE ENERGIES OF TRANSFER (ON THE MOLAR SCALE) FROM METHANOL TO AQUEOUS
METHANOL OF TETRAETHYLTIN, MERCURIC CHLORIDE, AND THE TETRAETHYLTIN-
MERCURIC CHLORIDE TRANSITION STATE(cal.mole^{-1}) AT 298 °K[24]

Mole fraction MeOH in solvent	$\delta\Delta G^{\ddagger}$	ΔG_t^0		
		Et_4Sn	$HgCl_2$	Tr
0.999	0	0	0	0
0.956	−184	150	34	0
0.914	−377	310	73	6
0.874	−546	475	110	39
0.800	−845	810	187	152
0.716	−1179	1220	294	335
0.640	−1464	1650	412	598
0.510	−1973	2450	619	1096
0.400	−2415	3345	859	1789
0.300	−2830	4250	1108	2528
0.200	−3100	5400	1356	3656
0.100	−3700	6600	1529	4429
0	−4150	7800	1623	5273

TABLE 13

FREE ENERGIES OF TRANSFER (ON THE MOLAR SCALE) FROM METHANOL TO AQUEOUS
METHANOL OF TETRAALKYLTINS AND TETRAALKYLTIN−MERCURIC CHLORIDE
TRANSITION STATES (cal.mole^{-1}) AT 298 °K[24]

Mole fraction MeOH in solvent	$\Delta G_t^0(R_4Sn)$				$\Delta G_t^0(R_4Sn/HgCl_2)$			
	$R = Me$	Et	Pr^n	Bu^n	$R = Me$	Et	Pr^n	Bu^n
0.999	0	0	0	0	0	0	0	0
0.956	105	150	185	230	−9	0	49	111
0.914	215	310	390	465	−15	6	114	225
0.874	320	475	590	700	−7	39	195	355
0.800	535	810	1010	1200	+50	152	417	683
0.716	810	1220	1580	1850	184	335	790	1160
0.640	1105	1650	2150	2500	377	598	1217	1692
0.510	1710	2450	3200	3750	790	1096	2009	2725
0	5120	7800	10,000	11,800	3500	5273		

effect of the more aqueous solvents on the transition state itself would tend to
lead to an *increase* in $\Delta G^{\ddagger}$.

Similar analyses were reported[24] for the substitution of other tetraalkyltins,
and the final values of $\Delta G_t^0(Tr)$ are given in Table 13 together with values of
$\Delta G_t^0(R_4Sn)$. The values of $\Delta G_t^0(R_4Sn)$ are themselves of interest since they show
that the effect of increasing size of a non-electrolyte is to increase the value of
ΔG_t^0 considerably. Exactly the same trend is shown in the $\Delta G_t^0(Tr)$ values, the larger

TABLE 14

COMPARISON OF FREE ENERGIES OF TRANSFER (ON THE MOLAR SCALE) FROM
METHANOL TO WATER OF TRANSITION STATES WITH SOME NON-POLAR AND IONIC
SOLUTES (kcal.mole^{-1}) AT 298 °K[24]

Non-polar solute	ΔG_t^0	Transition state	ΔG_t^0	Ion-pair	ΔG_t^0
Bu^{t}Cl	+4.1	[Bu^{t}Cl]‡	−2.2	Me$_4$NCl	−3.4
Bu^{t}Cl	+4.1	[Me$_3$N/MeI]‡	+1.7	Me$_4$NI	−1.7
Et$_4$Sn	+7.8	[Me$_4$Sn/HgCl$_2$]‡	+3.5	Me$_4$NClO$_4$	−0.1
Pr$^n_4$Sn	+10.0	[Et$_4$Sn/HgCl$_2$]‡	+5.3	Pr$^n_4$NClO$_4$	+2.9

the transition state the larger the (positive) value of ΔG_t^0(Tr). Since it is clear that
values of ΔG_t^0 depend markedly on the size of the solute species, Abraham[24]
compared values of ΔG_t^0 for various transition states with ΔG_t^0 values for non-
polar non-electrolytes and ΔG_t^0 values for ion-pairs, taking as far as was possible
non-electrolytes and ion-pairs of the same size as the transition state under
examination. Some of his comparisons are shown in Table 14. The value of ΔG_t^0
for the transition state in the solvolysis of *tert.*-butyl chloride is much nearer
that for the ion-pair than for the model non-polar solute; this is expected in view
of the pronounced polar character of this S_N1 transition state. The [R$_4$Sn/HgCl$_2$]‡
transition states appear to resemble the quoted ion-pairs rather more than the
model non-polar solutes, but the [Me$_3$N/MeI]‡ S_N2 transition state resembles
the non-polar solute slightly more than the ion-pair. Abraham[24] suggested that
charge separation in the transition states increased along the series

$$[Me_3N/MeI]^\ddagger < [R_4Sn/HgCl_2]^\ddagger < [Bu^tCl]^\ddagger$$

More recently, Abraham and Johnston[26] have attempted to put the above
comparisons on a quantitative basis. Following the procedure of Alfenaar and de
Ligny[27], they divided ΔG_t^0 values for transfer from methanol to aqueous methanol
into a non-electrostatic contribution and an electrostatic contribution, *viz.*

$$\Delta G_t^0 = \Delta G_e^0 + \Delta G_n^0 \qquad (20)$$

The value of ΔG_n^0 was taken as the free energy of transfer of a non-polar non-
electrolyte of the same size as the species under consideration; then for a given
species, knowing ΔG_t^0, it was possible to deduce a value for ΔG_e^0. Abraham and
Johnson[26] then showed that for uncharged transition states, ΔG_e^0 was related
directly to the degree of charge separation ($\delta\pm$) in the transition state. Some of
their calculated values of ΔG_e^0 for transfers from methanol to water, and corresponding deduced values of $\delta\pm$, are given in Table 15. Transition states in S_N1
solvolyses are characterised by high (> 0.8) values of $\delta\pm$, whereas S_N2 solvolysis
transition states are characterised by low ($\simeq$ 0.3) values of $\delta\pm$. Charge separations in the [Me$_4$Sn/HgCl$_2$]‡ transition state and the [Et$_4$Sn/HgCl$_2$]‡ transition

TABLE 15

VALUES OF ΔG_e^0 (kcal.mole^{-1}) FOR TRANSFER FROM METHANOL TO WATER OF SOME SOLUTES AND TRANSITION STATES, AND DERIVED VALUES OF $\delta\pm$, AT 298 °K[26]

Solute/transition state	ΔG_e^0	$\delta\pm$
CH$_4$	0	0
n-Hexane	0.1	$\simeq 0$
t-BuCl	0.1	$\simeq 0$
Et$_4$Sn	0	0
Me$_4$NCl[a]	−7.2	$\simeq 1.0$
Et$_4$NI[a]	−7.5	$\simeq 1.0$
Pr$^n_4$NClO$_4$[a]	−7.2	$\simeq 1.0$
[Bu^{t}Cl]‡[b]	−6.3	0.85
[Bu^{t}Br]‡[b]	−6.1	0.82
[Pr^{i}Br]‡[b]	−3.8	0.51
[Benzyl Cl]‡[b]	−3.0	0.40
[EtBr]‡[b]	−2.3	0.31
[MeI]‡[b]	−2.3	0.31
[Bu^{n}Br]‡[b]	−2.0	0.27
[Me$_3$N/MeI]‡[c]	−3.2	0.42
[Me$_4$Sn/HgCl$_2$]‡	−3.9	0.53
[Et$_4$Sn/HgCl$_2$]‡	−4.4	0.59

[a] Ion-pairs.
[b] Solvolysis transition states.
[c] Menschutkin reaction.

state are intermediate between those in S_N1 and S_N2 solvolyses.

Calculations were also made for transfers from methanol to aqueous methanol, and average values of $\delta\pm$ deduced are

Transition state	$\delta\pm$
[Me$_4$Sn/HgCl$_2$]‡	0.45–0.65
[Et$_4$Sn/HgCl$_2$]‡	0.60–0.70
[Pr$^n_4$Sn/HgCl$_2$]‡	0.50–0.60

These charge separations are rather lower than those deduced previously[20,21], although agreement is reasonable considering the varied methods used, but confirm that there is a high degree of charge separation in the tetraalkyltin–mercuric chloride transition states. Hence transition state (X) (p. 78), corresponding to mechanism S_E2(open), seems to be a good model for the transition state, at least in the methanol–water system.

Since both $\Delta G^{\ddagger}$ and $\Delta H^{\ddagger}$ have been dissected into initial-state and transition-state contributions for the tetraethyltin/mercuric chloride reaction, it is possible to achieve a similar separation in terms of the entropy function; some values given by Abraham[24] are in Table 16.

TABLE 16

ENTROPIES OF TRANSFER (ON THE MOLAR SCALE) FROM METHANOL TO AQUEOUS METHANOL OF TETRAETHYLTIN, MERCURIC CHLORIDE, AND THE TETRAETHYLTIN–MERCURIC CHLORIDE TRANSITION STATE (cal.mole^{-1}.deg^{-1}) AT 298 °K[24]

Mole fraction MeOH in solvent	$\delta\Delta S^{\ddagger}$	ΔS_t^0		
		Et_4Sn	$HgCl_2$	Tr
0.999	0	0	0	0
0.956	−0.6	0.1	1.3	0.8
0.914	−2.7	0.1	2.6	0
0.874	−3.6	0.5	3.6	0.5
0.800	−3.8	0.7	5.3	2.2
0.716	−2.3	0.6	7.2	5.5
0.640	−1.9	0.4	8.0	6.5
0.510	−0.4	−0.8	9.5	8.3
0.400[a]	+1.1	−3.3	10.2	8.0
0.300[a]	2.6	−5.9	10.6	7.3
0.200[a]	4.3	−9.5	10.7	5.5
0.100[a]	6.1	−13.6	10.1	2.6
0[a]	8.1	−17.8	8.8	−0.9

[a] Values at these compositions estimated.

The variation of $\delta\Delta S^{\ddagger}$($\delta\Delta S^{\ddagger} = \Delta S_2^{\ddagger} - \Delta S_1^{\ddagger}$) depends in a complicated way upon both initial-state effects and transition-state effects. The large negative values of ΔS_t^0(Et_4Sn) that arise on transfer from methanol to highly aqueous methanol may be due to a "structure-making" effect of the hydrocarbon-like solute on the solvent. Values of ΔS_t^0(Tr) in this composition region are much more positive than are values of ΔS_t^0(Et_4Sn); this behaviour is expected if the transition state carries a substantial charge separation (cf. values[28] of around +29 cal.mole^{-1}.deg^{-1} for molar entropies of transfer of alkali halides from methanol to water). It is possible to compare values of ΔS_t^0 for transition states with ΔS_t^0 values for various solutes of approximately the same size as the transition states. Details are given in Table 17 for transfers from methanol to water; values of ΔS_t^0 for the ionic species have been calculated from ΔH_t^0 values given by Choux and Benoit[29] and ΔG_t^0 values calculated by Abraham[24]. Although it would be preferable to use ΔS_t^0 values for ion pairs, rather than for the dissociated species in Table 17, the listed values show

TABLE 17

ENTROPIES OF TRANSFER (ON THE MOLAR SCALE) FROM METHANOL TO WATER OF
SOME SOLUTES AND TRANSITION STATES (cal.mole^{-1}.deg^{-1}) AT 298 °K

Species	ΔS_t^0	Species	ΔS_t^0	Species	ΔS_t^0
Me_4Sn	−10	Et_4Sn	−18	Bu^tCl	−6
$Me_4Sn+HgCl_2$	−1	$Et_4Sn+HgCl_2$	−9		
$[Me_4Sn/HgCl_2]^{\ddagger}$	+9	$[Et_4Sn/HgCl_2]^{\ddagger}$	−1	$[Bu^tCl]^{\ddagger}$	+9
$Me_4\overset{+}{N}+\overset{-}{I}$	+12	$Et_4\overset{+}{N}+\overset{-}{I}$	+1	$Me_4\overset{+}{N}+\overset{-}{Cl}$	+9

rather well that all three transition states behave as polar species and not as non-polar species. Hence from the $\Delta S_t^0(Tr)$ values, it may be concluded that charge separation is very high in all three transition states, for the particular solvent system (methanol–water) under consideration.

Abraham and Behbahany[30] have studied the kinetics of reaction (21) (X = Cl, I, and OAc)

$$Et_4Sn + HgX_2 \rightarrow EtHgX + Et_3SnX \tag{21}$$

TABLE 18

SECOND-ORDER RATE COEFFICIENTS (l.mole^{-1}.sec^{-1}) AND ACTIVATION PARAMETERS[a]
FOR THE SUBSTITUTION OF TETRAETHYLTIN BY MERCURIC CHLORIDE IN *tert.*-BUTANOL-
METHANOL MIXTURES AT 298 °K[30]

Mole fraction MeOH in solvent	$10^3 k_2$	$\Delta G^{\ddagger}$	$\Delta H^{\ddagger}$	$\Delta S^{\ddagger}$
1	3.33	20,832	14,350	−21.8
0.950	2.88	20,918		
0.900	2.55	20,991		
0.875	2.38	21,031		
0.840	2.13	21,097	14,850	−21.0
0.750	1.73	21,219		
0.700	1.60	21,266	15,150	−20.6
0.600	1.42	21,339		
0.500	1.26	21,408	14,400	−23.6
0.400	1.10	21,490	13,550	−26.6
0.300	0.973	21,562	12,400	−30.7
0.200	0.773	21,698	11,950	−32.7
0.100	0.500	21,956		
0.050	0.308	22,243	10,400	−39.7
0.025	0.183	22,550		
0	0.115	22,827	10,300	−42.0

[a] $\Delta G^{\ddagger}(\pm 9)$ and $\Delta H^{\ddagger}(\pm 250)$ in cal.mole^{-1}, $\Delta S^{\ddagger}(\pm 0.9)$ in cal.deg^{-1}.mole^{-1}.

using various *tert.*-butanol–methanol solvent mixtures ranging from methanol (dielectric constant, ε^{25}, = 32.6) to *tert.*-butanol ($\varepsilon^{25} = 12.5$). In Table 18 are listed rate coefficients and activation parameters for reaction (21) (X = Cl). The rate of reaction is considerably reduced as the solvent dielectric constant decreases, but the increase in $\Delta G^{\ddagger}$ is seen to be due to a large decrease in $\Delta S^{\ddagger}$. It was suggested[30] that for a reaction in which relatively non-polar reactants proceed to a polar transition state, a decrease in $\Delta S^{\ddagger}$ with decreasing solvent dielectric constant would be expected.

The effect of added tetra-*n*-butylammonium perchlorate on the rate of reaction (21) (X = Cl) was also studied. It was found[30] that the perchlorate greatly increased the value of the second-order rate coefficient, and that this positive kinetic salt effect was more marked the lower was the solvent dielectric constant. The salt effects were analysed in terms of the equation

$$\left\{ \frac{\log (k/k_0)}{I} \right\}_{I \to 0} = \frac{4\pi N e^4 Z^2 d}{2303(\varepsilon kT)^2} = JZ^2 d \tag{22}$$

where k_0 is the rate coefficient in the absence of salt, k is the coefficient in the presence of salt at an ionic strength I, N is the Avogadro number, e is the electronic charge, ε is the solvent dielectric constant, k is the Boltzmann constant, and T is the absolute temperature. The transition state is supposed to act as a dipole consisting of point charges $+Z$ and $-Z$ separated by a distance d. Details of the calculations are in Table 19; the value of Z has been deduced on the assumption that $d = 3.1$ Å.

For reactions taking place in methanol–water mixtures, values of Z calculated in this way (0.75–0.84) compare reasonably with previous estimates of $\delta\pm$, *viz.* 0.60–0.70 from ΔG_e^0 values. All these methods lead to the conclusion that charge

TABLE 19

ANALYSIS OF THE KINETIC SALT EFFECT OF TETRA-*n*-BUTYLAMMONIUM PERCHLORATE ON THE REACTION OF TETRAETHYLTIN WITH MERCURIC CHLORIDE[30]

Solvent	Temp. (°K)	ε	J	$\left\{\dfrac{\log (k/k_0)}{I}\right\}_{I \to 0}$	$Z^2 d$	Z^{a}
MeOH/H$_2$O[b]	313	32.2	0.903	1.56	1.73	0.75
MeOH	298	32.6	0.972	2.15	2.21	0.84
χ(MeOH) = 0.84[c]	298	27.2	1.396	3.73	2.67	0.93
χ(MeOH) = 0.40[c]	298	18.7	2.954	6.94	2.35	0.87
χ(MeOH) = 0.20[c]	298	15.9	4.086	9.20	2.25	0.85
t-BuOH	303	11.6	7.423	19.0	2.56	0.91

[a] Taking $d = 3.1$ A.
[b] From ref. 20; the mole fraction methanol is 0.914.
[c] Mole fraction methanol in *tert.*-butanol–methanol mixtures.

TABLE 20

SECOND-ORDER RATE COEFFICIENTS (l.mole^{-1}.sec^{-1}) AND RELATIVE RATE COEFFICIENTS
FOR THE SUBSTITUTION OF TETRAETHYLTIN BY MERCURIC SALTS AT 298 °K[30, 31]

| Solvent | Mole fraction MeOH in solvent | 10^3k_2 (HgI$_2$) | 10^3k_2 [Hg(OAc)$_2$] | Relative rate coefficients | | | |
				HgI$_3$$^-$	HgI$_2$	HgCl$_2$	Hg(OAc)$_2$
MeOH/H$_2$O	0.716	14.78	3330	0	0.60	1	137
MeOH/H$_2$O	0.914	4.50	1180	0	0.71	1	188
MeOH	1	2.53	850	0	0.76	1	255
t-BuOH/MeOH	0.840	1.63	297		0.76	1	139
t-BuOH/MeOH	0.400	0.72	98.0		0.65	1	89
t-BuOH/MeOH	0.200	0.35	45.8		0.45	1	59
t-BuOH	0	0.032	7.67	0	0.28	1	67
MeCN	0	34.0	83.7	0	1.82	1	4.5

separation in the transition state must be very high (for reaction in solvent methanol–water). The interesting feature of Table 19 is that values of Z^2d, and hence of Z, are even higher in tert.-butanol ($\varepsilon^{30} = 11.6$) than in methanol ($\varepsilon^{25} = 32.6$), thus indicating that the transition state still carries a high separation of charge even in a solvent of comparatively low dielectric constant. Abraham and Behbahany[30] suggested that the mechanism of reaction (21) (X = Cl) remained S$_E$2(open) along the entire solvent range from aqueous methanol through methanol to tert.-butanol.

Second-order rate coefficients for reaction (21) (X = I and OAc) were also reported by Abraham and Behbahany[30] and are given in Table 20. Kinetic salt effects of added tetra-n-butylammonium perchlorate were studied for reaction (21) (X = I and OAc) both with solvent methanol and solvent tert.-butanol. Reaction (21) (X = I) was accelerated in both solvents to about the same extent as was reaction (21) (X = Cl), and mechanism S$_E$2(open) was therefore suggested. The reaction of tetraethyltin with mercuric acetate was subject to very large positive salt effects in methanol, perhaps due to anion exchange, but was unaffected by the electrolyte in solvent tert.-butanol. Abraham and Behbahany[30] considered that it was not possible to deduce the mechanism of the acetate reaction and that further work was necessary to decide between mechanism S$_E$2(open) and mechanism S$_E$2(cyclic).

Reaction (21) (X = Cl, I, OAc) has been studied by Abraham and Hogarth[31] using acetonitrile as solvent. Rate coefficients and activation parameters are in Table 21. Tetra-n-butylammonium perchlorate and also lithium perchlorate accelerate reaction (21) (X = Cl); application of equation (22) yielded values for Z^2d of 0.652 (Bu$^n{}_4$NClO$_4$) and 0.674 (LiClO$_4$). If d is taken as 3.1 A, as before, these values correspond to 0.46 and 0.47 for Z. Although Z is much less than observed when hydroxylic solvents are used (see Table 19), Abraham and

TABLE 21

SECOND-ORDER RATE COEFFICIENTS (l.mole^{-1}.sec^{-1}) AND ACTIVATION PARAMETERS FOR THE SUBSTITUTION OF TETRAETHYLTIN BY MERCURIC SALTS IN SOLVENT ACETONITRILE[31]

Salt	$10^3 k_2$			$\Delta G^{\ddagger}$	$\Delta H^{\ddagger}$	$\Delta S^{\ddagger}$
	25 °C	40 °C	60 °C			
HgCl$_2$	18.67±0.55	45.8±0.3	154±4	19,812±17	11,300±200	−28.5±0.7
HgI$_2$	34.0±0.3	82.0±0.5	227±7	19,456±6	10,200±200	−31.0±0.7
Hg(OAc)$_2$	83.7					

Hogarth[31] considered the values of Z more compatible with mechanism S_E2(open) than with mechanism S_E2(cyclic). Added tetra-n-butylammonium perchlorate also accelerates reaction (21) (X = I), and mechanism S_E2(open) was again suggested to obtain.

Relative rate coefficients for reaction of the mercuric salts $HgI_3{}^-$, HgI_2, $HgCl_2$ and $Hg(OAc)_2$ with tetraethyltin are in Table 20. Abraham and Hogarth[31] drew attention to the fact that in methanol the sequence of reactivity is $HgCl_2 > HgI_2$ but in acetonitrile the sequence is $HgCl_2 < HgI_2$ even though all four substitutions proceed by mechanism S_E2(open), as judged from evidence of positive kinetic salt effects in all four cases. Hence, they suggest, reactivity sequences amongst the mercuric halides cannot be used as criteria of mechanism S_E2(open) or of mechanism S_E2(cyclic). The reactivity of mercuric acetate relative to mercuric chloride decreases by a factor of 57 on change of solvent from methanol to acetonitrile; however, little can be said about this large decrease in view of the difficulty in assigning a mechanism to the substitution of tetraethyltin by mercuric acetate (other than that it is some sort of S_E2 reaction).

1.6 SUBSTITUTION OF TETRAALKYLTINS BY DIMETHYLTIN DICHLORIDE

Tin-for-tin substitutions of the type

$$MeSnR_3 + Me_2SnCl_2 \xrightarrow{\text{MeOH}} Me_3SnCl + R_3SnCl \qquad (23)$$

have been studied by Tagliavini and co-workers[32]. Reactions were run using a large excess of the tetraalkyltin; under these conditions reaction (23) was first-order in dimethyltin dichloride. Second-order rate coefficients were calculated from the expression

$$k_2 = k_1/[MeSnR_3]$$

and in Table 22 are given values of k_1 and k_2 for the substitution of tetramethyltin

TABLE 22

FIRST-ORDER AND SECOND-ORDER RATE COEFFICIENTS FOR THE SUBSTITUTION OF
TETRAMETHYLTIN BY DIMETHYLTIN DICHLORIDE IN SOLVENT METHANOL[a] AT 30 °C[32]

Initial concentration		$10^5 k_1$ (sec^{-1})	$10^3 k_2$ $(l.mole^{-1}.sec^{-1})$
$10^2 [Me_4Sn]$	$10^4 [Me_2SnCl_2]$		
3.0	1.00	3.75	1.25
6.5	1.96	8.20	1.26
3.0	5.00	3.80	1.26
6.0	6.00	7.44	1.24
3.0	10.00	3.42, 3.35	1.14, 1.12

[a] An ionic strength of 0.01 M was maintained by the addition of sodium perchlorate; $\mu =$ [NaClO$_4$]+2[Me$_2$SnCl$_2$].

TABLE 23

SECOND-ORDER RATE COEFFICIENTS (l.mole^{-1}.sec^{-1}) FOR THE SUBSTITUTION OF
TETRAMETHYLTIN BY DIMETHYLTIN DICHLORIDE IN SOLVENT METHANOL IN
PRESENCE OF ADDED SALTS

(i) Rate coefficients at 30 °C: initial concentrations [Me$_4$Sn] = 3×10^{-2} M, [Me$_2$SnCl$_2$] = 5×10^{-4} M

(ii) Rate coefficients at 25 °C: initial concentrations [Me$_4$Sn] = 4.3×10^{-1} M, [Me$_2$SnCl$_2$] = 5×10^{-4} M

(i) Total salt[a] $(10^3 M)$	Total Cl^{-}[b] $(10^3 M)$	$10^4 k_2$	(ii) Total salt[a] $(10^3 M)$	Total Cl^{-}[b] $(10^3 M)$	$10^4 k_2$
10	1.0	12.6	1.0	1.0	7.5
10	2.0	6.6	11	1.0	8.1
10	4.0	3.7	51	1.0	11.0
10	6.0	2.4	101	1.0	12.5
10	7.0	2.0	161	1.0	15.1
10	10.0	1.4			

[a] Total [Cl$^-$]+total [NaClO$_4$].
[b] Added [NaCl]+2[Me$_2$SnCl$_2$].

by dimethyltin dichloride, reaction (23) (R = Me). Except at the highest concentration of dimethyltin dichloride, the second-order rate coefficient remains constant.

The effect of added salts on the rate coefficient for reaction (23) (R = Me) was also studied. Results are collected in Table 23 and show that, whereas sodium perchlorate has a marked accelerating effect (as it does also on the analogous substitution of dimethyldiethyltin), the addition of chloride ion strongly retards the substitution. It was suggested[32] that reaction (23) actually proceeds through the solvated Me$_2$Sn^{2+} ion, formed by the dissociation of dimethyltin dichloride

$$Me_2SnCl_2 \rightleftharpoons Me_2Sn^{2+}(solv)+2 Cl^-$$

In accord with this suggestion, it was observed[32] that reaction (23) did not take place when solvents in which dimethyltin dichloride is not dissociated were used (*e.g.* solvents such as acetonitrile, acetone, and dioxan).

Tagliavini and co-workers[32] wrote a mechanism for reaction (23) which involves coordination of the methanol solvent, S, to the tetraalkyltin, followed by a bimolecular substitution of the adduct by the species $Me_2SnS_3^{2+}$ through an "open" transition state, *viz.*

$$MeR_3Sn + S \rightleftharpoons MeR_3Sn \leftarrow S \tag{24}$$

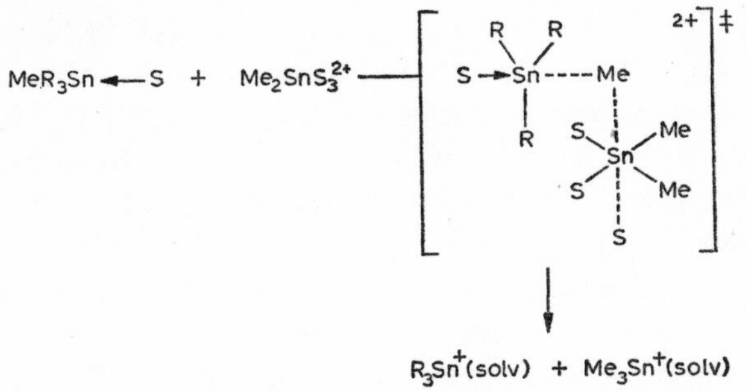

$$R_3\overset{+}{Sn}(solv) + Me_3Sn^+(solv)$$

Substitution of a number of tetraalkyltins was examined, and it was shown that only those tetraalkyls that contained at least one methyl–tin group underwent substitution. It was therefore deduced[32] that in mixed tetraalkyltins (such as Et_2SnMe_2) only the methyl group was cleaved from the tin atom. Rate coefficients for reaction of a number of tetraalkyltins with dimethyltin dichloride are given in Table 24, together with activation parameters where determined.

The reactivity sequence (see Table 24)

$$Me_4Sn > MeSnEt_3 > MeSnPr^n_3 > MeSnBu^n_3 > MeSnPr^i_3$$

TABLE 24

SECOND-ORDER RATE COEFFICIENTS (l.mole^{-1}.sec^{-1}) AND ACTIVATION PARAMETERS FOR THE SUBSTITUTION OF TETRAALKYLTINS BY DIMETHYLTIN DICHLORIDE IN SOLVENT METHANOL AT AN IONIC STRENGTH OF 0.01 M[32]

R_4Sn	$10^4k_2(25\,°C)$	E_a^a	$\Delta S^{\ddagger a}$
Me_4Sn	8.1	16.6	−17.0
$MeSnEt_3$	3.5	18.2	−13.3
$MeSnPr^n_3$	2.7	14.7	−25.6
$MeSnBu^n_3$	2.2	12.7	−32.5
$MeSnPr^i_3$	1.3	14.8	−26.8
Me_3SnEt	6.8		
Me_2SnEt_2	7.0		
Et_4Sn	No reaction		

^{a} $E_a(\pm 0.6)$ in kcal.mole^{-1} and $\Delta S^{\ddagger}(\pm 2)$ in cal.deg^{-1}.mole^{-1}. Errors estimated by the author.

was supposed to arise from a decreased tendency of methanol to coordinate to the tetraalkyltin, in reaction (24), along the above series[32].

There are a number of difficulties, however, in the proposed mechanism of substitution. First of all, it is not clear why tetraalkyltins should act as Lewis acids towards methanol. No co-ordination complexes of tetraalkyltins have ever been isolated[33] and it is generally assumed that compounds such as tetramethyltin are devoid of any power of forming co-ordination complexes[34]. It could be suggested that the position of equilibrium in (24) might lie well to the side of the uncomplexed tetraalkyltin, and that only a small fraction of the tetraalkyltin was actually present as $MeR_3Sn \leftarrow S$. But in this case it is not possible to deduce the percentage reaction taking place via the complex $MeR_3Sn \leftarrow S$ compared with the percentage reaction taking place via the uncomplexed substrate MeR_3Sn. Secondly, the rate coefficients given in Table 24 are not corrected for the statistical factors required when considering the substitution of tetraalkyltins that contain more than one methyl group. When these corrections are made, the rate coefficients for cleavage of the methyl–tin bond lead to sequences of rates (Table 25) that cannot be explained simply on the basis of the postulated equilibrium (24).

It seems to the author that the incorporation of the equilibrium (24) into the mechanism of reaction does not help in the interpretation of the results of Tagliavini and co-workers and that a direct reaction between the tetraalkyltin and the solvated Me_2Sn^{2+} cation cannot be ruled out. The leaving group effects shown in Table 25 are very small and can be accounted for by combinations of minor steric effects with polar effects in some transition state of the S_E2 type.

TABLE 25

STATISTICALLY CORRECTED SECOND-ORDER RATE COEFFICIENTS (l.mole^{-1}.sec^{-1}) FOR THE CLEAVAGE OF METHYL–TIN BONDS BY DIMETHYLTIN DICHLORIDE IN SOLVENT METHANOL AT 25 °C

Substrate	$10^4 k_2$	Substrate	$10^4 k_2$
Me–SnMe$_3$	2.0	Me–SnMe$_3$	2.0
Me–SnEt$_3$	3.5	Me–SnMe$_2$Et	2.3
Me–SnPr^n$_3$	2.7	Me–SnMeEt$_2$	3.5
Me–SnBu^n$_3$	2.2	Me–SnEt$_3$	3.5
Me–SnPr^i$_3$	1.3		

1.7 SUBSTITUTIONS OF TRIMETHYLANTIMONY BY ANTIMONY TRICHLORIDE

The substitution

$$Me_3Sb + SbCl_3 \xrightarrow{DMF} MeSbCl_2 + Me_2SbCl \qquad (26)$$

was shown by Weingarten and Van Wazer[35] to follow second-order kinetics, first-order in each reactant, in solvent dimethylformamide. Second-order rate coefficients were found to be $(6.7 \pm 0.5) \times 10^{-5}$ l.mole^{-1}.sec^{-1} at 72 °C and $(5.3 \pm 0.5) \times 10^{-4}$ l.mole^{-1}.sec^{-1} at 100 °C, and activation parameters were given[35] as $\Delta H^{\ddagger} = 18 \pm 0.6$ kcal.mole^{-1} and $\Delta S^{\ddagger} = -25 \pm 2$ cal.deg^{-1}.mole^{-1}. The large negative activation entropy was cited[35] as evidence that two sites of attachment were involved in the transition state, written as

However, for reaction (26), the change in translational entropy on forming a transition state of molecular weight 395 from the reactants of molecular weights 167 and 228 amounts to -33.5 cal.deg^{-1}.mole^{-1} at 25° and -34.0 cal.deg^{-1}. mole^{-1} at 100 °C, whatever is the actual structure of the transition state (these entropy changes have been calculated for a standard state of 1 mole/l). Entropy changes due to vibration and rotation, as well as to possible solvent involvement, all contribute to the overall observed value of $\Delta S^{\ddagger}$, but it seems clear that a large negative activation entropy cannot be used as evidence for a cyclic transition state. It might be more circumspect, therefore, merely to assign mechanism S$_E$2 to reaction (26) without detailing whether the "open" or "cyclic" varient obtains.

1.8 SUBSTITUTION OF PENTAAQUOPYRIDIOMETHYLCHROMIUM(III) IONS BY MERCURIC(II) SALTS

The substitution of the pentaaquo-4-pyridiomethylchromium(III) ion by mercuric chloride and mercuric bromide, proceeding according to equation (27) (X = Cl, Br),

was shown by Coombes and Johnson[36] to follow overall second-order kinetics, first-order in each reactant. Second-order rate coefficients for reaction (27) (X = Cl) were found to be 0.49 at 0.2°, 3.2 at 24.9°, and 10.0 at 40.0 °C (in l.mole^{-1}.sec^{-1}) and lead to the activation parameters[36] $\Delta H^{\ddagger} = 12.6$ kcal.mole^{-1} and $\Delta S^{\ddagger} = -13$ cal.deg^{-1}.mole^{-1}. Rate coefficients for the substitution of a number of other pyridiomethylchromium(III) ions by mercuric chloride are given in Table 26.

TABLE 26

SECOND-ORDER RATE COEFFICIENTS (l.mole^{-1}.sec^{-1}) FOR THE SUBSTITUTION OF
PENTAAQUO-Y-PYRIDIOMETHYLCHROMIUM(III) ION BY MERCURIC CHLORIDE IN
SOLVENT WATER AT 24.9 °C AND IONIC STRENGTH 0.02 M[36]

Y	2-	1-CH$_3$-2-	3-	1-CH$_3$-3-	4-	1-CH$_3$-4-
k_2	0.16	0.26	1.4	1.3	3.2	3.6

TABLE 27

SECOND-ORDER RATE COEFFICIENTS (l.mole^{-1}.sec^{-1}) FOR THE SUBSTITUTION OF
PENTAAQUO-4-PYRIDIOMETHYLCHROMIUM ION BY MERCURIC BROMIDE IN PRESENCE
OF BROMIDE ION IN SOLVENT WATER AT 24.9 °C AND IONIC STRENGTH 0.5 M[36]

Initial concentration of complex, 3×10^{-5} M; concentration of mercuric bromide (*i.e.* amount
added), 1×10^{-3} M.

[*LiBr*] (M)	Reactant composition (%)			k^{obs}
	$HgBr_2$	$HgBr_3^-$	$HgBr_4^{2-}$	
0	100	0	0	0.87
0.00384	58	36	6	0.56
0.0096	30	48	22	0.34
0.0192	13	43	44	0.19
0.0382	4.5	31	65	0.10
0.0576	2.2	23	74	0.065
0.0768	1.3	19	80	0.051
0.48	0.04	3.6	96.5	0.016

TABLE 28

SECOND-ORDER RATE COEFFICIENTS (l.mole^{-1}.sec^{-1}) FOR THE SUBSTITUTION OF
PENTAAQUO-Y-PYRIDIOMETHYLCHROMIUM(III) IONS BY VARIOUS MERCURIC
SPECIES IN SOLVENT WATER AT 24.9 °C AND IONIC STRENGTH 0.5 M[36, 38]

Mercuric species	Y		
	4	3	2
Hg(NO$_3$)$_2$[a]	125		
Hg(OAc)$_2$[a]	$\simeq 140$		
HgCl$_2$	3.2	1.5	0.17
HgCl$_3^-$	7.3	6.0	0.50
HgCl$_4^{2-}$	1.1	2.8	0.34
HgBr$_2$	0.87		
HgBr$_3^-$	0.17		
HgBr$_4^{2-}$	0.01		

[a] Ionic strength of the reaction medium not stated.

Addition of halide ion considerably affected the observed rate coefficients for reaction (27) (X = Cl, Br). The effect is more marked in the case of the addition of bromide ion to mercuric bromide than in the addition of chloride ion to mercuric chloride, and details of the bromide ion effect are given in Table 27. The concentrations of the various bromomercurate species as a function of added bromide ion are known[37] under the experimental conditions used (25 °C and $\mu = 0.5\ M$) and were calculated by Coombes and Johnson (see Table 27). These authors were thus able to evaluate the individual rate coefficients for substitution by the three possible species, $HgBr_2$, $HgBr_3^-$, and $HgBr_4^{2-}$. The rate coefficients are given in Table 28, together with those for the chloromercurate species deduced from similar experiments. Also given in Table 28 are rate coefficients for the substitution of the 4-pyridiomethylchromium(III) ion by mercuric nitrate and mercuric acetate[36], and rate coefficients for substitution of other pyridiomethylchromium(III) ions as given later by Johnson and co-workers[38].

The observed kinetic form suggests that all of the various substitutions listed in Tables 26 and 28 proceed by mechanism S_E2; there remains the problem of distinguishing between mechanism $S_E2(\text{open})$ and mechanism $S_E2(\text{cyclic})$. Mechanism $S_E2(\text{open})$ may be described as in equations (28)–(30), where R denotes the pyridio group, *viz.*

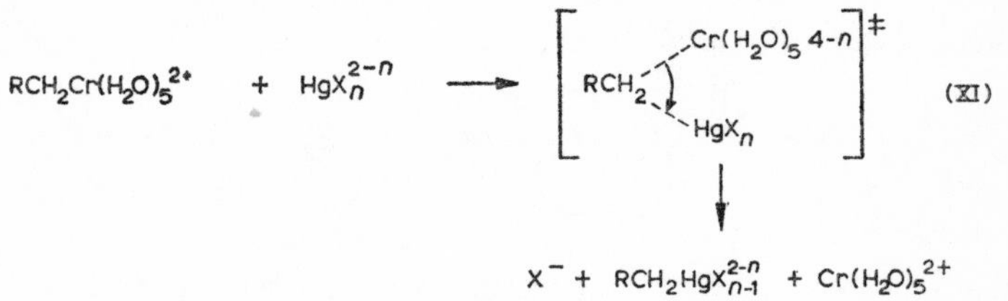

$$RCH_2Cr(H_2O)_5^{2+} \;+\; HgX_n^{2-n} \longrightarrow \left[RCH_2 \genfrac{}{}{0pt}{}{\nearrow Cr(H_2O)_5\,4-n}{\searrow HgX_n} \right]^{\ddagger} \qquad (\text{XI})$$

$$X^- + RCH_2HgX_{n-1}^{2-n} + Cr(H_2O)_5^{2+} \qquad (28)$$

$$Cr(H_2O)_5^{3+} + X^- \longrightarrow XCr(H_2O)_5^{2+} \qquad (29)$$

$$Cr(H_2O)_5^{3+} + H_2O \longrightarrow Cr(H_2O)_6^{3+} \qquad (30)$$

In the $S_E2(\text{cyclic})$ mechanism, it might be expected that a halogen atom could bridge the two metal centres, as shown in (31), where again R denotes the pyridio group.

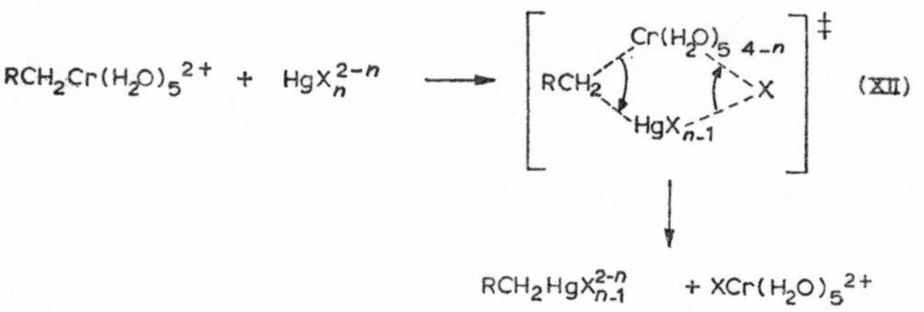

$$RCH_2Cr(H_2O)_5^{2+} \;+\; HgX_n^{2-n} \longrightarrow \left[RCH_2 \genfrac{}{}{0pt}{}{\nearrow Cr(H_2O)_5\,4-n}{\searrow HgX_{n-1}} X \right]^{\ddagger} \qquad (\text{XII})$$

$$RCH_2HgX_{n-1}^{2-n} \;+\; XCr(H_2O)_5^{2+} \qquad (31)$$

Since the complexes $[Cr(H_2O)_6]^{3+}$ and $[XCr(H_2O)_5]^{2+}$, where X = Cl or Br,

References pp. 105–106

TABLE 29

CHROMIUM SPECIES FORMED IN THE SUBSTITUTION OF THE PENTAAQUO-4-
PYRIDIOMETHYLCHROMIUM(III) ION BY MERCURIC HALIDES IN SOLVENT WATER[36]

Reactants (M)					Products (%)		
4-Pyridiomethyl-chromium(III)ion	$HgCl_2$	$LiCl$	$HgBr_2$	$LiBr$	$[ClCr(H_2O)_5]^{2+}$	$[Cr(H_2O)_6]^{3+}$	$[BrCr(H_2O)_5]^{2+}$
0.012	0.017	0			0	100	
0.012	0.017	0.1			<8	>92	
0.012	0.017	0.5			20	80	
0.012			0.02	0		100	0

are not interconvertable under the experimental conditions, it thus appears that the product analyses of the complexes formed could distinguish these two mechanisms[36]. Mechanism S_E2(cyclic) should give rise only to the complex $[XCr(H_2O)_5]^{2+}$, whereas mechanism S_E2(open) could give rise to a mixture of the complexes $[Cr(H_2O)_6]^{3+}$ and $[XCr(H_2O)_5]^{2+}$ (see equations (29) and (30)). If only the complex $[Cr(H_2O)_6]^{3+}$ were formed, then this would unambiguously indicate the operation of mechanism S_E2(open), through (28) and (30). Table 29 gives details of the product analyses under various conditions.

It is immediately clear that a halogen atom cannot bridge the two metal atoms in transition state (XII) when the electrophile is HgX_2, although a transition state with a bridging water molecule is a possibility. By comparison with other reactions in which the species $[Cr(H_2O)_5]^{3+}$ is liberated in the presence of halide ions, Coombes and Johnson[36] concluded that the product distribution in Table 29 was compatible with mechanism S_E2(open), reactions (28) to (30), even for the cases in which the electrophile is $HgCl_3^-$ and $HgCl_4^{2-}$. If, therefore, a chlorine atom in these electrophiles will not act as a bridge in transition state (XII) it is most unlikely that a water molecule will do so*. The transition state (XI) was hence suggested to obtain for all of the substitutions studied[36].

The order of reactivity of the halomercurate species, (see Table 28), $HgCl_2 < HgCl_3^- > HgCl_4^{2-}$, and $HgBr_2 > HgBr_3^- > HgBr_4^{2-}$ was suggested[36] to be due to the expected electrophilic power of the species, $HgX_2 \gg HgX_3^- \gg HgX_4^{2-}$, combined with the series resulting if coulombic attraction between the halomercurate species and the positively charged pentaaquopyridiomethylchromium(III) ions was the dominant factor, viz.: $HgX_2 < HgX_3^- < HgX_4^{2-}$.

---

* It has been suggested[39] that mercuric chloride in aqueous solution exists as the entity $[HgCl_2(H_2O)_2]$, and it is theoretically possible for a water molecule initially attached to mercury to be transferred to chromium in an S_E2(cyclic) transition state and thus to give rise to the species $[Cr(H_2O)_6]^{3+}$. Since chlorine will not act as a bridging group, even from $HgCl_4^{2-}$, it does not seem likely that H_2O will do so.

1.9 SUBSTITUTION OF PENTAAQUOPYRIDIOMETHYLCHROMIUM(III) IONS BY THALLIUM(III) SALTS

These substitutions proceed[38] in an analogous manner to those by the mercuric salts detailed in Section 1.8. Rate coefficients for substitutions by the species $TlCl_2^+$, $TlCl_3$, and $TlCl_4^-$ were calculated from experiments conducted at constant ionic strength (0.5 M) and with varying concentrations of perchloric acid and chloride ion. From the known proportions of these thallium(III) species under the experimental conditions, Johnson and co-workers[38] were able to dissect the observed second-order rate coefficients into the contributions from the individual species. Results are given in Table 30.

TABLE 30

SECOND-ORDER RATE COEFFICIENTS ($l.mole^{-1}.sec^{-1}$) FOR THE SUBSTITUTION OF PENTAAQUO-Y-PYRIDIOMETHYLCHROMIUM(III) IONS BY VARIOUS THALLIUM(III) SPECIES IN SOLVENT WATER AT 25.0 ± 0.2 °C AND IONIC STRENGTH 0.5 M[38]

Thallium(III) species	Y		
	4	3	2
$TlCl_2^+$	5.6	6.0	0.25
$TlCl_3$	6.1	8.4	0.40
$TlCl_4^-$	2.3	3.7	0.14

The only chromium species formed in the reaction of the 4-pyridiochromium(III) ion with $TlCl_3$ is $[Cr(H_2O)_6]^{3+}$, and even in presence of 0.4 M chloride ion, when some 70 % of the total reaction involves $TlCl_4^-$ as the reactant, the ratio of $[Cr(H_2O)_6]^{3+}$ to $[ClCr(H_2O)_5]^{2+}$ is about 9 : 1 (ref. 38). By the arguments advanced previously[36] (see Section 1.8) it may be deduced[38] that mechanism S_E2(open) is operative, and the substitutions therefore occur as in (32)–(34) where R denotes the pyridio group.

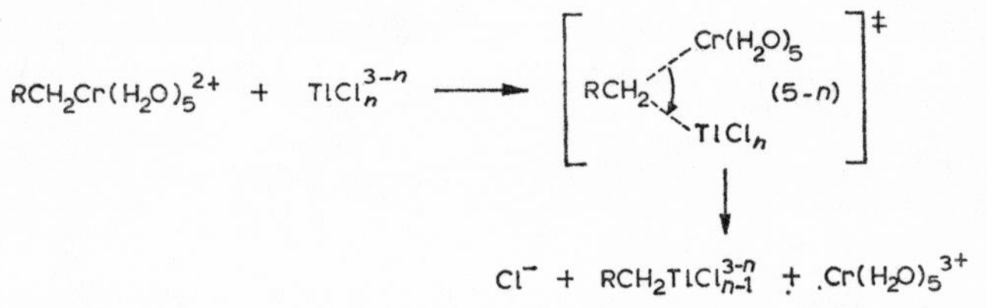

$$RCH_2Cr(H_2O)_5^{2+} + TlCl_n^{3-n} \longrightarrow \left[RCH_2 \underset{TlCl_n}{\overset{Cr(H_2O)_5}{\diagdown}} (5-n) \right]^{\ddagger}$$

$$Cl^- + RCH_2TlCl_{n-1}^{3-n} + \cdot Cr(H_2O)_5^{3+} \tag{32}$$

$$Cr(H_2O)_5^{3+} + Cl^- \longrightarrow ClCr(H_2O)_5^{2+} \tag{33}$$

$$Cr(H_2O)_5^{3+} + H_2O \longrightarrow Cr(H_2O)_6^{3+} \tag{34}$$

1.10 SUBSTITUTION OF PYRIDIOMETHYLDICARBONYL-π-CYCLOPENTA-DIENYLIRON BY MERCURY(II) AND THALLIUM(III) SALTS

Dodd and Johnson[40] have studied the displacement of iron from pyridiomethyl-iron species by mercury(II) and thallium(III) salts. Reactions were run using aqueous solutions at 25 °C and an ionic strength of 0.5 M, both with and without added chloride ion. Individual rate coefficients for attack of various mercury(II) and thallium(III) species were calculated by the method outlined in Section 1.8, and are given in Table 31.

TABLE 31

SECOND-ORDER RATE COEFFICIENTS (l.mole^{-1}.sec^{-1}) FOR THE SUBSTITUTION OF γ-PYRIDIOMETHYLDICARBONYL-π-CYCLOPENTADIENYLIRON IONS BY VARIOUS SPECIES IN SOLVENT WATER AT 25 °C AND IONIC STRENGTH 0.5 M^{40}

Electrophile	Y	k_2	Electrophile	Y	k_2
Hg^{2+}	3	300	Tl^{3+}	3	2.1
$HgCl^+$	3	35	$TlOH^{2+}$	3	22.0
$HgCl_2$	3	5×10^{-3}	$TlCl^{2+}$	3	2.8
$HgCl_3$	3	6×10^{-4}	$TlCl_2^+$	3	0.12
$HgCl_4^-$	3	3×10^{-5}	$TlCl_3$	3	4.3×10^{-2}
			$TlCl_4^-$	3	1.6×10^{-3}
Hg^{2+}	4	155			
			Tl^{3+}	4	0.29
			$TlOH^{2+}$	4	0.18

It was suggested that the rate coefficients in Table 31 represent electrophilic reactivities of the various species towards the pyridiomethyliron compound by mechanism S_E2(open); the reaction scheme shown in equations (35)–(37) was proposed, *viz.*

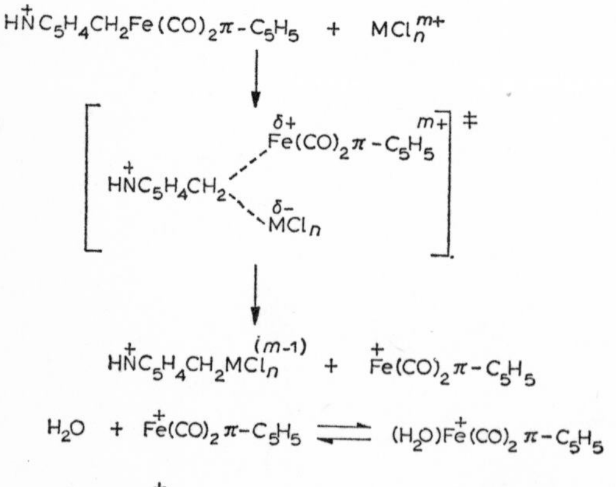

$$\overset{+}{H}NC_5H_4CH_2Fe(CO)_2\pi\text{-}C_5H_5 \quad + \quad MCl_n^{m+}$$

$$\left[\overset{+}{H}NC_5H_4CH_2 \underset{\delta^-}{\overset{\delta^+}{\cdots}} \begin{matrix} Fe(CO)_2\pi\text{-}C_5H_5 \\ \cdots \\ MCl_n \end{matrix} \right]^{m\mp\ddagger}$$

$$\overset{+}{H}NC_5H_4CH_2MCl_n^{(m\text{-}1)} \quad + \quad \overset{+}{Fe}(CO)_2\pi\text{-}C_5H_5 \tag{35}$$

$$H_2O \quad + \quad \overset{+}{Fe}(CO)_2\pi\text{-}C_5H_5 \rightleftharpoons (H_2O)\overset{+}{Fe}(CO)_2\pi\text{-}C_5H_5 \tag{36}$$

$$Cl^- \quad + \quad \overset{+}{Fe}(CO)_2\pi\text{-}C_5H_5 \rightleftharpoons ClFe(CO)_2\pi\text{-}C_5H_5 \tag{37}$$

1.11 SUBSTITUTION OF PYRIDIOMETHYLPENTACYANOCOBALTATE(III) IONS BY MERCURY(II), THALLIUM(III), INDIUM(III), AND GALLIUM(III) SALTS

The electrophilic displacement of cobalt by various species has been investigated by Bartlett and Johnson[41]. Using aqueous solutions at 25 °C with an ionic strength 0.5 M, they showed that the displacements were bimolecular (S_E2) processes, being first-order in the cobalt species and first-order in mercury(II) or thallium(III). Indium(III) and gallium(III) salts in the presence of chloride ion were found not to displace cobalt, so that both mercury(II) and thallium(III) are much stronger electrophiles. Individual rate coefficients for displacement of cobalt by various species are given in Table 32. The electrophilic displacements were suggested[41] to proceed by mechanism S_E2(open), and the substitution written as

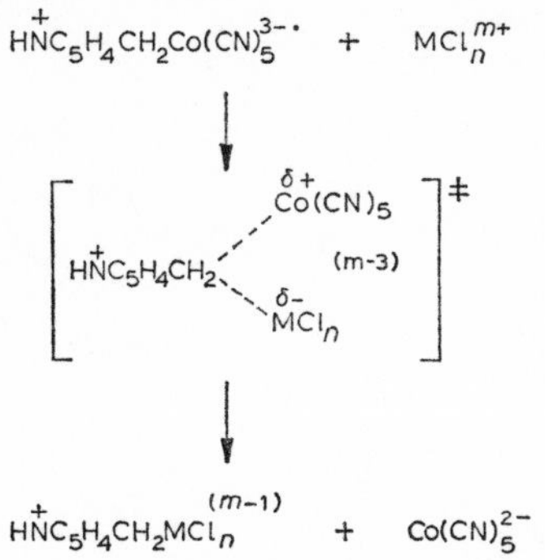

$$H\overset{+}{N}C_5H_4CH_2Co(CN)_5^{3-\cdot} \quad + \quad MCl_n^{m+}$$

$$\left[H\overset{+}{N}C_5H_4CH_2 \overset{\overset{\delta+}{Co(CN)_5}}{\underset{\underset{\delta-}{MCl_n}}{\diagdown}} \quad (m-3) \right]^{\ddagger}$$

$$H\overset{+}{N}C_5H_4CH_2MCl_n^{(m-1)} \quad + \quad Co(CN)_5^{2-} \tag{38}$$

$$Cl^- + Co(CN)_5^{2-} \longrightarrow ClCo(CN)_5^{3-} \tag{39}$$

$$H_2O + Co(CN)_5^{2-} \longrightarrow (H_2O)Co(CN)_5^{2-} \tag{40}$$

TABLE 32

SECOND-ORDER RATE COEFFICIENTS (l.mole^{-1}.sec^{-1}) FOR SUBSTITUTION OF γ-PYRIDIO-METHYLPENTACYANOCOBALTATE(III) IONS BY VARIOUS SPECIES IN SOLVENT WATER AT 25 °C AND IONIC STRENGTH 0.5 M[41]

Electrophile	Y	k_2	Electrophile	Y	k_2
$HgCl_2$	2	0.57	$TlCl_2^+$	2	5.7
$HgCl_3^-$	2	0.17	$TlCl_3$	2	3.3
$HgCl_4^{2-}$	2	Small	$TlCl_4^-$	2	0.31
$HgCl_2$	3	$\simeq 5$	$TlCl_2^+$	3	
$HgCl_3^-$	3	$\leqslant 1$	$TlCl_3$	3	$\simeq 150$
$HgCl_4^{2-}$	3	Small	$TlCl_4^-$	3	$\simeq 7$
$HgCl_2$	4	8.5	$TlCl_2^+$	4	157
$HgCl_3^-$	4	2.4	$TlCl_3$	4	58.2
$HgCl_4^{2-}$	4	0.06	$TlCl_4^-$	4	2.8

TABLE 33

SECOND-ORDER RATE COEFFICIENTS ($l.mole^{-1}.sec^{-1}$) FOR THE ELECTROPHILIC SUBSTITUTION OF Y-PYRIDIOMETHYL–METAL SPECIES BY MERCURY(II) AND THALLIUM(III) SPECIES IN SOLVENT WATER AT 25 °C AND IONIC STRENGTH 0.5 M

	$\overset{+}{H}NC_5H_4CH_2Fe(CO)_2\pi\text{-}C_5H_5$	$\overset{+}{H}NC_5H_4CH_2Co(CN)_5^{3-}$	$\overset{+}{H}NC_5H_4CH_2Cr(H_2O)_5^{2+}$		$\overset{+}{H}NC_5H_4CH_2Fe(CO)_2\pi\text{-}C_5H_5$	$\overset{+}{H}NC_5H_4CH_2Co(CN)_5^{3-}$	$\overset{+}{H}NC_5H_4CH_2Cr(H_2O)_5^{2+}$
(a) $Y = 3$				(a) $Y = 3$			
Hg^{2+}	300		205	Tl^{3+}	2.1		
$HgCl^+$	35			$TlCl^{2+}$	2.8		
$HgCl_2$	5×10^{-3}	5	1.5	$TlCl_2^+$	0.12		6.0
$HgCl_3^-$	6×10^{-4}	$\leqslant 1$	6.0	$TlCl_3$	4.3×10^{-2}	150	8.4
$HgCl_4^{2-}$	3×10^{-5}	Small	2.8	$TlCl_4^-$	1.6×10^{-3}	7	3.7
(b) $Y = 4$				(b) $Y = 4$			
$HgCl_2$		8.5	3.2	$TlCl_2^+$		157	5.6
$HgCl_3^-$		2.4	7.3	$TlCl_3$		58.2	6.1
$HgCl_4^{2-}$		0.06	1.1	$TlCl_4^-$		2.8	2.3
(c) $Y = 2$				(c) $Y = 2$			
$HgCl_2$		0.57	0.17	$TlCl_2^+$		5.7	0.25
$HgCl_3^-$		0.17	0.50	$TlCl_3$		3.3	0.40
$HgCl_4^{2-}$		Small	0.34	$TlCl_4^-$		0.31	0.14

It is of some interest to compare reactivities of the mercury(II) and thallium(III) species in the displacement of cobalt, reaction (38), with reactivities in displacement of chromium and iron. Second-order rate coefficients, obtained by Johnson *et al.*, are collected in Table 33. They show, first of all, that the position of the $-CH_2M$ group in the pyridine ring has little influence on relative reactivities of either the mercury or thallium species. The order of electrophilic power towards the chromium compound is always $HgCl_2 < HgCl_3^- > HgCl_4^{2-}$ and $TlCl_2^+ < TlCl_3 > TlCl_4^-$ no matter whether the $-CH_2M$ group is in position 2, 3, or 4.

The reactivities in the 3-pyridiomethyl series, for which most information is available, show clearly, however, that the charge of the substrate does influence the relative reactivity of the electrophilic series. Towards the iron compound, in which the metal atom carries zero formal charge, the reactivity sequences parallel the ability of the electrophilic species to complex with, for example, halide ion, *viz.*

$$Hg^{2+} > HgCl^+ > HgCl_2 > HgCl_3^- > HgCl_4^{2-}$$

$$Tl^{3+} \simeq TlCl^{2+} > TlCl_2^+ > TlCl_3 > TlCl_4^-$$

The only exception is that Tl^{3+} reacts rather more slowly than expected. The above sequences are also obeyed, as far as one can tell, in reaction with the cobalt compound; in this compound the metal atom carries a formal charge of -3. Towards the chromium complex, in which the chromium carries a formal charge of $+2$, the reactivities of the mercury(II) species (and the thallium(III) species) tend to be similar. Johnson[42] suggests that the sequences towards the iron compound are to be regarded as the normal sequence of electrophilic power of the various species, but that towards the chromium complex these sequences are modified by coulombic interaction between the charged electrophiles and the positively charged chromium atom. Hence the rate coefficients for attack of the species $HgCl_4^{2-}$, $HgCl_3^-$, and $TlCl_4^-$ tend to be increased in value, and the rate coefficients for attack of $TlCl_2^+$ tend to be reduced in value, the net result being that all three species $HgCl_2$, $HgCl_3^-$, and $HgCl_4^{2-}$ (and all three species $TlCl_2^+$, $TlCl_3$, and $TlCl_4^-$) are of about the same electrophilic power towards the chromium complex.

1.12 SUBSTITUTION OF PENTAAQUOPYRIDIOMETHYLCHROMIUM(III) IONS BY MERCURY(I)

The displacement of chromium(III) from pentaaquo-Y-pyridiomethylchromium(III) ions by mercury(I) follows second-order kinetics, and has been suggested[43] to involve a direct attack of the binuclear Hg_2^{2+} ion at saturated carbon through mechanism S_E2(open), *viz.*

$$HNC_5H_4CH_2Cr(H_2O)_5^{2+} + Hg_2^{2+} \xrightarrow{H_2O} HNC_5H_4CH_2Hg^+$$
$$+ Hg^0 + Cr(H_2O)_6^{3+} \qquad (41)$$

Second-order rate coefficients ($l.mole^{-1}.sec^{-1}$) for reaction (41) were reported to be

Y	2	3	4
$k_2(Hg_2^{2+})$	12	135	240
$k_2(Hg^{2+})$	30	205	500

Reactions were run using solvent water and at an ionic strength of 0.5 M; also given above are corresponding rate coefficients for attack of the mercury(II) ion, Hg^{2+}.

1.13 SUBSTITUTION OF ALKYLGOLD(I) AND ALKYLGOLD(III) COMPLEXES BY MERCURIC SALTS AND BY ALKYLMERCURIC SALTS

A number of substitutions of alkylgold complexes by mercuric compounds have been examined by Gregory and Ingold[44]. The S_E1 substitution of 1-cyano-1-carbethoxypentyl(triphenylphosphine) gold(I) has been described in Chapter 4, Section 2.3 (p. 32) and in the present section those substitutions proceeding by bimolecular mechanisms are discussed.

1.13.1 Substitution of alkyl(triphenylphosphine)gold(I) complexes by mercuric salts

The stoichiometry of these substitutions in solvents dioxan, dioxan–water, acetone, and dimethylformamide (DMF) was reported[44] to be

$$RAuPPh_3 + HgX_2 = RHgX + XAuPPh_3 \qquad (42)$$

In the above solvents, reaction (42) follows second-order kinetics, first-order in the gold complex and first-order in mercuric salt. The determined second-order rate coefficients are assembled in Table 34; the relative standard deviation in these coefficients is usually about 5 %.

Gregory and Ingold[44] note that for reaction (42) (R = Me, Et, and n-

TABLE 34

SECOND-ORDER RATE COEFFICIENTS (l.mole^{-1}.sec^{-1}) FOR THE SUBSTITUTION OF ALKYL(TRIPHENYLPHOSPHINE)GOLD(I) COMPLEXES, RAuPPh$_3$, BY MERCURIC SALTS AT 24.7 °C[44]

Solvent[a]	HgI_2	$HgBr_2$	$HgCl_2$	$Hg(OAc)_2$
$R = Me$				
Dioxan[b]	159	144	74	
Dioxan (19) : AcOH(1)		155		~4500
Dioxan (9) : water (1)		430		
Dioxan (4) : water (1)[b]		730		
Acetone (at 0.0 °C)[b]		1320	1120	
Dimethylformamide		200		
$R = Et$				
Dioxan	700	610	250	
Dioxan (19) : AcOH (1)				$>10^4$
$R = n\text{-}BuC(CN)CO_2Et$				
Dioxan[c]	43	45	22	255[d]

[a] Mixed solvents given as v/v.
[b] k_2 for substitution by HgBr$_3^-$ = 0.
[c] k_2 for substitution by Hg(NO$_3$)$_2$ > 5000.
[d] In the presence of 0.001 M acetic acid.

BuC(CN)CO$_2$Et) the sequence of reactivity of the electrophile is

$$Hg(NO_3)_2 > Hg(OAc)_2 > HgCl_2 \sim HgBr_2 \sim HgI_2 > HgBr_3^- = 0 \quad (43)$$

and suggest that this sequence indicates that reaction (42) proceeds by mechanism S$_E$2(open). As can be seen from Table 34, sequence (43) was found only for reactions run in solvent dioxan (or dioxan with added acetic acid). Assignment of mechanism on the basis of reactivity sequences such as (43) has been criticised by Jensen and Rickborn and in Chapter 5, Section 2 (p. 48) are presented various arguments to suggest that mechanisms S$_E$2(open) and S$_E$2(cyclic) cannot, in general, be differentiated in this way. There is little additional evidence to support the assignment of mechanism S$_E$2(open) to reaction (42), although it is worth noting that reaction (42) (R = Me, X = Br) in solvent dioxan is considerably accelerated on addition of water (see Table 34). This is, of course the expected behaviour if mechanism S$_E$2(open) obtained, but it is possible that such a gross change in solvent could itself change the mechanism of substitution. Since mechanisms S$_E$2(open) and S$_E$2(cyclic) can be regarded as the two extremes of a whole spectrum of mechanisms involving internal co-ordination to a lesser or greater extent, a change in solvent from dioxan to dioxan : water (4 : 1, v/v) might well be expected to shift the transition state towards an open transition state with charge separation from a transition state with more cyclic character.

Salt effects on reaction (42) (R = Me, X = Br) were also studied using the solvents dioxan(9) : water(1), dioxan(4) : water(1), and acetone. In all three cases it was shown that added bromide ion strongly depressed the rate and hence that the species HgBr$_3^-$ was inoperative as an electrophile[44]. The effect of added lithium perchlorate was also investigated, but at the low concentrations used (0.09×10^{-4} M to 3.25×10^{-4} M) no salt effect could be observed.

For reaction (42) in solvent dioxan, the evidence for mechanism S$_E$2(open) rests on the interpretation of sequence (29), and for reaction (42) in solvents acetone and dimethylformamide there is no available evidence for or against mechanism S$_E$2(open). It seems to the author that some substantial evidence is required in order to support the hypothesis that a bimolecular reaction between two neutral molecules in a solvent of as low a dielectric constant as dioxan ($\varepsilon = 2.1$) proceeds through a transition state with a degree of charge separation, when an alternative pathway is available for the reaction to proceed through an S$_E$2(cyclic) transition state in which little or no charge separation occurs.

The effect of the alkyl group, R, on the rate of reaction (42) in solvent dioxan may be seen from Table 34. Relative rate coefficients are given in Table 35, and show that for all the mercuric salts used, the order of reactivity of the gold complexes is that of increasing electron release by the group R, viz.

$$n\text{-BuC(CN)CO}_2\text{Et} < \text{Me} < \text{Et}$$

TABLE 35

RELATIVE SECOND-ORDER RATE COEFFICIENTS FOR THE SUBSTITUTION OF ALKYL (TRIPHENYLPHOSPHINE)GOLD(I) COMPLEXES, RAuPPh$_3$, BY MERCURIC SALTS IN SOLVENT DIOXAN AT 24.7 °C[44]

	R		
HgX_2	Me	Et	$n\text{-}BuC(CN)CO_2Et$
HgI$_2$	100	440	27
HgBr$_2$	100	420	31
HgCl$_2$	100	340	30
Hg(OAc)$_2$	100[a]	>220[a]	$\simeq$6[b]

[a] In solvent dioxan (19) : acetic acid (1) v/v.
[b] In solvent dioxan+0.001 M acetic acid.

1.13.2 Substitution of alkyl(triphenylphosphine)gold(I) complexes by methylmercuric salts

Reaction (44) (R = Me and Et)

$$RAuPPh_3 + MeHgX = RHgMe + XAuPPh_3 \qquad (44)$$

was also studied by Gregory and Ingold[44], who used dioxan–water and dimethyl-sulphoxide as solvents. The complex n-BuC(CN)CO$_2$Et·AuPPh$_3$ was also used as the substrate in one particular case. The substitutions were found to follow second-order kinetics; rate coefficients are listed in Table 36. Gregory and Ingold[44] suggest that the sequence of reactivity in the electrophile

$$MeHgNO_3 > MeHgOAc > MeHgCl \simeq MeHgBr \simeq MeHgI \qquad (45)$$

indicates that mechanism S_E2(open) is operative. As has been mentioned before, the use of such reactivity sequences in the deduction of mechanism is open to criticism.

TABLE 36

SECOND-ORDER RATE COEFFICIENTS (l.mole^{-1}.sec^{-1}) FOR THE SUBSTITUTION OF ALKYL(TRIPHENYLPHOSPHINE)GOLD(I) COMPLEXES, RAuPPh$_3$ BY METHYLMERCURIC SALTS AT 24.7 °C[44]

Solvent	R	$MeHgI$	$MeHgBr$	$MeHgCl$	$MeHgOAc$	$MeHgNO_3$
Dioxan (4) : H$_2$O (1) (v/v)	Me	0.59	0.41	0.47	5.2	$\simeq$8500
Dimethylsulphoxide	Me	2.9	1.16	0.77	0.84	>4000
Dimethylsulphoxide	Et	2.7	0.78	0.19	0.16	$\simeq$8000
Dioxan	n-BuC(CN)CO$_2$Et			16.2		

TABLE 37

SECOND-ORDER RATE COEFFICIENTS (l.mole^{-1}.sec^{-1}) FOR THE SUBSTITUTION OF
METHYL(TRIPHENYLPHOSPHINE)GOLD(I)[a] BY METHYLMERCURIC BROMIDE[a] AT 24.7 °C
IN THE PRESENCE OF ADDED SALTS[44]

Solvent	Added salt (M)	k_2
Dioxan (4) : water (1) (v/v)	None	0.41 ± 0.015[b]
Dioxan (4) : water (1) (v/v)	LiClO$_4$, 1.29×10^{-3}	0.43
Dioxan (4) : water (1) (v/v)	LiClO$_4$, 2.15×10^{-3}	0.47
Dioxan (4) : water (1) (v/v)	NaOAc, 1.96×10^{-3}	0.46
Dioxan (4) : water (1) (v/v)	LiBr, 1.94×10^{-3}	0.66
Dimethylsulphoxide	None	1.16 ± 0.09[b]
Dimethylsulphoxide	LiClO$_4$, 1.18×10^{-3}	1.20
Dimethylsulphoxide	LiBr, 0.67×10^{-3}	2.3
Dimethylsulphoxide	LiBr, 1.33×10^{-3}	3.4

[a] Initial concentration 5.00×10^{-4} M.
[b] Standard deviations calculated by the author.

Salt effects on reaction (44) (R = Me, X = Br) were also studied[44] and details
are given in Table 37. The acceleration caused by added lithium perchlorate and
added sodium acetate to solvent dioxan(4) : water(1) is of considerable interest,
since these are the only kinetic salt effects (other than the special salt effects due
to added halides) reported by Gregory and Ingold in which the increase in the rate
coefficient is outside experimental error. The equation used by Abraham and
Behbahany[30], (22) p. 85, can be applied in the present case; for dioxan(4) :
water(1) at 25 °C, J is 9.02 and the value of the term $[\log(k/k_0)]/I$ as $I \to 0$ is
approximately 8.7. Hence Z^2d is easily calculated to be 0.96 A. A reasonable
value for the mercury-to-gold distance in the transition state for reaction (44)
(R = Me, X = Br) is 3.0 A, so that if the value of d is taken as this distance, then
Z is about 0.57 units of charge. Because of the very limited data in Table 37, the
final value of Z is approximate only. Nevertheless, the degree of charge separation
in the transition state for reaction (44) (R = Me, X = Br) must be high enough
to suggest that reaction (44) (R = Me, X = Br) in solvent dioxan(4) : water(1)
proceeds by mechanism S$_E$2(open).

The effect of lithium perchlorate on the rate of reaction (44) (R = Me, X = Br)
in solvent dimethylsulphoxide is unfortunately within experimental error, so that
a calculation on the above lines would be meaningless. Since, however, mechanism S$_E$2(open) is in force when dioxan(4) : water(1), of $\varepsilon^{25} = 10.5$, is the
solvent, it would not be unreasonable if a similar mechanism applied also to
reaction (44) (R = Me, X = Br) when dimethylsulphoxide, of $\varepsilon^{25} = 46.5$, is
the solvent.

1.13.3 Substitution of trimethyl(triphenylphosphine)gold(III) by mercuric salts

Only one methyl group is cleaved from the above gold(III) complex by mercuric salts, with the substitution proceeding as

$$Me_3AuPPh_3 + HgX_2 \rightarrow MeHgX + Me_2AuPPh_3X \qquad (46)$$

Second-order kinetics were found for substitution by mercuric salts using a number

TABLE 38

SECOND-ORDER RATE COEFFICIENTS ($l.mole^{-1}.sec^{-1}$) FOR THE SUBSTITUTION OF TRIMETHYL(TRIPHENYLPHOSPHINE)GOLD(III) BY MERCURIC SALTS AT 24.7 °C[44]

Solvent	Mercuric salt	k_2
Dioxan	$HgBr_2$	130
Dioxan (19) : AcOH (1) (v/v)	$HgBr_2$	184
Dioxan (19) : AcOH (1) (v/v)	$Hg(OAc)_2$	$\simeq 9500$
Dioxan (4) : water (1) (v/v)	$HgBr_2$	1450
Acetone (at 0.0 °C)	$HgBr_2$	2000

of solvents, and the reported[44] rate coefficients are given in Table 38. It was observed that added lithium bromide (in acetone) quenches the reaction almost quantitatively and hence that in this solvent the species $HgBr_3^-$ is inoperative as an electrophile. On the basis of the relative rates $HgBr_3^- < HgBr_2$ (solvent acetone) and $HgBr_2 < Hg(OAc)_2$ (solvent dioxan–acetic acid), Gregory and Ingold[44] suggested that mechanism S_E2(open) was operative. The reservations noted before about deduction of mechanism from such sequences apply here also.

It is of interest that change of solvent from dioxan (dielectric constant 2.1), to dioxan(4) : water(1), (dielectric constant 11), increases the rate of reaction (46) (X = Br) by a factor of 11. Since mechanisms S_E2(open) and S_E2(cyclic) may be regarded as the two extremes of a whole range of bimolecular mechanisms with varying amounts of charge separation in the transition state, it is evident that such a profound change of solvent could lead to a change in the character of the transition state, with the more polar solvent inducing a transition state nearer the S_E2(open) end of the range and the less polar solvent supporting a transition state nearer the S_E2(cyclic) end.

Without further experimental evidence, however, nothing definite can be said about the mechanism of reaction (46) other than that some bimolecular mechanism is in force under the conditions given in Table 38.

2. Alkyl exchanges between compounds R_nM and R_mM'

A number of exchanges of alkyl groups between fully-alkylated metals have been studied by proton magnetic resonance. The mechanism of these exchanges is not entirely clear, but Oliver[45] has suggested that the fast exchange reactions between organometallic compounds of Groups I, II, and III proceed through bridged transition states. Since such alkyl exchange reactions have been well reviewed recently[45,46] they will not be considered here. In any case, it is a moot point whether such exchanges can be considered to proceed by the mechanism of electrophilic substitution at saturated carbon.

REFERENCES

1 A. F. REID AND C. J. WILKINS, *J. Chem. Soc.*, (1955) 4029.
2 E. C. ASHBY, *Quart. Rev.*, 21 (1967) 259.
3 M. H. ABRAHAM AND P. H. ROLFE, *J. Organometal. Chem.*, 8 (1967) 395.
4 D. S. MATTESON AND J. O. WALDBILLIG, *J. Am. Chem. Soc.*, 85 (1963). 1019; 86 (1964) 3778.
5 D. S. MATTESON, J. O. WALDBILLIG AND S. W. PETERSON, *J. Am. Chem. Soc.*, 86 (1964) 3781.
6 D. S. MATTESON AND M. L. TALBOT, *J. Am. Chem. Soc.*, 89 (1967) 1119.
7 D. S. MATTESON AND M. L. TALBOT, *J. Am. Chem. Soc.*, 89 (1967) 1123.
8 D. S. MATTESON AND R. A. BOWIE, *J. Am. Chem. Soc.*, 87 (1965) 2587.
9 F. R. JENSEN AND K. L. NAKAMAYE, *J. Am. Chem. Soc.*, 88 (1966) 3437.
10 D. S. MATTESON, R. A. BOWIE AND G. SRIVASTAVA, *J. Organometal. Chem.*, 16 (1969) 33.
11 D. S. MATTESON AND E. KRAMER, *J. Am. Chem. Soc.*, 90 (1968) 7261.
12 M. GIELEN AND J. NASIELSKI, *Ind. Chim. Belge*, (1961) 1393.
13 D. S. MATTESON AND P. G. ALLIES, *J. Am. Chem. Soc.*, 92 (1970) 1801.
14 C. R. HART AND C. K. INGOLD, *J. Chem. Soc.*, (1964) 4372.
15 F. R. JENSEN AND D. HEYMAN, *J. Am. Chem. Soc.*, 88 (1966) 3438.
16 M. H. ABRAHAM AND T. R. SPALDING, *J. Chem. Soc. A*, (1968) 2530.
17 M. H. ABRAHAM, P. L. GRELLIER, M. J. HOGARTH, T. R. SPALDING, M. FOX AND G. R. WICKHAM, *J. Chem. Soc. A*, (1971) 2972.
18 M. H. ABRAHAM AND T. R. SPALDING, *J. Chem. Soc. A*, (1969) 399.
19 M. H. ABRAHAM AND T. R. SPALDING, *J. Chem. Soc. A*, (1969) 784.
20 M. H. ABRAHAM AND G. F. JOHNSTON, *J. Chem. Soc. A*, (1970) 188.
21 M. H. ABRAHAM AND G. F. JOHNSTON, *J. Chem. Soc. A*, (1970) 193.
22 M. H. ABRAHAM, R. J. IRVING AND G. F. JOHNSTON, *J. Chem. Soc. A*, (1970) 199.
23 C. M. SLANSKY, *J. Am. Chem. Soc.*, 62 (1940) 2430.
24 M. H. ABRAHAM, *J. Chem. Soc. A*, (1971) 1061.
25 M. H. ABRAHAM, J. F. C. OLIVER AND J. A. RICHARDS, *J. Chem. Soc. A*, (1970) 203.
26 M. H. ABRAHAM AND G. F. JOHNSTON, *J. Chem. Soc. A*, (1971) 1610.
27 M. ALFENAAR AND C. L. DE LIGNY, *Rec. Trav. Chim.*, 86 (1967) 929.
28 C. M. CRISS, R. P. HELD AND E. LUKSHA, *J. Phys. Chem.*, 72 (1968) 2970.
29 G. CHOUX AND R. L. BENOIT, *J. Am. Chem. Soc.*, 91 (1969) 6221.
30 M. H. ABRAHAM AND F. BEHBAHANY, *J. Chem. Soc. A*, (1971) 1469.
31 M. H. ABRAHAM AND M. J. HOGARTH, *J. Chem. Soc. A*, (1971) 1474.
32 G. PLAZZOGNA, S. BRESADOLA AND G. TAGLIAVINI, *Inorg. Chim. Acta*, 2 (1968) 333.
33 I. R. BEATTIE, *Quart. Rev.*, 17 (1963) 382.
34 G. E. COATES AND K. WADE, *Organometallic Compounds*, Methuen, London, 3rd edn., 1967, Vol. 1, p. 416.
35 H. WEINGARTEN AND J. R. VAN WAZER, *J. Am. Chem. Soc.*, 88 (1966) 2700.
36 R. G. COOMBES AND M. D. JOHNSON, *J. Chem. Soc. (A)*, (1966) 1805.

37 Y. MARCUS, *Acta Chem. Scand.*, 11 (1957) 599.
38 R. G. COOMBES, M. D. JOHNSON AND D. VAMPLEW, *J. Chem. Soc. A*, (1968) 2297.
39 V. JEDINÁKOVÁ AND J. CELEDA, *Collection Czech. Chem. Commun.*, 32 (1967) 1679.
40 D. DODD AND M. D. JOHNSON, *J. Chem. Soc. B*, (1971) 662.
41 E. H. BARTLETT AND M. D. JOHNSON, *J. Chem. Soc. A*, (1970) 517.
42 M. D. JOHNSON, *Record Chem. Progr.*, 31 (1970) 143.
43 D. DODD AND M. D. JOHNSON, *Chem. Commun.*, (1970) 460.
44 B. J. GREGORY AND C. K. INGOLD, *J. Chem. Soc. B*, (1969) 276.
45 J. P. OLIVER, *Advances in Organometallic Chemistry*, 8 (1970) 167.
46 N. S. HAM AND T. MOLE, *Progr. NMR Spectry.*, 4 (1969) 91.

Chapter 7

Acidolysis of Organometallic Compounds

Cleavage of the alkyl–metal bond by acids is a characteristic reaction of organometallic compounds; the general equation

$$R-MX_n + HA = R-H + M(A)X_n \tag{1}$$

may be written, where HA = HCl, HOH, etc. Even very weak acids react in this way with the alkyls of the more electropositive elements. From the data of Dessy and Kim[1] and the more recent work of Robinson[2], an estimate of the relative reactivity of metal alkyls and aryls towards acids may be made.

$$R_2Cd(20,000) : R_2Hg(110) : R_4Pb(60) : R_4Sn(1) \tag{2}$$

In general, the ease of acidolysis of metal alkyls parallels the Pauling-scale electronegativity[3] of the metal atom, with the alkyls of the Group IA metals reacting explosively with water, whereas the alkyls of mercury, boron, and the Group IVA metals are inert towards water. Kinetic studies of reaction (1) have been restricted, in the main, to alkyls of the less reactive metals. The following discussion is drawn up in the order of the Periodic Group of the metal. Additionally, the acidolysis of the 4-pyridiomethylmercuric ion by mechanism S_E1 has been discussed in Chapter 4, Section 1.2 (p. 25), and acidolyses of allyl compounds are described in Chapter 10, Section 3 (p. 201).

1. Stereochemical studies

The acidolysis of optically active di-*sec.*-butylmercury, and of the *cis* and *trans* isomers of di-4-methylcyclohexylmercury, proceeds with predominant retention of configuration at the carbon atom undergoing substitution[4].

Cleavage of the optically active cyclopropyl compounds (I; M' = Li)[5] by methanol in ether or ether/1,2-dimethoxyethane and (I; M' = SnMe₃)[6] by concentrated hydrochloric acid in carbon tetrachloride and by anhydrous hydrogen bromide in CCl₄ also take place with retention of optical activity and configuration, *viz.*

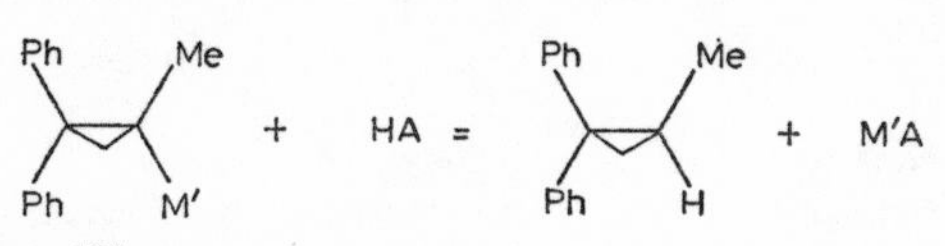

$$\tag{3}$$

Davies and Roberts[7] have shown that the dibutyl ester of optically active 1-phenylethylboronic acid is cleaved by $C_7H_{15}CO_2D$ in boiling diglyme to yield $(-)$-1-deuterioethylbenzene with 95 % retention of configuration. The acidolysis was suggested[7] to proceed by mechanism S_E2(cyclic), reaction (4),

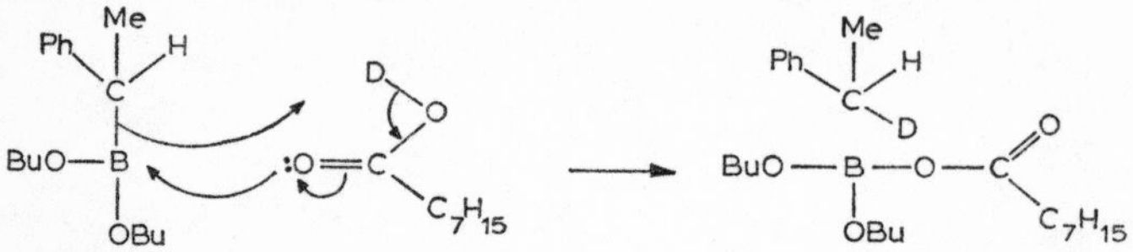

$$(4)$$

in accord with previous work[8] on the acidolysis of tri-*exo*-norbornylboron. When the diethanolamine ester of optically active 1-phenylethylboronic acid was cleaved by OD^- in solvent boiling D_2O, the 1-deuterioethylbenzene was formed with 54 % net inversion of configuration, and mechanism S_E1-OD^- was suggested[7] for this cleavage. It may be noted that Weinheimer and Marsico[9] had previously observed retention of configuration when a diasterioisomer of $PhMeCH \cdot C(BH_2)MePh$ was subjected to alkali cleavage, and had suggested that mechanism S_E2(co-ord) obtained (5) ($R = PhMeCH \cdot CMePh$)*, *viz.*

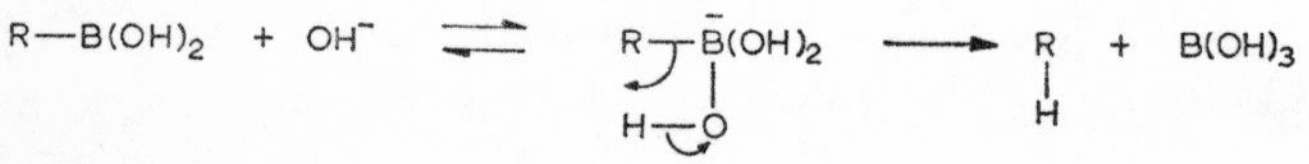

$$(5)$$

Although the stereochemical consequences of reaction (1) are not entirely clear-cut, the general conclusion may be reached that in acidolyses proceeding by mechanisms S_E2 (especially by mechanisms S_E2(cyclic) and S_E2(co-ord)) retention of configuration at the carbon atom undergoing substitution is observed.

2. Acidolysis of organomagnesium compounds

2.1 THE CLEAVAGE OF GRIGNARD REAGENTS** BY 1-ALKYNES

A long series of papers on the acidolysis of alkyl Grignard reagents by 1-alkynes has been published by Wotiz *et al.* (for Part XII of this series, see ref. 10). Most of these papers deal with the relative reactivities of various Grignard reagents, as determined[11] by evolution of the volatile gaseous hydrocarbon, *e.g.*

$$\text{"RMgX"} + BuC \equiv CH = RH\uparrow + BuC \equiv CMgX \qquad (6)$$

Rate coefficients for the cleavage of alkyl Grignard reagents were recorded in one paper, and it was shown[12] that the reaction between 1.0 M ethylmagnesium

---

* The compound RBH_2 is of course rapidly hydrolysed to $RB(OH)_2$.
** The symbol "RMgX" will be used to indicate the Grignard reagent, although the constitution of this reagent is more correctly described by equations (8) and (11).

bromide and 1.0 M hexyne-1 in diethylether at 35 °C followed overall second-order kinetics with $k_2 = 1.7 \times 10^{-3}$ l.mole^{-1}.sec^{-1}. From earlier work[13] on the kinetics of the reaction between ethylmagnesium bromide and a large excess of acetylene, it is probable that reaction (6) (RMgX = EtMgBr) is first-order in each reactant in solvent diethylether. A reinvestigation of the kinetics of reaction (6) (RMgX = EtMgBr) yielded[14] a value for k_2 of 1.5×10^{-3} l.mole^{-1}.sec^{-1} at 35 °C in solvent diethylether, in good agreement with the earlier value[12]. Wotiz et al.[12] also showed that for the reaction of ethylmagnesium bromide with BuC≡CH and BuC≡CD, in diethylether, $k_H/k_D = 4.3$. Relative reactivities of various Grignard reagents towards hexyne-1 in diethylether were found[11] to be*

$$\text{``MeMgBr''} \ (6) < \text{``Pr}^n\text{MgBr''} \ (59) < \text{``EtMgBr''} \ (100) < \text{``Pr}^i\text{MgBr''} \ (210) \tag{7}$$

There appears to be no correlation[15] between the kinetic acidities of a number of 1-alkynes and their relative rates of acidolysis[16] of ethylmagnesium bromide in solvent diethylether.

All of the above information is compatible with the suggestion[12, 15] that the acidolysis (6) involves transition state (II), S_E2(cyclic) type, but the actual reaction

(II)

mechanism cannot be deduced, since the organometallic species taking part in the reaction is not known. At the concentrations used, 1.0 M, ethylmagnesium bromide in diethylether has an association factor of about 2, and is thus, on the average, dimeric[18]. There are at least two dimeric species that may exist in ethereal solutions of Grignard reagents and these species are in equilibrium with various other species in the equilibrium[18]

$$\tag{8}$$

The actual species in (8) that reacts with the 1-alkyne is not known, and it therefore seems to the author that further mechanistic speculation is not profitable**.

---

* Some twenty years previously, Ivanoff et al.[17] had observed an identical sequence (qualitatively) for the acidolysis of Grignard reagents by weak acids such as indene.
** It may be noted that the kinetics of acidolysis of phenylmagnesium bromide by hexyne-1 in solvent diethylether have been described as overall second-order (first-order in each reactant)[19], and as overall first-order[20]. Again, constitutional difficulties may be responsible.

Hashimoto *et al.*[14] have concluded that reaction (6) (RMgX = EtMgBr) follows third-order kinetics in solvent tetrahydrofuran (THF) and write the mechanism as

$$EtMgBr + BuC{\equiv}CH \rightleftharpoons EtMgBr{\cdot}BuC{\equiv}CH \tag{9}$$

$$EtMgBr{\cdot}BuC{\equiv}CH + EtMgBr \rightarrow EtH + BuC{\equiv}CMgBr + EtMgBr \tag{10}$$

It is thus assumed[14] that the reactive entity is the species EtMgBr. However, in THF the constitution of ethylmagnesium bromide is best described[18] by the equilibrium

$$Et_2Mg + MgBr_2 \rightleftharpoons 2\,EtMgBr \tag{11}$$

in which the value of the equilibrium constant is 4. Thus at a concentration 0.5 M in Grignard reagent (the conditions mainly used in the kinetic experiments), the actual concentration of EtMgBr will be 0.25 M and that of Et_2Mg will be 0.12 M. It seems to the author that difficulties over the constitution of the reactive species preclude, again, any detailed mechanistic conjectures.

Pocker and Exner[59] have determined deuterium isotope effects in the acidolysis of a number of Grignard reagents by oxygen acids (such as water and alcohols) and by carbon acids (such as alkynes), using as solvents ether and THF. They find that values of k_H/k_D are generally only slightly greater than unity for the oxygen acids and suggest that in these cases the mechanism of acidolysis consists of an initial stage in which the oxygen acid co-ordinates to magnesium, followed by a fast stage in which the actual proton transfer takes place. On the other hand, values of k_H/k_D are quite large (from 3 to 6) for the carbon acids, suggesting that proton transfer is now involved in the rate-determining step.

2.2 THE CLEAVAGE OF DIALKYLMAGNESIUMS BY HEXYNE-1

It was early reported[12] that the action of hexyne-1 on diethylmagnesium in solvent diethylether proceeded by the two competitive consecutive second-order reactions (12) (R = Et) and (13) (R = Et).

$$R_2Mg + BuC{\equiv}CH \rightarrow RH + BuC{\equiv}CMgR \tag{12}$$

$$BuC{\equiv}CMgR + BuC{\equiv}CH \rightarrow RH + (BuC{\equiv}C)_2Mg \tag{13}$$

For the reaction of diethylmagnesium with $BuC{\equiv}CH$ and $BuC{\equiv}CD$ a value for k_H/k_D of 2.8 was recorded[12]. Later workers[21] also interpreted the action of hexyne-1 on dialkylmagnesiums in terms of reactions (12) and (13); the various rate coefficients are collected in Table 1, and lead[21] to a reactivity sequence

$$Pr''_2Mg\ (59) < Et_2Mg\ (100) < Pr^i_2Mg\ (180) \tag{14}$$

for reaction (12) very close that for the cleavage of Grignard reagents by hexyne-1 (see sequence (7)).

TABLE 1

SECOND-ORDER RATE COEFFICIENTS ($l.mole^{-1}.sec^{-1}$) FOR THE ACIDOLYSIS OF
DIALKYLMAGNESIUMS BY HEXYNE-1 IN SOLVENT DIETHYLETHER AT 33 °C
REACTIONS (12) AND (13)

R_2Mg	$k_{(12)}$	$k_{(13)}$	Ref.
Et_2Mg	1.3×10^{-2}	2.5×10^{-3}	12
Et_2Mg	2.2×10^{-2}	3.0×10^{-3}	21
$Pr^n{}_2Mg$	1.3×10^{-2}	2.0×10^{-3}	21
$Pr^i{}_2Mg$	4.0×10^{-2}	8.0×10^{-3}	21

Under the concentrations, about 0.5 M, used in the kinetic studies association factors are 1.6 for dimethylmagnesium and 1.2 for diethylmagnesium in diethylether[18]; the dipropylmagnesiums might well have association factors close to unity. It is thus quite probable that the reactive species in these acidolyses is monomeric R_2Mg, and that reaction (12) proceeds through a cyclic transition state such as

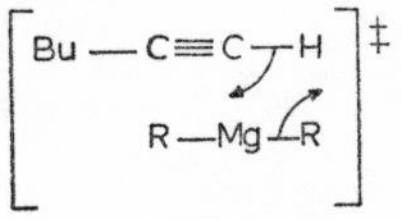

$$\left[\begin{array}{c} Bu - C \equiv C - H \\ R - Mg - R \end{array} \right]^{\ddagger}$$

(III)

3. Acidolysis of dialkylzincs and of dialkylcadmiums

The difficulties over the constitution of organomagnesium compounds mentioned above, prompted Abraham and Hill[22] to study the kinetics of acidolysis of reactive organometallic compounds of well-defined constitution. Dialkylzincs are monomeric substances[23] which may be purified by distillation, and should therefore be more suitable substrates. Acidolysis of di-n-propylzinc by the weak acids p-toluidine and cyclohexylamine at 76 °C in solvent diisopropyl ether was shown to follow kinetics compatible with the two competitive consecutive second-order reactions (15) (R = Pr^n, R' = p-tolyl or cyclohexyl) and (16) (R = Pr^n, R' = p-tolyl or cyclohexyl),

$$R_2Zn + H_2NR' \rightarrow RH + RZnNHR' \tag{15}$$

$$RZnNHR' + H_2NR' \rightarrow RH + Zn(NHR')_2 \tag{16}$$

in which the rate coefficient for the first step, reaction (15), is about six times the value of that for the second step, reaction (16). Values of the second-order rate coefficient at 76 °C in solvent diisopropyl ether were 2.1×10^{-2} $l.mole^{-1}.sec^{-1}$ for reaction (15) (R = Pr^n, R' = p-tolyl) and 1.8×10^{-3} $l.mole^{-1}.sec^{-1}$ for reaction (15) (R = Pr^n, R' = cyclohexyl).

TABLE 2

RELATIVE SECOND-ORDER RATE COEFFICIENTS FOR THE ACIDOLYSIS OF
DIALKYLZINCS BY p-TOLUIDINE, REACTION (15) (R' = p-tolyl)[22]

Me_2Zn	Et_2Zn	Pr^n_2Zn	Bu^n_2Zn	Pr^i_2Zn	Solvent
41	100	42	27	61	Ether, 35 °C
	100	33	35	67	i-Pr$_2$O, 68 °C
	100			71	Hexane, 69 °C
	100			82	THF, 66 °C

Relative reactivities of various dialkylzincs towards p-toluidine were obtained by competitive experiments (see ref. 24) in which two dialkylzincs were allowed to react with a deficiency of p-toluidine; results are given in Table 2.

For the acidolysis of unsymmetrical dialkylzincs by p-toluidine at 68 °C in solvent diisopropyl ether, relative rate coefficients were[22] k(Et) 100 to k(Pr^n) 52 and k(Pr^n) 52 to k(Bu^n) 41 with the compounds EtZnPr^n and Pr^{n}ZnBu^n respectively.

Abraham and Hill[22] suggested that the mechanism of acidolysis was that of S_E2(cyclic), in which electrophilic attack at the carbon atom undergoing substitution was an important feature; the reactivity sequence p-toluidine > cyclohexylamine is that of acid strength. A transition state such as (IV) is thus indicated.

(IV)

The acidolysis of dialkylzincs by diphenylamine in the presence of a number of complexing agents was later studied by Inoue and Yamada[57] using solvent benzene at 40 °C. Only one alkyl group was cleaved from the dialkylzinc, and rate coefficients for this cleavage were reported as shown in Table 3. Under the given conditions, more of the dialkylzinc is converted into the $R_2Zn\cdot$TMED complex than into the $R_2Zn\cdot$bipy complex, so that the enhanced reactivity of the dialkylzincs in presence of the three ligands with conjugated structures cannot be ascribed to differences in complexing ability. Inoue and Yamada[57] suggest that in these three cases, back donation from the zinc atom to the ligand reduces the electron density at zinc and facilitates co-ordination of diphenylamine to zinc in an S_E2(cyclic) transition state analogous to (IV).

The acidolysis of diethylcadmium by a number of nuclear-substituted benzyl alcohols has been studied[56] using solvent ether at 34.5 °C. Both ethyl groups were

TABLE 3

SECOND-ORDER RATE COEFFICIENTS (l.mole^{-1}.sec^{-1}) FOR THE ACIDOLYSIS OF
DIALKYLZINCS (0.30 M) BY DIPHENYLAMINE (0.30 M) IN THE PRESENCE OF
COMPLEXING AGENTS (0.30 M) IN SOLVENT BENZENE AT 40 °C[57]

Complexing agent	$10^6 k_2$		
	Et_2Zn	Bu^n_2Zn	Bu^s_2Zn
None	4.4	4.2	Very slow
1 2-Dimethoxyethane	1.9	2.2	
Triethylamine	3.1		
N, N, N', N'-Tetramethylethylenediamine	3.3	4.2	
Pyridine	27		
8-Dimethylaminoquinoline	89		
Bipyridyl	110	97	

cleaved from cadmium, the overall reaction being

$$Et_2Cd + 2\ ArCH_2OH = 2\ EtH + (ArCH_2O)_2Cd$$

The observed kinetic form was given by the equation

$$dx/dt = k_2(A_0 - x/2)(B_0 - x)$$

where x is the concentration of ethane at time t, and A_0 and B_0 are the initial
concentrations of diethylcadmium and benzyl alcohol respectively. If the acidolysis
proceeded, as might be expected, by two competitive consecutive second-order
reactions, then the above rate equation would correspond (*cf.* ref. 58) to the case
of a slow first step followed by a rapid second step. Although the first cleavage of
an alkyl group from an organometallic of type R_nM is usually faster than the
second cleavage, at least one other case is known in which a slow first cleavage
is followed by a more rapid second cleavage (see p. 126).

Second-order rate coefficients (l.mole^{-1}.sec^{-1}) for reaction of diethylcadmium
with the alcohols X–C$_6$H$_4$CH$_2$OH were recorded[56] as

X	p-OMe	p-Me	m-Me	H	m-OMe	p-Cl	m-Cl	m-CF$_3$	p-CF$_3$
$10^3 k_2$	1.8	1.5	1.4	1.3	1.2	1.1	1.0	0.98	0.85

Hammett's equation is obeyed, with $\rho = -0.35$, and it was therefore suggested[56]
that nucleophilic attack at the cadmium atom was more important than electrophilic attack at the ethyl group. A kinetic isotope effect, $k(PhCH_2OH)/k(PhCH_2OD)$
$= 2.5$, indicates that the O–H bond is broken in the rate-determining step, so that
a mechanism in which co-ordination of oxygen to cadmium slightly precedes
electrophilic attack at the ethyl group may be put forward, *viz.*

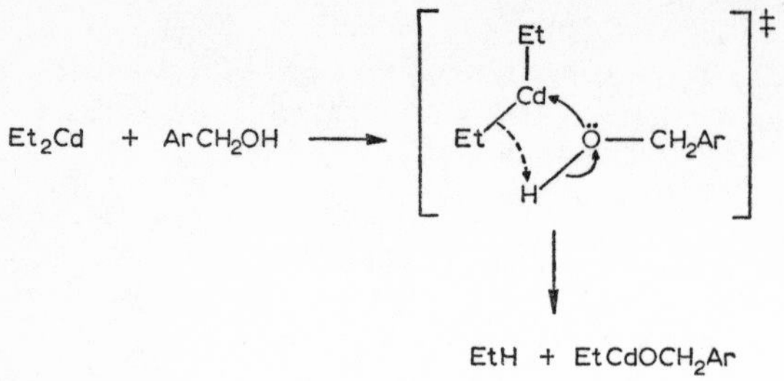

$$\text{Et}_2\text{Cd} + \text{ArCH}_2\text{OH} \longrightarrow \left[\begin{array}{c} \text{Et} \\ | \\ \text{Cd} \\ \text{Et} \diagdown \quad \diagup \ddot{\text{O}} - \text{CH}_2\text{Ar} \\ \text{H} \end{array} \right]^{\ddagger}$$

$$\downarrow$$

$$\text{EtH} + \text{EtCdOCH}_2\text{Ar}$$

4. Acidolyses of organomercury compounds

4.1 ACIDOLYSIS OF ALKYLMERCURIC IODIDES BY AQUEOUS SULPHURIC AND PERCHLORIC ACIDS

Kreevoy[25] studied the action of aqueous sulphuric and perchloric acids on methylmercuric iodide, and showed that reaction took place through the two steps shown in equations (17) and (18) (R = Me).

$$H^+ + RHgI \xrightarrow{\text{slow}} RH + HgI^+ \qquad (17)$$

$$HgI^+ + RHgI \xrightarrow{\text{fast}} RHg^+ + HgI_2 \qquad (18)$$

Kinetics were followed using a large excess of acid, and the rate-determining step (17) (R = Me) was shown to be first-order. At equal molar concentrations of the two acids, the first-order rate coefficient for reaction (17) (R = Me), k_1, using sulphuric acid was 1.71 times that for perchloric acid. The quantities $k_1/[\text{H}_2\text{SO}_4]$ and $k_1/[\text{HClO}_4]$ were constant over the range 1–6 M in acid, thus

TABLE 4

SECOND-ORDER RATE COEFFICIENTS ($1.\text{mole}^{-1}.\text{sec}^{-1}$) AT 110 °C AND ACTIVATION
PARAMETERS FOR REACTION (17)[26]

Solvent water plus 2 % methanol

R	$k_2{}^a$	$\Delta H^{\ddagger}$ ($kcal.mole^{-1}$)	$\Delta S^{\ddagger}$ ($cal.deg^{-1}.mole^{-1}$)
Me	1.72×10^{-5}	22.3 ± 0.3	-23.3 ± 1
Et	6.8×10^{-6}	22.8 ± 1.7	-25 ± 4
Pr^n	3.8×10^{-6}	23 ± 3	-23 ± 9
Hex^n	$\sim 2 \times 10^{-6}$		
Pr^i	2.2×10^{-6}	25.0 ± 2.4	-21 ± 6
cyclo-Pr	1.3×10^{-2}	20.9 ± 1.4	-11 ± 4
Bu^t	$\sim 1.4 \times 10^{-7b}$		

^{a} Rate coefficients with 1 M sulphuric acid.
^{b} Solvent water plus 6 % methanol.

suggesting[25] that a water molecule is incorporated in the transition state, probably as $\overset{+}{H-OH_2}$. No solvent deuterium isotope effect was observed and it was considered that the Me–Hg bond was almost completely broken in the transition state, whilst the $\overset{+}{H-OH_2}$ bond was only slightly weakened.

In a subsequent paper, Kreevoy and Hansen[26] reported rate coefficients and activation parameters for reaction (17) for a number of alkylmercuric iodides; details are in Table 4.

Experimental difficulties were encountered in following the kinetics of acidolysis of the isopropyl- and *tert.*-butyl-mercuric iodides, since these compounds were susceptible to atmospheric oxidation.

The transition state

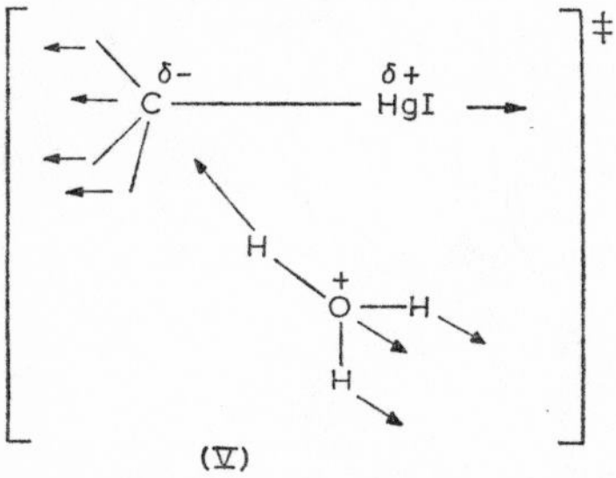

(Ⅴ)

was advanced[26] for reaction (17); in the terminology used at present this corresponds to mechanism S_E2(open).

4.2 ACIDOLYSIS OF DIALKYLMERCURYS BY ACETIC ACID AND BY PERCHLORIC ACID IN SOLVENT ACETIC ACID

When dialkylmercurys are allowed to react with acetic acid, the latter also acting as the solvent, one alkyl group only is cleaved from the mercury atom, *viz.*

$$R_2Hg + CH_3CO_2H \rightarrow RH + RHgOCOCH_3 \tag{19}$$

Winstein and Traylor[27] showed that under the above conditions, that is with an excess of acetic acid, reaction (19) follows kinetics first-order in dialkylmercury; their reported first-order rate coefficients are collected in Table 5. Relative rates of acetolyses at 25 °C are thus:

$$Bu_2{}^sHg\ (640),\ Bu_2{}''Hg\ (67),\ 4\text{-camphyl}_2Hg\ (14),\ neophyl_2Hg(1) \tag{20}$$

Addition of sodium acetate does not affect the observed first-order rate coefficient, and it was suggested that reaction (19) is an elementary reaction with acetic acid acting as the electrophile in an S_E2(cyclic) reaction, proceeding through the transition state

TABLE 5

FIRST-ORDER RATE COEFFICIENTS (sec^{-1}) FOR THE ACETOLYSIS OF DIALKYLMERCURYS, REACTION (19)[27]

Compound	Temp. (°C)	k_1
Bu^s_2Hg	25.00	2.30×10^{-5}
	50.01	1.47×10^{-4}
Bu^n_2Hg	25.0	2.4×10^{-6a}
	50.06	3.76×10^{-5}
	75.05	3.91×10^{-4}
Neophyl$_2$Hg	25.0	3.6×10^{-8a}
	75.07	1.02×10^{-5}
	74.97	0.95×10^{-5b}
	99.61	9.35×10^{-5}
4-Camphyl$_2$Hg	25.0	5.2×10^{-7a}
	49.6	5.1×10^{-6}
	75.0	3.75×10^{-5}

[a] Extrapolated value.
[b] In the presence of 0.0492 M added sodium acetate.

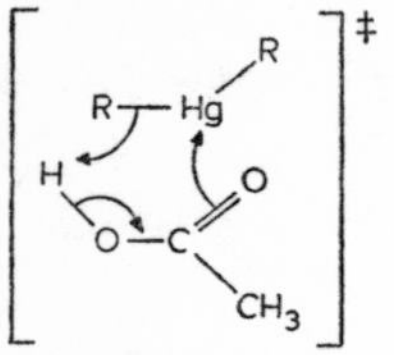

(VI)

The fact that di-4-camphylmercury (VII) is not especially unreactive (see sequence (20)) indicates that the acetolysis involves retention of configuration at the carbon atom undergoing substitution. If inversion of configuration was the preferred mode, (VII) would be expected to be extremely unreactive.

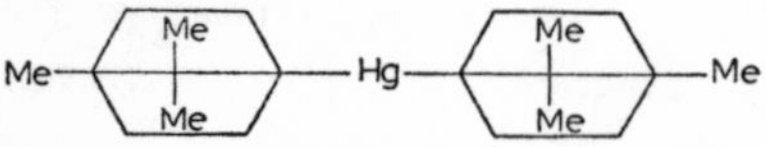

(VII)

Activation parameters calculated by Winstein and Traylor[27] for the first-order acetolyses are Bu^n_2Hg ($\Delta H^{\ddagger} = 20.4$ kcal.mole^{-1}, $\Delta S^{\ddagger} = -16$ cal.deg^{-1}.mole^{-1}), neophyl$_2$Hg ($\Delta H^{\ddagger} = 22.6$, $\Delta S^{\ddagger} = -17$), and 4-camphyl$_2$Hg ($\Delta H^{\ddagger} = 17$, $\Delta S^{\ddagger} = -12$).

Addition of perchloric acid to the solvent acetic acid leads to a second-order acidolysis, with the rate proportional to $[R_2Hg][HClO_4]$. The relative reactivity of di-4-camphylmercury to di-neophylmercury is $\simeq 5:1$ for this second-order component, and mechanism S_E2(open) was now suggested to obtain, by way of transition state (VIII), the electrophile being $CH_3CO_2H_2{}^+$.

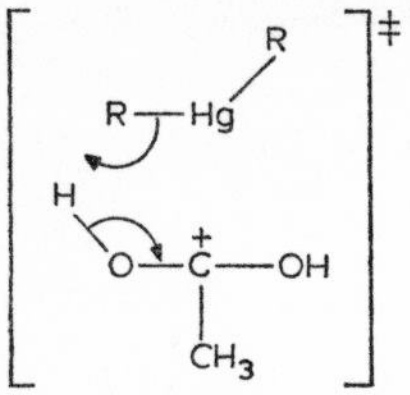

(VIII)

4.3 ACIDOLYSIS OF DIALKYLMERCURYS BY HYDROGEN CHLORIDE

The acidolysis of dialkylmercurys by hydrogen chloride in solvent DMSO-(10) : dioxan (1) was shown by Dessy et al.[28], to proceed by the simple reaction

$$R_2Hg + HCl \rightarrow RH + RHgCl \tag{21}$$

Rate coefficients and activation parameters for the substitution of a number of dialkylmercurys are in Table 6.

The rate of substitution of diethylmercury by hydrogen chloride in DMSO was increased on addition of dioxan and on addition of sodium chloride, but was unaffected by added sodium sulphate. The corresponding substitution of diphenylmercury was retarded by the addition of water to the DMSO solvent, and was unaffected by added sulphuric acid. All of these observations indicate[28] that the electrophile is undissociated hydrogen chloride (i.e. unionised hydrogen chloride plus ionised but not dissociated hydrogen chloride; in other words, covalent

TABLE 6

SECOND-ORDER RATE COEFFICIENTS $(l.mole^{-1}.sec^{-1})$ AND ACTIVATION PARAMETERS FOR THE SUBSTITUTION OF DIALKYLMERCURYS BY HYDROGEN CHLORIDE IN SOLVENT DMSO(10) : DIOXAN(1), REACTION (21)[28,29]

E_a in kcal.mole^{-1}(± 0.5), and $\Delta S^{\ddagger}$ in cal.deg^{-1}.mole^{-1}(± 2).

R_2Hg	$k_2(25\,°C)$	$k_2(40\,°C)$	$k_2(50\,°C)$	E_a	$\Delta S^{\ddagger}$
Ph_2Hg	9.3×10^{-3}	2.5×10^{-2}		12.2	-29
cyclo-Pr_2Hg	6.6×10^{-2}	2.5×10^{-1}		16.5	-11
Me_2Hg			1.3×10^{-4}		
Et_2Hg	0.82×10^{-4a}	3.8×10^{-4}	8.2×10^{-4}	15.5	-27
Pr^n_2Hg	0.51×10^{-4a}	2.3×10^{-4}	5.1×10^{-4}	16.5	-25
Pr^i_2Hg	0.56×10^{-4a}	2.6×10^{-4}	5.6×10^{-4}	15.4	-28

^{a} Values of 0.82×10^{-3}, 0.51×10^{-3}, and 0.56×10^{-3} were given[29], but these are clearly misprints.

H–Cl plus intimate ion-pairs, H^+Cl^-). A mechanism involving concerted attack by the proton and by the chloride ion was proposed. Since the deuterium isotope effect of DCl was very small, it was suggested[28] that the H–Cl bond in the transition state was only slightly weakened. In the present terminology, the mechanism would be regarded as S_E2(cyclic), perhaps verging towards S_E2(co-ord), *viz.*

(IX)

Dessy *et al.*[28] concluded by stating that nucleophilic attack on mercury was as important as electrophilic attack on carbon. In a subsequent paper, Dessy and Kim[29] state, "It is obvious that the important rate-determining attack lies with the H, and not with the Cl" but the earlier conclusion would seem to be more in keeping with the experimental observations*.

Reaction (21) has been studied for a few other miscellaneous cases; details are given in Table 7.

TABLE 7

SECOND-ORDER RATE COEFFICIENTS ($l.mole^{-1}.sec^{-1}$) FOR REACTION (21)

R_2Hg	k_2	Solvent	Ref.
Et$_2$Hg	1.95×10^{-3}(50 °C)	Dioxan(9) : water(1)	30
	7.52×10^{-3}(60 °C)		
	2.57×10^{-2}(70 °C)		
Et$_2$Hg	1.9×10^{-3}(40 °C)	DMF(10) : dioxan(1)	1
	6.4×10^{-3}(50 °C)		
Et$_2$Hg	3.2×10^{-5}(20 °C)	THF+15 % water	31
Bu$_2$Hg	1.18×10^{-2}(70 °C)	Dioxan(9) : water(1)	30

Jensen and Rickborn[60] have severely criticised the work of Dessy on the acidolysis of dialkylmercurys by hydrogen chloride. They state that, in their view, "no firm mechanistic conclusions can be drawn from the published results", although a four-centred mechanism at present seems to be reasonable.

It is not possible to extrapolate the mechanistic findings of Dessy *et al.*[28] to reactions in other solvents, as will be evident from the activation parameters for reaction (21) (R = Et), *viz.*

* Dessy *et al.*[28, 29] do not distinguish between aromatic substitution and aliphatic substitution and their conclusions as to the mechanism of reaction (21) apparently include both of these types of substitution.

Solvent	E_a $(kcal.mole^{-1})$	$\Delta S^{\ddagger}$ $(cal.deg^{-1}.mole^{-1})$	Ref.
DMSO(10) : dioxan(1)	15.5	-27	28
DMF(10) : dioxan(1)	24.4	$+5$	1
Dioxan(9) : water(1)	28.4	$+15$	30

A variation of some 42 cal.deg^{-1}.mole^{-1} in $\Delta S^{\ddagger}$ must indicate considerable changes in either the reactants or the transition state (or both) on transfer from one solvent to another.

4.4 A NOTE ON THE KHARASCH "ELECTRONEGATIVITY" SERIES

Over a period of years, Kharasch et al.[32, 33], and also Whitmore and Bernstein[34], studied the cleavage of unsymmetrical dialkylmercurys by hydrogen chloride in solvent ethanol, viz.

$$\text{RHgR}' + \text{HCl} \underset{\text{EtOH}}{\overset{\text{EtOH}}{\lessgtr}} \begin{array}{l} \text{RH} + \text{R'HgCl} \\ \text{R'H} + \text{RHgCl} \end{array} \tag{22}$$

Relative rates of cleavage of the groups R and R' were obtained, and the qualitative sequence

$$\text{Me} > \text{Et} > \text{Pr}^n > \text{Bu}^n > \text{Pr}^i > \text{Heptyl}^n > \text{Bu}^s > \text{Pe}^{neo}, \text{Bu}^t \tag{23}$$

may be constructed for the tendency of alkyl groups to be cleaved from mercury under the above conditions. Kharasch and Flenner[33] probably viewed* the mechanism of reaction (22) as S_E1, for which we may write

$$\text{R}_2\text{Hg} \xrightarrow{\text{slow}} \text{R}^- + \text{HgR}^+ \tag{24}$$

$$\text{R}^- + \text{HCl} \xrightarrow{\text{fast}} \text{RH} + \text{Cl}^- \tag{25}$$

In line with a mechanism involving a rate-determining ionisation (24), sequence (23) was considered to be that of the electronegativity of alkyl groups. The methyl

---

* "We consider the radical which dissociates more readily from the mercury and then combines with the hydrogen ion in solution to be the more electronegative of the two radicals"[33]. It is not clear if this statement refers to dissociation of a negatively charged radical (using the term "radical" in the sense of "group") or to dissociation of a free radical. The latter would appear unlikely since it is difficult to imagine Kharasch and Flenner envisaging a simple combination of a free radical with a hydrogen ion.

group was thus taken as the most electronegative group (of those in sequence (23)) and hence most able to support a negative charge on the corresponding carbanion.

The S_E1 mechanism, equations (24) and (25), implies that dialkylmercurys should ionise according to equation (24) in solvent ethanol, even in the absence of hydrogen chloride. But dialkylmercurys are quite stable in ethanolic solution and thus reaction (24) would seem to be unrealistic; Winstein and Traylor[27] have already pointed out that the typical dialkylmercurys used in the studies of Kharasch do not follow an S_E1 mechanism. There is thus very little theoretical basis for regarding (23) as an electronegativity series.

Although mechanism S_E1 is ruled out, the electrophilic substitution (22) might well follow mechanism S_E2(cyclic). If electrophilic attack at the carbon atom was rather more important than nucleophilic attack at the mercury atom, as in transition state (X), application of the theory of Abraham and Hill[35] would lead to the

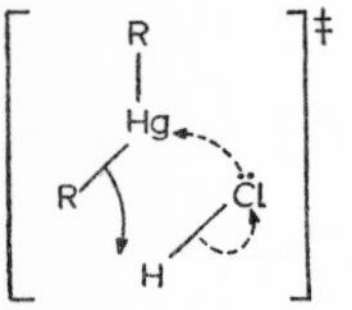

(X)

conclusion that steric effects of the alkyl groups undergoing substitution would dominate the observed reactivity sequence. In the author's opinion, sequence (23) may indeed be interpreted as a steric sequence, and there is but little basis for supposing that (23) represents an order of electronegativity*.

4.5 ACIDOLYSIS OF BENZYLMERCURIC COMPOUNDS

Reaction (26) has been reported[30,36] to follow second-order kinetics with a value[30] of 3.54×10^{-3} l.mole^{-1}.sec^{-1} for the rate coefficient at 70 °C in solvent dioxan(9) : water(1) and of[36] 0.45×10^{-3} l.mole^{-1}.sec^{-1} at 69 °C in solvent dioxan**.

$$(C_6H_5CH_2)_2Hg + HCl = C_6H_5CH_3 + C_6H_5CH_2HgCl \tag{26}$$

Reutov et al.[37] claimed that the reaction proceeded by mechanism S_E1 in solvents

---

* Whether or not reaction (22) may be used to deduce electronegativity sequences amongst aryl groups is another matter.
** Second-order rate coefficients (l.mole^{-1}.sec^{-1}) are reported[36] for $(PhCH_2CH_2CH_2)_2Hg$, 1.88×10^{-3}; $(PhCH_2CH_2)_2Hg$, 1.25×10^{-3}; $(PhCH_2)_2Hg$, 0.45×10^{-3}: solvent dioxan, temperature 69 °C.

DMSO, DMF, THF, aqueous acetonitrile, and n-butanol, since the overall kinetic form was first-order in dibenzylmercury[37]. Later work[38] indicated that the situation was more complex than first thought; dibenzylmercury and DCl in solvents DMSO, DMF, acetonitrile, and dioxan afforded not only the compound $C_6H_5CH_2D$, but other deuterated species such as (XI) and (XII).

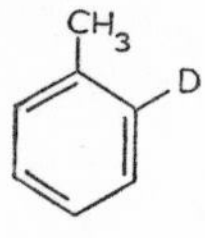

(XI) 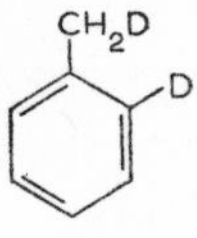   (XII) 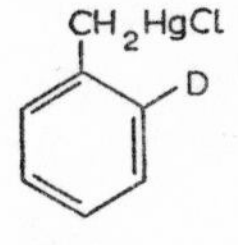   (XIII)

It was thought[38] that the mechanism was intermediate between S_E1 and S_E2. Other workers[39] have suggested that the kinetic results of Reutov *et al.*[37] suffer from irregularities arising from atmospheric oxidation of dibenzylmercury. When conducted under a nitrogen atmosphere, reaction (26) followed second-order kinetics in solvent aqueous acetonitrile[39].

Benzylmercuric chloride is also cleaved by hydrogen chloride with the stoichiometry[40]

$$C_6H_5CH_2HgCl + HCl = C_6H_5CH_3 + HgCl_2 \tag{27}$$

Using equimolar quantities of the two reactants, the observed kinetics are second-order[40]. Reported values of the rate coefficient ($l.mole^{-1}.sec^{-1}$) are 0.357×10^{-2} (60 °C), 1.00×10^{-2} (70 °C), 2.48×10^{-2} (80 °C), and the activation parameters recorded[40] are $E_a = 21$ kcal.mole^{-1} and $\Delta S^{\ddagger} = -7.4$ cal.deg^{-1}.mole^{-1}. The solvent used was slightly aqueous dioxan, made up from 10 ml dioxan and 0.04–0.08 ml of concentrated hydrochloric acid[40]. From the above values of the rate coefficient, reaction (27) would appear to be faster than reaction (26), under comparable conditions — an extraordinary situation in view of the reported stoichiometry of (26). The action of DCl on benzylmercuric chloride in solvent dioxan also leads to introduction of deuterium into the aromatic ring, compounds (XII) and (XIII) being formed[41].

4.6 THE ANION-CATALYSED ACIDOLYSIS OF α-CARBETHOXYBENZYL-MERCURIC CHLORIDE

Although α-carbethoxybenzylmercuric chloride is quite stable to perchloric acid in solvent dioxan–water (7 : 3 v/v), it undergoes demercuration in the presence of added chloride ion according to the stoichiometry

$$PhCH(CO_2Et)HgCl + H^+ + 2\,Cl^- = PhCH_2CO_2Et + HgCl_3{}^-$$

Kinetics of the demercuration were very complicated, but Coad and Ingold[55] were able to fit their data to the rate equation

$$\text{Rate} = \frac{k[\text{RHgCl}][\text{H}_3\text{O}^+]^2[\text{Cl}^-]^2}{[\text{HgCl}_3{}^-]}$$

where RHgCl refers to the α-carbethoxybenzylmercuric chloride. A mechanism consistent with this expression was put forward by Coad and Ingold[55] and is shown below. The actual cleavage of the carbon–mercury bond occurs in step (iii), where a bimolecular protolysis of the substrate takes place with rearrangement. The overall mechanism may be described[55] as an anion-catalysed bimolecular protolysis with rearrangement, S_E2'–2Cl^-.

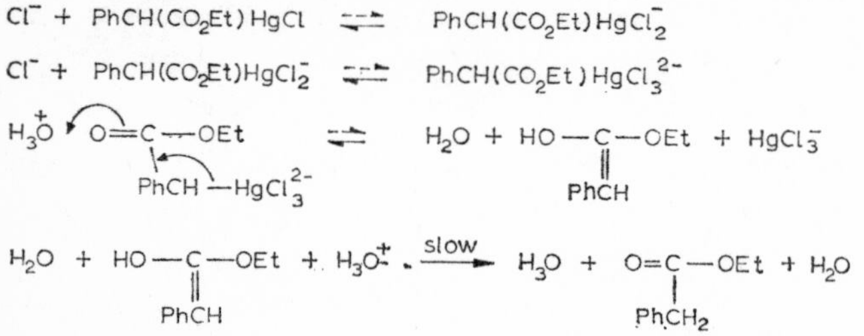

$$\text{Cl}^- + \text{PhCH(CO}_2\text{Et)HgCl} \; \rightleftharpoons \; \text{PhCH(CO}_2\text{Et)HgCl}_2^- \qquad (i)$$

$$\text{Cl}^- + \text{PhCH(CO}_2\text{Et)HgCl}_2^- \; \rightleftharpoons \; \text{PhCH(CO}_2\text{Et)HgCl}_3^{2-} \qquad (ii)$$

(iii)

(iv)

5. Acidolysis of triethylboron by carboxylic acids

An extensive investigation by Dessy and co-workers[42] showed that the acidolysis

$$\text{Et}_3\text{B} + \text{RCO}_2\text{H} \xrightarrow{\text{diglyme}} \text{EtH} + \text{RCO}_2\text{BEt}_2 \qquad (28)$$

was first-order in triethylboron and first-order in carboxylic acid. Second-order rate coefficients for the acidolysis were obtained using a number of aliphatic and aromatic carboxylic acids, as well as for the related acidolyses with several non-carboxylic acids. Table 8 records results for the aliphatic carboxylic acids together with their pK values[43]. As pointed out by Dessy and co-workers[42], there is a linear relationship between $\log k_2$ and pK for these acids, the weaker the acid the faster being the acidolysis (28). A suggested mechanism involved the sequence

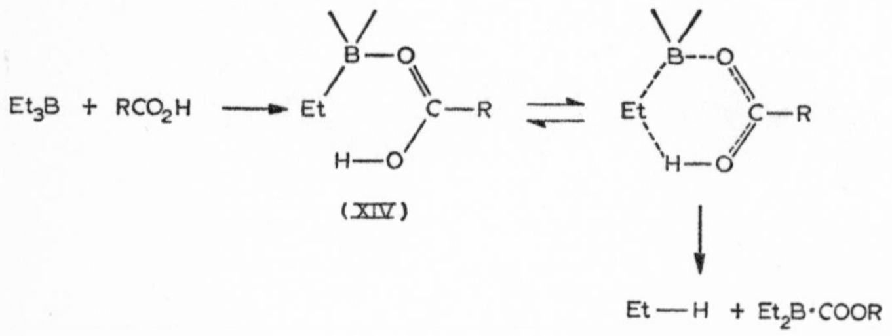

(29)

TABLE 8

SECOND-ORDER RATE COEFFICIENTS ($\text{l.mole}^{-1}.\text{sec}^{-1}$) FOR THE ACIDOLYSIS OF TRIETHYLBORON BY CARBOXYLIC ACIDS IN SOLVENT DIGLYME AT 31 °C, REACTION $(28)^{42}$

Acid	k_2	Acid pK
$(CH_3)_3CCO_2H$	8.1×10^{-3}	5.03
$CH_3(CH_2)_6CO_2H$	5.9×10^{-3}	4.89
$CH_3CH_2CO_2H$	4.8×10^{-3}	4.87
CH_3CO_2H	4.7×10^{-3}	4.75
$PhCO_2H$	4.2×10^{-3}	4.19
$PhCH_2CO_2H$	3.6×10^{-3}	4.28
Ph_2CHCO_2H	2.4×10^{-3}	3.94
$ClCH_2CO_2H$	5.1×10^{-4}	2.85
Cl_2CHCO_2H	5.9×10^{-5}	1.48
Cl_3CCO_2H	Very slow	0.70
$CH_3CO_2D^a$	1.4×10^{-4} $(1.4 \times 10^{-3})^a$	

[a] See footnote below.

Sequence (29) was described[42] as a pre-rate-determining nucleophilic coordination of the carboxylic acid, followed by a proton transfer to carbon. That proton transfer is involved in a rate-determining step as indicated by the ratio k_H/k_D of 3.3 (see Table 8)*. As written in sequence (29), if formation of (XIV) is held to be pre-rate-determining, then (XIV) must be formed very rapidly and must decompose slowly to the products. But if this were so, the observed kinetic order would be unity (see Chapter 3, Section 1.4, p. 13). A more correct sequence would appear to be that shown in (30), where the complex (XIV) is formed rapidly and reversibly.

$$Et_3B + RCO_2H \underset{\text{fast}}{\overset{\text{fast}}{\rightleftharpoons}} (XIV) \xrightarrow{\text{slow}} \text{products} \tag{30}$$

Unlike the trialkyls of the other Group IIIA elements, those of boron are remarkably inert towards water and aqueous strong acids. Consideration of Pauling-scale electronegativities[3], *viz.* B(2.0), Al(1.5), Ga(1.6), In(1.7), Tl(1.8), Hg(1.9), Sn^{IV}(1.9), and C(2.5), suggests that a carbon atom attached to boron should indeed be rather inert towards electrophilic reagents since the percentage ionic character[44a] of the B–C bond is only some 6 %. In this respect boron alkyls might be expected to resemble those of mercury and tin; significantly, the alkyls of type R_2Hg and R_4Sn are also inert towards water and aqueous strong acids. We may consider that two requirements are necessary for an acid to cleave the B–C bond, (*i*) the acid should be a good electrophile, *i.e.* should be a strong acid,

---

* Values for k_2 are given[42] as 4.7×10^{-3} (CH_3CO_2H) and 1.4×10^{-4} (CH_3CO_2D). Since it is stated[42] that $k_H/k_D = 3.3$ for these two acids, it is possible that the value for CH_3CO_2D is a misprint for 1.4×10^{-3}.

and (ii) it should be a good nucleophile and thus able to weaken the B–C bond by coordination to the boron atom. Hence aqueous strong acids are not effective reagents since although H_3O^+ is a good electrophile, it is a very poor nucleophile. Very weak acids such as water, alcohols, and amines are also not effective reagents since they are very poor electrophiles, although good nucleophiles. Carboxylic acids, and also some other weak acids, seem to possess just the right amount of electrophilic and nucleophilic character to be effective as acidolysis reagents.

6. Acidolysis of the alkyls of metals in Group IVA

The cleavage of a methyl group from $Me_3Si(CH_2)_2CO_2H$ by sulphuric acid at high concentration in water has been shown to be first-order in the organometallic substrate and first-order in "unionised H_2SO_4". More complex kinetics were found[45] for the analogous cleavage of $Me_3Si(CH_2)_3CO_2H$. The acidolysis of Me_4Si and Me_3SiCH_2Cl by an excess of fluorosulphonic acid at $-47\,°C$ yields methane[61]; the acid, diluted 1 : 1 with sulphur dioxide, also acted as the solvent. Reported first-order rate coefficients are 4.3×10^{-3} sec^{-1} (Me_4Si) and 4.6×10^{-5} sec^{-1} (Me_3SiCH_2Cl). Similarly, Me_3SiOSO_2F and $Me_2Si(CH_2Cl)OSO_2F$ were cleaved by an excess of neat fluorosulphonic acid at $57.5\,°C$ to yield methane with first-order rate coefficients 1.92×10^{-5} sec^{-1} and 1.90×10^{-6} sec^{-1} respectively. It was suggested[61] that the electron density at the silicon atom was an important factor in the acidolysis and hence that the electron-withdrawing CH_2Cl group was responsible for lower rates of methane production in the two sets of acidolyses.

It has been reported[46] that the relative rates of cleavage of tetraalkyltins by hydrogen chloride are $Me_4Sn(1) : Et_4Sn(7.5) : Pr^n_4Sn(3) : Pr^i_4Sn(3)$ in solvent benzene, and $Me_4Sn(1) : Et_4Sn(3)$ in solvent dioxan.

The acetolysis of a number of tetraalkylleads was studied by Robinson[2] who used an excess of acetic acid as the solvent. He found that the acetolyses were first-order in tetraalkyllead; rate coefficients and activation parameters for reaction (31) are given in Table 9. Robinson also studied the cleavage of tetraalkylleads by perchloric acid in solvent acetic acid.

$$R_4Pb + CH_3CO_2H \rightarrow RH + R_3PbO\cdot COCH_3 \tag{31}$$

With initial concentrations of tetraalkyllead and perchloric acid at about 0.01 M, the cleavage proved to be first-order in tetraalkyllead and first-order in perchloric acid; the second-order rate coefficients are also given in Table 9. Apart from suggesting that both the acetolysis and the perchloric acid cleavage were examples of mechanism S_E2, Robinson[2] refrained from mechanistic speculation. For the perchloric acid cleavage, however, it seems reasonable to postulate a transition state such as (XV).

TABLE 9

FIRST-ORDER RATE COEFFICIENTS (sec^{-1}) AND ACTIVATION PARAMETERS FOR THE ACETOLYSIS OF TETRAALKYLLEADS, REACTION (31)[2]

$\Delta H^{\ddagger}$ in kcal.mole^{-1}($\pm$0.3) and $\Delta S^{\ddagger}$ in cal.deg^{-1}.mole^{-1}($\pm$1)

R_4Pb	$10^5 k_1$			$\Delta H^{\ddagger}$	$\Delta S^{\ddagger}$	$10^2 k_2$[a] (25 °C)
	24.9 °C	49.8 °C	60.0 °C			
Me$_4$Pb	1.16	17.1	41.2	20.8	-12	22
Et$_4$Pb	0.80	10.9	28.2	20.1	-15	2.4
Pr$^n_4$Pb	0.225	3.1	8.2	20.7	-15	0.89
Bu$^n_4$Pb	0.31	4.3	10.6	21.2	-13	0.68
iso-Amyl$_4$Pb	0.30	4.2	12.2	21.2	-13	0.59

[a] Second-order rate coefficients, in l.mole^{-1}.sec^{-1}, for acidolysis by perchloric acid in solvent acetic acid.

(XV)

Horn and Huber[47] have comprehensively studied the acetolysis of tetraethyllead by acetic acid in solvent anhydrous toluene. In contrast to Robinson[2] who observed only reaction (32) when acetic acid was used as solvent, Horn and Huber showed that the two competitive consecutive reactions (32) and (33)

$$Et_4Pb + CH_3CO_2H \rightarrow EtH + Et_3PbO \cdot COCH_3 \tag{32}$$

$$Et_3PbO \cdot COCH_3 + CH_3CO_2H \rightarrow EtH + Et_2Pb(O \cdot COCH_3)_2 \tag{33}$$

took place, and they were able to determine the second-order rate coefficient for reaction (33) simply by allowing triethyllead acetate to react with acetic acid in

TABLE 10

SECOND-ORDER RATE COEFFICIENTS (l.mole^{-1}.sec^{-1}) FOR THE ACETOLYSIS OF TETRAETHYLLEAD (REACTION (32)) AND TRIETHYLLEAD ACETATE (REACTION (33)) IN SOLVENT TOLUENE[47]

Temp. (°C)	$k_{(32)}$	$k_{(33)}$	$k_{(33)}/k_{(32)}$
60.0	0.44×10^{-5}	0.21×10^{-3a}	48
80.0	2.18×10^{-5}	1.37×10^{-3}	63
100.0	5.83×10^{-5}	12.5×10^{-3}	214
E_a(kcal.mole^{-1})	15	22.1	

[a] Extrapolated value, using the reported activation energy[47].

separate experiments. Both (32) and (33) were shown[47] to follow second-order kinetics, first-order in each reactant, and the rate coefficients and activation energies are given in Table 10. The activation energy for reaction (32) is reduced by almost 6 kcal.mole^{-1} as the solvent is changed from acetic acid to toluene; part of this reduction, of course, may be due to the solvent effect on the reactants, but since acetic acid is probably dimerised in toluene as well as in neat acetic acid[44b] there is no obvious specific solvent effect to invoke.

The activation energies for reactions (32) and (33) in solvent toluene differ by 7 kcal.mole^{-1} and hence the ratio of the rate coefficients k_{33}/k_{32} is temperature-dependent, being 48 at 60 °C and 214 at 100 °C. A normal second-order plot of $x/a(a-x)$ versus t for the case of equal initial concentrations* (0.2 M) of tetraethyllead and acetic acid at 60 °C gives a good straight line for the first 13 % reaction, but a similar plot at 100 °C deviates from a straight line even at 5 % reaction, due to the much faster relative rate of reaction (33) at 100 °C compared with that at 60 °C.

It is remarkable that reaction (33) is much faster than reaction (32) in solvent toluene, whereas reaction (33) was not detected[2] when acetic acid was the solvent.

Bade and Huber[62] have reported second-order rate coefficients for the acidolysis of Me$_4$Pb and Me$_3$PbO·COCD$_3$ by the acid CD$_3$CO$_2$D. At 100 °C in solvent nitrobenzene containing 0.135 mole.l^{-1} of toluene, the coefficients were 4.08×10^{-4} l.mole^{-1}.sec^{-1} (for Me$_4$Pb) and 0.938×10^{-4} l.mole^{-1}.sec^{-1} (for Me$_3$PbO·COCD$_3$).

Second-order rate coefficients for acidolysis by hydrochloric acid in solvent 20 % methanol : 80 % benzene at 50 °C were found[63] to be

Compound:	Me$_4$Pb	Ph$_4$Pb	(p-MeOC$_6$H$_4$)$_4$Pb
k_2(l.mole^{-1}.sec^{-1}):	0.0473	0.717	3.28

The acidolysis of a series of tetraalkyltins and tetraalkylleads by hydrochloric acid in solvent methanol has also been studied by Bade and Huber[63]; second-order rate coefficients are given in Table 11. The value of K_a for hydrochloric acid in methanol at 25 °C is[64] 0.059 l^{-1}.mole, so that at the acid concentrations used by Bade and Huber (typically 0.126 l^{-1}.mole) about half the acid must be present as unionised HCl and about half present as the ionised species (MeOH$_2^+$ + Cl$^-$)**.

---

* This particular case is quoted only because the author found it more convenient to calculate than the general second-order expression. Full numerical details are given by Horn and Huber[47] for various combinations of initial concentrations.

** Although K_a refers to 25 °C, the kinetic experiments were carried out at 50 °C so that a complete calculation would require the value of K_a at 50 °C. Nevertheless, it is clear that significant amounts of both unionised and ionised forms will be present under the kinetic conditions.

TABLE 11

SECOND-ORDER RATE COEFFICIENTS ($l.mole^{-1}.sec^{-1}$) FOR THE ACIDOLYSIS OF ALKYL–TIN AND ALKYL–LEAD COMPOUNDS BY HYDROCHLORIC ACID IN METHANOL AT 50 °C[63]

| R | 10^4k_2 | | | |
	R_4Sn	R_3SnCl	R_4Pb	R_3PbCl
Me	2.30	0.173	187.7	0.320
Et			40.8	0.370
Pr^n			23.0	1.892
Bu^n	0.82	0.400	21.0	1.138
Pr^i	0.68	0.192		

Since tetraalkyltins and tetraalkylleads are mostly inert towards strong (*i.e.* ionised) acids, it seems to the author quite possible that unionised HCl is the actual electrophilic species. Bade and Huber[63] have discussed the sequences for acidolysis of the tetraalkyls shown in Table 11. They suggest that from considerations of bond strength, the rate of acidolysis should be in the order $Me_4M < Et_4M$ because the Me–M bond is much stronger than the Et–M bond (see Table 12). On the other hand, the products of acidolysis are solvated in the sense $Me_3M > Et_3M$, so that if the energy of solvation (Me_3M less Et_3M) were great enough to counteract the bond energy differences, then the reactivity sequence could shift to the order $Me_4M > Et_4M$, as observed.

It seems to the author, however, that other factors besides the ones considered by Bade and Huber must be involved. For example, in the cleavage of tetraalkyltins by mercuric chloride in methanol, bond energy differences are the same as in the acidolysis reaction and so are solvation differences because in both reactions the products are the compounds R_3SnCl. Hence if only bond energy differences and R_3SnCl solvation differences were important, the same reactivity sequences would be expected. But this is not observed at all as can be seen from the series of relative reactivity ($Me_4Sn = 100$) in Table 13. The reactivity sequence for

TABLE 12

MEAN BOND DISSOCIATION ENERGIES, $\bar{D}$, AND BOND DISSOCIATION ENERGIES, D, IN KCAL.MOLE^{-1} AT 25 °C

| | R | | | | | | |
	Me	*Et*	Pr^n	Bu^n	Pr^i	Bu^t	*Ref.*
$\bar{D}$(Sn–R)	52	46	47	47	42[a]		65
$\bar{D}$(Pb–R)	37	31					65
$D(Me_3Sn–R)$	69	64	63[a]		60[a]	54[a]	66

[a] Calculated by the author from data in refs. 65–67.

TABLE 13

REACTIVITY SEQUENCES IN THE SUBSTITUTION OF TETRAALKYLTINS

Reactants	Solvent	R					Ref.
		Me	Et	Pr^n	Bu^n	Pr^i	
$R_4Sn+HgCl_2$	MeOH	100	0.22	0.04	0.04	$<10^{-6}$	68
$R_4Sn+HCl$	MeOH	100			36	30	63
$R_3SnCl+HCl$	MeOH	100			230	110	63
$R_4Sn+HCl$	Benzene	100	750	300		300	46

acidolysis in methanol is nearer, if anything, to that for acidolysis in benzene than to the cleavage by mercuric chloride. Yet solvation of R_3SnCl in benzene must be very small compared with solvation in methanol, and solvation differences in benzene (with respect to the group R) must also be small. Hence, on the hypothesis of Bade and Huber, in benzene the sequence should follow that predicted by bond strengths, *viz.* $Me_4Sn \ll Et_4Sn \simeq Pr^n_4Sn \ll Pr^i_4Sn$. To the author, it is evident that one additional factor that must be introduced is the mechanism of reaction; whether reaction follows mechanism S_E2(open) or S_E2(cyclic) is clearly going to affect reactivity sequences. For the reactions in Table 13, it has been shown that cleavage by mercuric chloride in methanol follows mechanism S_E2(open) and on intuitive grounds it is likely that the acidolysis in benzene follows mechanism S_E2(cylic). The exact mechanism of acidolysis in methanol is not known, but if reaction proceeded through a cyclic transition state with a degree of charge separation (*i.e.* a transition-state intermediate between an "open" and a "cyclic non-polar" state), then the observed sequence is comprehendable.

Bade and Huber[63] also give rate coefficients for the cleavage of the second alkyl group in the acidolysis reaction. The species present in the second stage could be either R_3MCl or R_3M^+ or both. Now Riccoboni[69] has reported that K_D for Et_3SnCl in methanol is 6×10^{-4} l^{-1}.mole, so that it may be calculated that at a (formal) concentration of 0.063 l^{-1}.mole, the concentration used typically by Bade and Huber, nearly all the tin is present as Et_3SnCl. Hence in Table 11, rate coefficients for the second cleavage are given as rate coefficients for acidolysis of R_3MCl. It was suggested[63] that in a reaction going from R_3MCl to R_2MCl_2, solvation differences would largely cancel out and the sequence of reactivity would then follow that expected on bond energy values. Although the reactivity sequence does not correlate well with bond energies (or the inverse of bond energies), support for Bade and Huber's contention that solvation differences would tend to cancel is obtained from a comparison of the reactivity sequence for acidolysis of R_3SnCl in methanol with R_4Sn in benzene (Table 13). Clearly solvation in benzene must be minimal, yet the two sequences are extremely close.

7. Acidolysis of the alkyls of metals in Group VIII and Group VI B

7.1 ACIDOLYSIS OF TRANS-BIS(TRIETHYLPHOSPHINE)HALO(METHYL) PLATINUM(II)

Kinetics of reaction (34) (L = PEt$_3$, X = Cl and I)

$$trans\text{-}(PtL_2MeX) + H^+ + Cl^- = MeH + trans\text{-}(PtL_2ClX) \tag{34}$$

have been studied by Belluco et al.[48]. With an excess of p-toluene sulphonic acid and an excess of chloride ion in solvent methanol, reaction (34) was found to follow first-order kinetics with respect to the complex trans-(PtL$_2$MeX). The observed first-order rate coefficients, k_1^{obs}, are given in Table 14 for various experiments run under conditions of constant ionic strength, and from the values of k_1^{obs} it may be deduced[48] that the overall rate law is of the form shown in equation (35) ((XVI) = trans-(PtL$_2$MeX)), viz.

$$-d[(XVI)]/dt = k_2[H^+][(XVI)] + k_3[H^+][Cl^-][(XVI)] \tag{35}$$

TABLE 14

FIRST-ORDER RATE COEFFICIENTS (sec^{-1}) FOR THE ACIDOLYSIS OF trans-(PtL$_2$MeX) IN SOLVENT METHANOL AT 40 °C[48]

Ionic strength 0.058 M

X	[Cl$^-$] (M)	[H$^+$] (M)	$10^4 k^{obs}$
Cl	0.01	0.020	9.4
Cl	0.01	0.015	7.65
Cl	0.01	0.0125	6.3
Cl	0.01	0.010	4.42
Cl	0.01	0.008	4.4
Cl	0.01	0.005	2.5
Cl	0.01	0.003	1.54
I	0.005	0.025	1020
I	0.005	0.020	790
I	0.005	0.014	550
I	0.005	0.011	420
I	0.005	0.008	335
I	0.005	0.005	210
I	0.005	0.003	130
Cl	0.020	0.008	6.2
Cl	0.015	0.008	4.6
Cl	0.010	0.008	4.4
Cl	0.008	0.008	3.35
Cl	0.005	0.008	2.5
Cl	0.003	0.008	2.14
I	0.025	0.011	390
I	0.0126	0.011	392
I	0.005	0.011	420

The observed first-order rate coefficient is thus given by

$$k_1^{obs} = k_2[H^+] + k_3[H^+][Cl^-] \tag{36}$$

and it follows that a plot of k_1^{obs} against $[H^+]$ should be a straight line of slope $(k_2 + k_3[Cl^-])$ and of zero intercept at constant ionic strength and constant $[Cl^-]$. At constant ionic strength and constant $[H^+]$, a plot of k_1^{obs} against $[Cl^-]$ should yield a straight line of slope $k_3[H^+]$ and intercept $k_2[H^+]$. In this way, the values of k_2 and k_3, below,

Complex	$k_2(l.mole^{-2}.sec^{-1})$	$k_3(l^2.mole^{-2}.sec^{-1})$
trans-(PtL$_2$MeCl)	1.78×10^{-1}	3.0
trans-(PtL$_2$MeI)	4.1	0

were obtained[48] at ionic strength 0.058 M and 40 °C for the two complexes studied.

It was suggested[48] that the second-order term in equation (35) corresponds to a mechanism in which the complex trans-(PtL$_2$MeX) is converted to a six-coordinate platinum(IV) complex by attack of H^+ and a solvent molecule, S. The six-coordinate complex then slowly eliminates methane to yield trans-(PtL$_2$XS) and finally this latter complex presumably is converted to trans-(PtL$_2$XCl) in a rapid substitution reaction, viz.

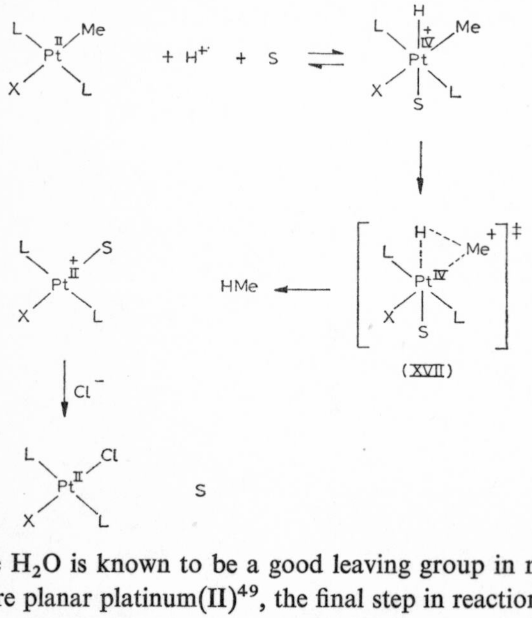

$$(37)$$

Since H_2O is known to be a good leaving group in nucleophilic substitutions at square planar platinum(II)[49], the final step in reaction (37) would be expected to be very rapid with MeOH as the leaving group.

For the iodo complex, reaction (37) is the only pathway available, k_3 being zero, but the complex trans-(PtL$_2$MeCl) is also acidolysed by a mechanism corresponding to the third-order term in equation (35). Mechanism (38) was suggested[48] for this reaction path.

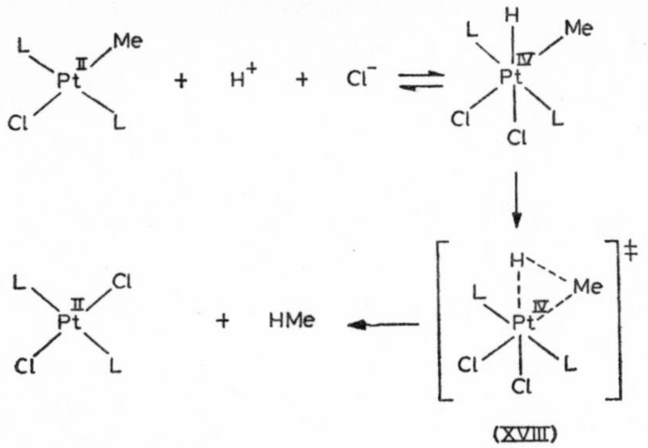

$$(38)$$

(XVIII)

Further experiments revealed that reaction (34) (X = Cl) was retarded by an increase in ionic strength and by addition of water to the methanol solvent. The latter experiments[48] (see Table 15) were carried out with [Cl$^-$] = 0.1 M and, under these conditions, the third-order term in equation (35) must account for about 95 % of the total rate. Belluco et al.[48] suggested that the rate decrease on addition of water could be due to increased solvation of the reactants H$^+$ and Cl$^-$, and it is of interest to calculate the rate decrease caused by the reduction in free energy of H$^+$ and Cl$^-$ on addition of water. The free energy of transfer of (H$^+$ plus Cl$^-$) from methanol to aqueous methanol, ΔG_t^0, may be obtained by interpolation from data given by Feakins and co-workers[50] and the rate decrease corresponding to this free energy term may then be calculated as $\exp(\Delta G_t^0/RT)$*. Details are given in Table 15, and it can be seen that the reduction in free energy

TABLE 15

THE EFFECT OF ADDED WATER ON THE RATE OF ACIDOLYSIS OF trans-(PtL$_2$MeCl) IN SOLVENT METHANOL AT 30 °C

[H$^+$] = 0.1 M, [Cl$^-$] = 0.1 M

Water (vol. %)	$10^4 k_1^{obs}$ (sec^{-1})	$k_1^{obs}/k_1^{obs}(MeOH)$	$\Delta G_t^0(H^+, Cl^-)$ (cal.mole^{-1})	$\exp(\Delta G_t^0/RT)$
0	57	1	0	1
1.5	16.4	0.29	−700	0.31
2.5	8.1	0.14	−1050	0.17

* Strictly speaking, the term ΔG_t^0 should be corrected for the secondary medium effect to take into account the fact that the activity coefficient of 0.1 M HCl in 100 % methanol will not be the same, necessarily, as that 0.1 M HCl in 98.5 % methanol–1.5 % water and in 97.5 % methanol–2.5 % water.

of H^+ and Cl^- accounts almost exactly for the observed rate decrease. Any solvent effect on *trans*-(PtL_2MeCl) must therefore be counterbalanced by an exactly similar solvent effect on the transition state (XVIII). Since the charge separation in (XVIII) would not be expected to be very different to the charge separation in *trans*-(PtL_2MeCl), the results of the calculations in Table 15 are thus in agreement with the postulated transition state (XVIII).

7.2 ACIDOLYSIS OF THE BENZYLCHROMIUM ION

Anet and Leblanc[51] in 1957 reported the preparation of aqueous solutions of benzylpentaaquochromium(III) perchlorate, formulated as containing the benzyl-pentaaquochromium(III) ion (XIV),

$$[C_6H_5CH_2Cr(H_2O)_5]^{2+}$$
$$(XIV)$$

They showed that the ion reacted with aqueous mercuric chloride to yield benzyl-mercuric chloride and the hexaaquochromium(III) ion with no change in pH, reaction (39)

$$[C_6H_5CH_2Cr(H_2O)_5]^{2+} + HgCl_2 = C_6H_5CH_2HgCl + [Cr(H_2O)_6]^{3+} + Cl^-$$
$$(39)$$

and suggested[51] that this was evidence that (XIV) was a derivative of chromium (III) and not of chromium(II). Furthermore, Anet and Leblanc[51] also suggested that the hydrolytic stability of (XIV) was due, at least in part, to the known inertness of chromium(III) complexes to nucleophilic substitution. Later workers[52] also clearly referred to (XIV) as a benzylchromium(III) species.

The acidolysis of the ion (XIV) was studied by Kochi and Buchanan[53], using aqueous ethanol as solvent and buffers such as acetic acid–acetate. They postulated a two-step process in which (XIV) first undergoes a reversible nucleophilic substitution by some nucleophilic species X, and the resulting intermediate then is subject to a rate-determining acidolysis. Kochi and Buchanan[53] wrote these steps as

$$C_6H_5CH_2Cr^{2+} + X^- \rightleftharpoons C_6H_5CH_2CrX^+ \tag{40}$$

$$C_6H_5CH_2CrX^+ + H^+ \rightarrow C_6H_5CH_3 + CrX^{2+} \tag{41}$$

omitting the solvent molecules coordinated to chromium. It is again clear from the stoichiometry of equations (40) and (41) that the species (XIV), *i.e.* $C_6H_5CH_2Cr^{2+}$, is a derivative of chromium(III). However(XIV), or $C_6H_5CH_2Cr^{2+}$, is specifically stated by Kochi and Buchanan[53] to be a derivative of chromium(II), and is referred to as the benzylchromium(II) ion. Since the kinetic interpretations* of Kochi and Buchanan appear to depend, at least in part, on the assumption that (XIV) is a derivative of chromium(II), any further detailed discussion would seem to be unwarranted.

REFERENCES

1 R. E. Dessy and J.-Y. Kim, *J. Am. Chem. Soc.*, 83 (1961) 1167.
2 G. C. Robinson, *J. Org. Chem.*, 28 (1963) 843.
3 L. Pauling, *The Nature of the Chemical Bond*, Cornell University Press, New York, 3rd edn., 1960, p. 93.
4 L. H. Gale, F. R. Jensen and J. A. Landgrebe, *Chem. Ind. (London)*, (1960) 118.
5 H. M. Walborsky, F. J. Impastato and A. E. Young, *J. Am. Chem. Soc.*, 86 (1964) 3283.
6 K. Sisido, S. Kozima and K. Takizawa, *Tetrahedron Letters*, (1967) 33; K. Sisido, T. Miyanisi and T. Isida, *J. Organometal. Chem.*, 23 (1970) 117.
7 A. G. Davies and P. B. Roberts, *J. Chem. Soc. C*, (1968) 1474.
8 H. C. Brown and K. J. Murray, *J. Org. Chem.*, 26 (1961) 631.
9 A. J. Weinheimer and W. E. Marsico, *J. Org. Chem.*, 27 (1962) 1926.
10 J. H. Wotiz and G. L. Proffitt, *J. Org. Chem.*, 30 (1965) 1240.
11 J. H. Wotiz, C. A. Hollingsworth and R. E. Dessy, *J. Am. Chem. Soc.*, 77 (1955) 103.
12 J. H. Wotiz, C. A. Hollingsworth and R. E. Dessy, *J. Am. Chem. Soc.*, 79 (1957) 358.
13 H. Kleinfeller and H. Lohmann, *Chem. Ber.*, 71 (1938) 2608.
14 H. Hashimoto, T. Nakano and H. Okado, *J. Org. Chem.*, 30 (1965) 1234.
15 R. E. Dessy, Y. Okuzumi and A. Chen, *J. Am. Chem. Soc.*, 84 (1962) 2899.
16 J. H. Wotiz, C. A. Hollingsworth and R. E. Dessy, *J. Org. Chem.*, 20 (1955) 1545.
17 D. Ivanoff and A. Spassoff, *Bull. Soc. Chim. France*, 51 (1932) 619; D. Ivanoff and I. Abdouloff, *Compt. Rend.*, 196 (1933) 491; D. Ivanoff and I. Ibdulov, *Godishnik Sofiskiya Univ.: Fiz.-Mat. Fak.*, 30 (1934) 53.
18 E. C. Ashby, *Quart. Rev. (London)*, 21 (1967) 259.
19 R. E. Dessy and R. M. Salinger, *J. Org. Chem.*, 26 (1961) 3519.
20 L. V. Guild, C. A. Hollingsworth, D. M. McDaniel, S. K. Podder and J. H. Wotiz, *J. Org. Chem.*, 27 (1962) 762.
21 S. K. Podder, E. W. Smalley and C. A. Hollingsworth, *J. Org. Chem.*, 28 (1963) 1435.
22 M. H. Abraham and J. A. Hill, *Proc. Chem. Soc.*, (1964) 175; *J. Organometal. Chem.*, 7 (1967) 23.
23 G. E. Coates and K. Wade, *Organometallic Compounds*, Vol. I, Methuen, London, 3rd edn., 1967.
24 T. S. Lee, in *Technique of Organic Chemistry*, S. L. Fiess and A. Weissberger (Eds.), Vol. VIII, Interscience, New York, 1953, p. 100 *et seq.*
25 M. M. Kreevoy, *J. Am. Chem. Soc.*, 79 (1957) 5927.
26 M. M. Kreevoy and R. L. Hansen, *J. Am. Chem. Soc.*, 83 (1961) 626.
27 S. Winstein and T. G. Traylor, *J. Am. Chem. Soc.*, 77 (1955) 3747; 78 (1956) 2597.
28 R. E. Dessy, G. F. Reynolds and J.-Y.Kim, *J. Am. Chem. Soc.*, 81 (1959) 2683.
29 R. E. Dessy and J.-Y. Kim, *J. Am. Chem. Soc.*, 82 (1960) 686.
30 A. N. Nesmeyanov, A. E. Borisov and I. S. Savel'eva, *Dokl. Akad. Nauk SSSR*, 155 (1964) 603; *Proc. Acad. Sci. USSR*, 155 (1964) 280.

---

* Kochi and Buchanan[53] state that reaction (41) is rate-determining, and hence equilibrium (40) must be set up rapidly. Although this might be feasible for an octahedral complex of chromium(II), it does not seem feasible if (XIV) is a derivative of chromium(III) since octahedral complexes of chromium(III) are invariably kinetically inert towards nucleophilic substitution (see ref. 54).

31 R. N. STERLIN, V.-G. LI AND I. L. KNUNYANTS, *Dokl. Akad. Nauk SSSR*, 140 (1961) 137; *Proc. Acad. Sci. USSR*, 140 (1961) 893.

32 M. S. KHARASCH AND R. MARKER, *J. Am. Chem. Soc.*, 48 (1926) 3130; M. S. KHARASCH AND S. SWARTZ, *J. Org. Chem.*, 3 (1938) 405.

33 M. S. KHARASCH AND A. L. FLENNER, *J. Am. Chem. Soc.*, 54 (1932) 674.

34 F. C. WHITMORE AND H. BERNSTEIN, *J. Am. Chem. Soc.*, 60 (1938) 2626.

35 M. H. ABRAHAM AND J. A. HILL, *J. Organometal. Chem.*, 7 (1967) 11.

36 F. NERDEL AND S. MAKOWER, *Naturwissenschaften*, 45 (1958) 490.

37 O. A. REUTOV, I. P. BELETSKAYA AND L. A. FEDOROV, *Dokl. Akad. Nauk SSSR*, 163 (1965) 1381; *Proc. Acad. Sci. USSR*, 163 (1965) 794.

38 O. A. REUTOV, I. P. BELETSKAYA AND L. A. FEDOROV, *Zh. Org. Khim.*, 3 (1967) 225; *J. Org. Chem. USSR*, 3 (1967) 213.

39 B. F. HEGARTY, W. KITCHING AND P. R. WELLS, *J. Am. Chem. Soc.*, 89 (1967) 4816.

40 O. A. REUTOV, I. P. BELETSKAYA AND M. YA. ALEINIKOVA, *Zh. Fiz. Khim.*, 36 (1962) 489; *Russ. J. Phys. Chem. (Engl. Transl.)*, 36 (1962) 256.

41 O. A. REUTOV, YU. G. BUNDEL' AND N. D. ANTONOVA, *Dokl. Akad. Nauk SSSR*, 166 (1966) 1103; *Proc. Acad. Sci. USSR*, 166 (1966) 191; YU. G. BUNDEL, V. I. ROZENBERG, I. N. KROKHINA AND O. A. REUTOV, *Zh. Org. Khim.*, 6 (1970) 1519; *J. Org. Chem. USSR*, 6 (1970) 1531.

42 L. H. TOPORCER, R. E. DESSY AND S. I. E. GREEN, *J. Am. Chem. Soc.*, 87 (1965) 1236.

43 *Handbook of Chemistry and Physics*, Chemical Rubber Co., Cleveland, 48th edn., 1967–1968, p. D-90.

44 Ref. 3, (a) see p. 98, (b) pp. 477–478.

45 L. H. SOMMER, W. P. BARIE AND J. R. GOULD, *J. Am. Chem. Soc.*, 75 (1953) 3765; L. M. SHORR, H. FREISER AND J. L. SPEIER, *J. Am. Chem. Soc.*, 77 (1955) 547.

46 R. WALRAEVENS, quoted by M. GIELEN AND J. NASIELSKI, *Rec. Trav. Chim.*, 82 (1963) 228.

47 H. HORN AND F. HUBER, *Monatsh. Chem.*, 98 (1967) 771.

48 U. BELLUCO, M. GIUSTINIANI AND M. GRAZIANI, *J. Am. Chem. Soc.*, 89 (1967) 6494.

49 C. H. LANGFORD AND H. B. GRAY, *Ligand Substitution Processes*, Benjamin, New York, 1965, p. 33.

50 A. L. ANDREWS, H. P. BENNETTO, D. FEAKINS, K. G. LAWRENCE AND R. P. T. TOMKINS, *J. Chem. Soc., A*, (1968) 1486.

51 F. A. L. ANET AND E. LEBLANC, *J. Am. Chem. Soc.*, 79 (1957) 2649.

52 F. GLOCKLING, R. P. A. SNEEDEN AND H. ZEISS, *J. Organometal. Chem.*, 2 (1964) 109 (see especially p. 115).

53 J. K. KOCHI AND D. BUCHANAN, *J. Am. Chem. Soc.*, 87 (1965) 853.

54 D. R. STRANKS, in *Modern Coordination Chemistry*, S. LEWIS AND R. G. WILKINS (Eds.), Interscience, New York, 1960.

55 J. R. COAD AND C. K. INGOLD, *J. Chem. Soc. B*, (1968) 1455.

56 A. JUBIER, E. HENRY-BASCH AND P. FRÉON, *Compt. Rend. Ser. C*, 267 (1968) 842; A. JUBIER, G. EMPTOZ, E. HENRY-BASCH AND P. FRÉON, *Bull. Soc. Chim. France*, (1969) 2032.

57 S. INOUE AND T. YAMADA, *J. Organometal. Chem.*, 25 (1970) 1.

58 A. A. FROST AND R. G. PEARSON, *Kinetics and Mechanism*, Wiley, New York, 1953, p. 165.

59 Y. POCKER AND J. H. EXNER, *J. Am. Chem. Soc.*, 90 (1968) 6764.

60 F. R. JENSEN AND B. RICKBORN, *Electrophilic Substitution of Organomercurials*, McGraw-Hill, New York, 1968.

61 D. H. O'BRIEN AND C. M. HARBORDT, *J. Organometal Chem.*, 21 (1970) 321.

62 V. BADE AND F. HUBER, *J. Organometal. Chem.*, 24 (1970) 691.

63 V. BADE AND F. HUBER, *J. Organometal. Chem.*, 24 (1970) 387.

64 G. J. JANZ AND S. S. DANYLUK, *Chem. Rev.*, 60 (1960) 209.

65 H. A. SKINNER, *Advan. Organometal. Chem.*, 2 (1964) 49.

66 D. B. CHAMBERS AND F. GLOCKLING, *Inorg. Chim. Acta*, 4 (1970) 150.

67 D. J. COLEMAN AND H. A. SKINNER, *Trans. Faraday Soc.*, 62 (1966) 1721.

68 M. H. ABRAHAM AND G. F. JOHNSTON, *J. Chem. Soc. A*, (1970) 193.

69 L. RICCOBONI, *Gazz. Chim. Ital.*, 71 (1941) 696.

Chapter 8

Halogenolysis (Halogenodemetallation) of Organometallic Compounds

In addition to cleavage by acids, (Chapter 7), the carbon–metal bond is also broken more-or-less readily by halogens (Y_2), *viz.*

$$RMX_n + Y_2 = RY + YMX_n \tag{1}$$

to yield the corresponding alkyl halide. Halogenolyses of the alkyls of mercury and tin have been the subject of intensive kinetic studies, and, indeed, the work of Gielen and Nasielski on the halogenolysis of tetraalkyltins is one of the classic investigations in the field of kinetics and mechanism of electrophilic substitution.

1. Stereochemical studies

The stereochemical course of the brominolysis of *cis*- and *trans*-4-methyl-cyclohexylmercuric bromide (I) and of optically active *sec.*-butylmercuric bromide (II) has been comprehensively studied by Jensen *et al.*[1,2]. Some of their results

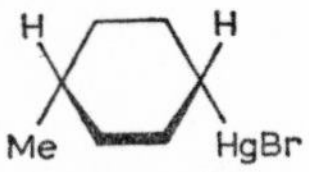

(*trans* - I)

are given in Table 1, in terms of the percentage retention of configuration at the carbon atom undergoing substitution. These results may be interpreted[1,2] as follows: in non-polar solvents such as CS_2, brominolysis takes place by a non-stereospecific free-radical mechanism which appears to be inhibited by oxygen (air). In most other solvents a combination of the free-radical mechanism and a polar mechanism occurs, and in brominolysis by Br_2/pyridine the polar mechanism occurs exclusively. The stereochemical course of the polar mechanism is that of retention of configuration.

More recent results[3,4] on the brominolysis of optically active *sec.*-butylmercuric bromide are in accord with the results obtained by Jensen *et al.*; details are given in Table 1.

A number of conflicting reports exist on the stereochemical course of halogenolysis of carbon–lithium and carbon–tin bonds when the carbon atom undergoing substitution is part of an alicyclic ring. Halogenolysis of optically active 1-methyl-

TABLE 1

STEREOCHEMICAL COURSE OF BROMINOLYSIS OF 4-METHYLCYCLOHEXYMERCURIC
BROMIDE (I) AND sec-BUTYLMERCURIC BROMIDE (II)[1,2]

RHgBr	Reagent	Atmosphere	Solvent	Temp. (°C)	Retention (%)
Trans-I	Br$_2$	N$_2$	CS$_2$	0	0
Trans-I	Br$_2$	Air	CS$_2$	0	0
Trans-I	Br$_2$	N$_2$	{CHCl$_3$+	0	3
Trans-I	Br$_2$	Air	0.75 % EtOH}	0	92
Trans-I	Br$_2$	N$_2$	CH$_3$CO$_2$H	25	18
Trans-I	Br$_2$	Air	CH$_3$CO$_2$H	25	8
Cis-I	Br$_2$/ZnBr$_2$	N$_2$	CH$_3$CO$_2$H	25	18
Trans-I	Br$_2$/ZnBr$_2$	Air	CH$_3$CO$_2$H	25	92
Trans-I	Br$_2$	Air	CH$_3$OH	25	85
Trans-I	Br$_2$/pyridine	Air or N$_2$	Pyridine	25	100
Cis-I	Br$_2$/pyridine	Air or N$_2$	Pyridine	25	100
(+)-II	Br$_2$	Air	CS$_2$	25	0
(+)-II	Br$_2$	Air	CS$_2$(10 % MeOH)	25	5
(+)-II	Br$_2$	Air	CH$_2$Cl$_2$(10 % MeOH)	25	11
(−)-II	Br$_2$/pyridine	Air	Pyridine	0	86
(+)-II	Br$_2$/pyridine	Air	Pyridine	−45	99.7
(−)-II	Br$_2$/base	Air	γ-Collidine/ pyridine	−65	100
II[a]	Br$_2$	N$_2$	CCl$_4$	25	0
II[b]	Br$_2$	Air	CCl$_4$	25	30
II[b]	Br$_2$	Air	CCl$_4$(4 % MeOH)	25	80

[a] Ref. 4.
[b] Ref. 3.

2,2-diphenylcyclopropyllithium by bromine in ether, or iodine in ether, yields[5] the corresponding cyclopropyl halide with 95 % retention of optical activity and configuration, whereas brominolysis of cis-2-methylcyclopropyllithium by bromine in 94 % pentane–6 % ether is non-stereospecific[6] at −70 °C or at 0 °C, although use of bromine in pentane alone[6] yields 93 % of cis-2-methylcyclopropylbromide at 30°. Brominolysis of exo-2-norbornyllithium by bromine in pentane at −70° yields[7] mainly the endo-bromide, a case of preferential inversion of configuration. (+)-1-Methyl-2, 2-diphenylcyclopropyltintrimethyl is reported[8] to be cleaved by halogens in carbon tetrachloride to (±)-1-methyl-2, 2-diphenylcyclopropylhalide and trimethyltin halide, but this has been suggested[8, 9] to be due to a free-radical mechanism (cf. the results of Jensen et al., Table 1). Halogenolysis of the cis and trans isomers of 2-methylcyclopropyltintrimethyl to 2-methylcyclopropylhalides proceeds[9] with complete retention of configuration for the reagents Br$_2$/CH$_3$CO$_2$H, Br$_2$/C$_6$H$_5$Cl, I$_2$/CH$_3$OH, and I$_2$/CH$_3$CO$_2$H. An experiment involving chlorinolysis has been reported[10], compound (III) yielding the corresponding exo-chloride with Cl$_2$/CH$_3$CO$_2$H and Cl$_2$/CHCl$_3$, i.e. with retention of configuration.

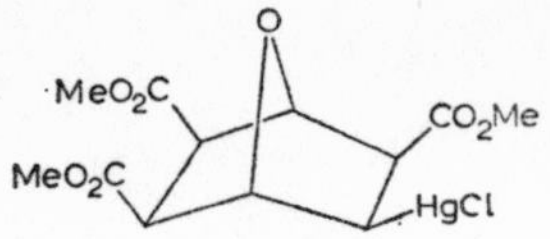

(III)

The MeCHCO$_2$Et group is cleaved from manganese or iridium by bromine in tetrahydrofuran at $-78\,^{\circ}$C with retention of optical activity and configuration[60].

A number of halogenolyses has recently been shown to proceed with inversion of configuration at the carbon atom undergoing substitution, *viz.* Bu^s-cobaltoxime with I$_2$ and Br$_2$ in CH$_2$Cl$_2$ (ref. 63), *threo*-Me$_3$CCHDCHD–Fe(CO)$_2$C$_p$ with Br$_2$ in CDCl$_3$ (ref. 64), and Bn^s–SnPe$_3^{neo}$ with Br$_2$ in MeOH (ref. 65). There is, there-fore, no general rule for the stereochemical course of halogenolyses of alkyl–metal compounds[1–4,63–65].

2. Halogenolysis of organomercury compounds

2.1 IODINOLYSIS OF ALKYLMERCURIC IODIDES BY I$_2$/I$^-$

Early studies by Keller[11] showed that iodinolysis of alkylmercuric iodides by iodine in solvent dioxan was free-radical in nature; in the presence of air or oxygen the kinetic form was second-order in iodine and zero-order in alkylmercuric iodide. Addition of iodide ion altered the kinetic form to first-order in alkyl-mercuric iodide and first-order in I$_3^-$, that is second-order overall. Winstein and Traylor[12] studied the iodinolysis of 4-camphylmercuric iodide (IV) by I$_2$/I$^-$ in

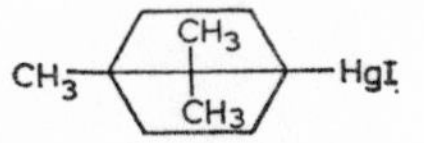

(IV)

solvent 95 % dioxan–5 % water and reported the second-order rate coefficient to be 1.27×10^{-3} l.mole^{-1}.sec^{-1} at 55 $^{\circ}$C. Relative rates for the reaction

$$RHgI + I_2 \rightleftharpoons RI + HgI_2 \tag{2}$$

were R = Bu^n (66): 4-camphyl (3.1): neophyl (1) at 55 $^{\circ}$C in solvent 90 % dioxan–10 % water, and these iodinolyses in the presence of iodide ion were sug-gested[12] to proceed through the transition state (V).

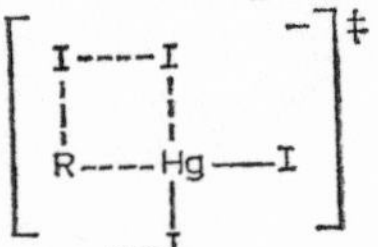

(V)‡

2.2 GENERAL KINETIC EXPRESSIONS FOR IODINOLYSES IN THE PRESENCE OF IODIDE ION

It is convenient to consider at this point the kinetic form expected for the general reaction

$$RMX_n + I_2 \overset{I^-}{=} RI + IMX_n \tag{3}$$

In the presence of iodide ion, iodine is converted into the species I_3^- in accord with the equilibrium

$$I_2 + I^- \rightleftharpoons I_3^- \tag{4}$$

Values of K, the equilibrium constant for reaction (4), are given in Table 2 and it can be seen that in the solvents commonly used for kinetic studies, values of K are around 10^4–10^7 l.mole^{-1}. Since kinetic studies are usually carried out with a large excess of iodide ion, such values of K result in almost complete conversion of iodine into I_3^-, and at any time during a kinetic run $[I_2] \simeq [I_3^-]$. Hence for reaction (3) the decrease in the concentration of organometallic substrate RMX_n, denoted by R, will equal the decrease in $[I_3^-]$, and the velocity, v, of the iodinolysis may thus be expressed as

$$v = -\frac{d[R]}{dt} = -\frac{d[I_3^-]}{dt} = k_2^{obs} \cdot [R][I_3^-] \tag{5}$$

where k_2^{obs} is the observed second-order rate coefficient for reaction (3)*. Following the procedure of Gielen and Nasielski[13], we may consider a number of possible terms that may contribute to the overall kinetic expression, viz.

$$v' = k_2^a \cdot [R][I_2] \qquad \text{term 2a}$$
$$v'' = k_2^b \cdot [R][I_3^-] \qquad \text{term 2b} \qquad (6)$$
$$v''' = k_3 \cdot [R][I_2][I^-] \qquad \text{term 3}$$

The overall rate is then given by

$$v = k_2^a \cdot [R][I_2] + k_2^b \cdot [R][I_3^-] + k_3 \cdot [R][I_2][I^-] \tag{7}$$

Combining equations (5) and (7) leads to the result that

$$k_2^{obs} = k_2^a \cdot \frac{[I_2]}{[I_3^-]} + k_2^b + k_3 \cdot \frac{[I_2][I^-]}{[I_3^-]} \tag{8}$$

Then introducing K, the equilibrium constant for reaction (4), we have that

$$k_2^{obs} = \frac{k_2^a}{K[I^-]} + k_2^b + \frac{k_3}{K} \tag{9}$$

---

* We are considering only the case of observed second-order kinetics for reaction (3).

TABLE 2

EQUILIBRIUM CONSTANTS (l.mole^{-1}) FOR TRIHALIDE ION FORMATION

$$X_2(soln) + X^-(soln) \rightleftharpoons X_3^-(soln)$$

Solvent	K	Temp. (°C)	μ	Refs.
(a) X = I				
H_2O	7.4×10^2	25	0	14
MeOH	1.89×10^4	20	10^{-4}	13
MeOH	2.04×10^4	20	0.1–0.2	13
EtOH	3.32×10^4	25	10^{-4}–10^{-3}	15
Pr^{n}OH	2.48×10^4	25	10^{-4}–10^{-3}	15
Pr^{i}OH	2.09×10^5	25		16
DMSO	2.5×10^5	22	0.1	See 17
DMF	1.6×10^7	22	0.1	See 17
CH$_3$CN^a	2.5×10^7	25	0.1	18
Sulfolane	3×10^7	22	0.1	17, 19
Acetone	2×10^8	25	0.1	20
(b) X = Br				
H_2O	1.6×10^1	25	0.2	21
MeOH	1.8×10^2	25	0.2	21
CH$_3$CO$_2$H	5.2×10^1	25	0.1	22
DMF	2×10^6	22	0.1	17
CH$_3$CN	1×10^7	25	0.1	20
Sulfolane	6×10^6	22	0.1	17, 19
Acetone	2×10^9	25	0.1	20

^{a} The lower values of K that have been reported[20,58] may be due to traces of water in the acetonitrile[58].

Gielen and Nasielski[13] point out that, at constant ionic strength, a plot of k_2^{obs} versus $1/[I^-]$ should thus be a straight line of slope k_2^a/K and of intercept $(k_2^b + k_3/K)$. In this way the overall rate may be partitioned between term 2a on the one hand, and terms 2b and 3 on the other. There is no kinetic treatment that can be used to determine the individual contributions of terms 2b and 3.

A similar procedure to the above may also be carried out for brominolyses by bromine in the presence of bromide ion; some values of the equilibrium constant for formation of Br_3^- are given in Table 2.

2.3 IODINOLYSIS OF BENZYLMERCURIC CHLORIDE BY I_2/I^-

Reaction (10)

$$C_6H_5CH_2HgCl + I_2 \overset{I^-}{=\!=} C_6H_5CH_2I + HgClI \tag{10}$$

has been studied by Reutov and co-workers[23] who used cadmium iodide as the source of iodide ion and 70 % dioxan–30 % water as the solvent. With initial concentrations of benzylmercuric chloride and iodine at $(2.4–4.8) \times 10^{-3}$ M and a concentration of 5×10^{-2} M cadmium iodide, the reaction followed second-order kinetics (first-order in benzylmercuric chloride and first-order in iodine (or I_3^-, since the concentrations of iodine and I_3^- are essentially equal)). The second-order rate coefficient found under the above conditions was identical in value to that obtained with initial reactant concentrations $(4.1–8.3) \times 10^{-4}$ M and cadmium iodide 5×10^{-3} M. Since k_2^{obs} thus appears not to depend upon $[I^-]$, it may be deduced from equation (9) that the contributing terms to the total kinetic law can only be 2b and 3, in equation (6). Rate coefficients and activation parameters for reaction (10) are given in Table 3. The relative standard deviation of k_2^{obs} is about 3 %, and hence the standard deviations in E_a and $\Delta S^{\ddagger}$ must be about 0.8 kcal. mole^{-1} and 3 cal.deg^{-1}.mole^{-1} respectively (estimated by the author).

In a subsequent paper[24], rate coefficients and activation parameters were given for reaction (10) in solvents methanol, ethanol, and dimethylformamide (DMF); details are also given in Table 3. The source of iodide ion was cadmium iodide, present in a ten-fold molar excess.

The dependence of the value of k_2^{obs}, for reaction (10), on the iodide ion concentration has briefly been investigated. Using cadmium iodide as the source of iodide ion and methanol as the solvent, it was shown[25] that k_2^{obs} was independent of iodide ion concentration when the latter was present in two-fold excess or more. In addition, k_2^{obs} was also found to be independent of the source of iodide ion. A mechanism involving attack by I_3^- on the benzylmercuric halide was proposed, possibly through transition states (VI) or (VII)[25].

(VI)   (VII)

TABLE 3

SECOND-ORDER RATE COEFFICIENTS (l.mole^{-1}.sec^{-1}) AND ACTIVATION PARAMETERS FOR THE IODINOLYSIS OF BENZYLMERCURIC CHLORIDE BY IODINE IN THE PRESENCE OF CADMIUM IODIDE[23, 24]

Solvent	E_a (kcal.mole^{-1})	log A	$\Delta S^{\ddagger}$ (cal.deg^{-1}.mole^{-1})	k_2^{obs} 288 °K	293 °K	298 °K
Aq. dioxan[a]	10.8	7.7	−23	0.30	0.366	0.58
DMF	15.3	11.1	−8	0.142	0.215	0.347
Methanol	9.6	7.1	−26	0.575	0.807	0.951
Ethanol	11.6	8.6	−20	0.189	0.279	0.375

[a] 70 % Dioxan–30 % water; k_2^{obs} at 297 °K was 0.48 l.mole^{-1}.sec^{-1}.

Such a mechanism corresponds to term 2b in equation (6), p. 138, and implies that the observed second-order rate coefficient, k_2^{obs}, is the rate coefficient, k_2^b, for attack by I_3^- on benzylmercuric chloride. Since k_2^{obs} is independent of $[I^-]$, term 2a in equation (6) cannot contribute to the overall rate, but as mentioned before it is not possible to distinguish between term 2b and term 3, in equation (6). A reasonable mechanism corresponding to term 3 in equation (6) is given in equations (11) and (12) ($R = PhCH_2$).

$$RHgCl + I^- \overset{K_{(11)}}{\rightleftharpoons} RHgClI^- \tag{11}$$

$$RHgClI^- + I_2 \overset{k_{(12)}}{\longrightarrow} products \tag{12}$$

Provided that the complex $RHgClI^-$ is present only in very low concentration at any time, and provided, as always, that an excess of iodide ion sufficient to convert essentially all iodine to I_3^- is present, then it may be deduced that equations (11) and (12) lead to an expression*

$$v''' = \frac{K_{(11)} \cdot k_{(12)}}{K} \cdot [RHgCl][I_2][I^-] \left. \vphantom{\begin{array}{c} \\ \\ \\ \\ \end{array}} \right\}$$

or

$$v''' = k_3 \cdot [RHgCl][I_2][I^-] \tag{13}$$

Thus the kinetic observations of Reutov et al.[25] demonstrate that in the presence of an excess of iodide ion, reaction (10) may proceed via mechanisms corresponding to terms 2b and 3 in equation 6. The mechanism corresponding to term 3 is given in equations (11) and (12) ($R = PhCH_2$), and the mechanism corresponding to term 2b is the simple bimolecular reaction

$$C_6H_5CH_2HgCl + I_3^- \rightarrow products \tag{14}$$

for which the transition states (VI) and (VII) have been proposed.

The influence of various solvents and solvent mixtures on the rate of reaction (10) has also been investigated. Graphs showing the dependence of k_2^{obs}, E_a and $\Delta S^{\ddagger}$ on solvent composition are given[26] for the binary mixtures DMF/ethanol, DMSO/ethanol, DMF/methanol, DMF/dioxan, DMSO/dioxan, and DMF/DMSO. The above parameters do not vary in any orderly way with solvent composition, and no detailed explanation of the variations was attempted. Numerical data on the activation parameters of reaction (10) were given[26] for a number of solvents, and are recorded in Table 4. It may be noted that the values for E_a and $\Delta S^{\ddagger}$ for solvent DMF given in Table 4 ($E_a = 13.8$ kcal.mole^{-1}, $\Delta S^{\ddagger} = -15$ cal.deg^{-1}.mole^{-1}) do not correspond to the values given in Table 3

---

* The derivation of equation (13) contains the assumption that the complex $RHgClI^-$ is always in equilibrium with the reactants RHgCl and I^-.

TABLE 4

SECOND-ORDER RATE COEFFICIENTS ($l.mole^{-1}.sec^{-1}$) AND ACTIVATION PARAMETERS
FOR THE IODINOLYSIS OF BENZYLMERCURIC CHLORIDE BY IODINE IN THE PRESENCE
OF IODIDE ION[26]

Solvent	k_2^{obs} (293 °K)	E_a (kcal.mole^{-1})	$\Delta S^{\ddagger}$ (cal.deg^{-1}. mole^{-1})
Aq. dioxan[a]	0.366	11.5	−23
DMF[b]	0.215	13.8	−15
Methanol	0.807	9.2	−30
Ethanol	0.279	12.4	−20
Butanol	0.072		
CH$_3$CN	0.590	10.1	−27
DMSO	0.145	15.2	−10.5
Benzene	6.6	9.7	−24

[a] 70 % Dioxan–30 % water.
[b] Ref. 27 gives k_2^{obs} = 0.18 (288 °K), 0.25 (293 °K), 2.47 (323 °K), E_a = 13.9, and $\Delta S^{\ddagger}$ = −16.

(E_a = 15.3, $\Delta S^{\ddagger}$ = −8). In another paper[27], the values given are E_a = 13.9 and $\Delta S^{\ddagger}$ = −16, i.e. close to those in Table 4.

The rates of iodinolysis of several nuclear-substituted benzylmercuric chlorides by iodine in the presence of iodide ion were determined[27] at 20 °C using methanol and DMF as solvents; relative rate coefficients are given in Table 5. It is clear that Hammett's equation is not followed, since rate accelerations are found for the electron-attracting substituent, p-NO$_2$, and for the electron-donating substituent, p-Me. It was suggested[27] that the mechanism approached S$_E$1 for the p-NO$_2$ substituted compound and approached S$_E$2 for the methyl and halogen substituted compounds. Activation parameters were also given for the various substitutions in solvent DMF (except for the p-nitrocompound). For the remaining twelve

TABLE 5

RELATIVE SECOND-ORDER RATE COEFFICIENTS AT 293 °K FOR IODINOLYSIS OF
NUCLEAR-SUBSTITUTED BENZYLMERCURIC CHLORIDES BY IODINE IN THE PRESENCE
OF CADMIUM IODIDE[27]

| Substituent | Solvent | | Substituent | Solvent | |
	MeOH	DMF		MeOH	DMF
p-NO$_2$	Very fast		p-Br		0.76
p-MeO	11.1	11.1	p-Cl	0.95	0.64
p-Me	2.57	2.04	m-Br	0.61	0.48
o-Me	2.33	1.80	m-F	0.56	0.48
p-F	1.12	0.75	o-Cl	0.31	0.32
m-Me	1.11	0.92	o-F	0.29	0.32
H	1	1			

TABLE 6

RELATIVE SECOND-ORDER RATE COEFFICIENTS FOR IODINOLYSIS OF NUCLEAR-
SUBSTITUTED α-CARBETHOXYBENZYLMERCURIC BROMIDES BY IODINE IN THE
PRESENCE OF CADMIUM IODIDE[28]

Solvent toluene+1.5 % methanol

Substituent	Relative $k_2^{obs\,a}$	Substituent	Relative $k_2^{obs\,a}$
p-NO$_2$	61.5	H	1
m-Br	7.98	m-Me	0.7
p-I	4.36	p-Pr^i	0.457
p-Cl	3.34	p-Bu^t	0.353
p-F	1.4		

^{a} The temperature at which kinetic runs were set up is not given in ref. 28 but it is probable
that the experiments were performed at 20 °C.

compounds given in Table 5, E_a was approximately constant at 14 ± 0.2 kcal.
mole^{-1}. $\Delta S^{\ddagger}$ varied between -15 and -18 cal.deg^{-1}.mole^{-1} except for the p-
methoxycompound for which $\Delta S^{\ddagger} = -12$ cal.deg^{-1}.mole^{-1}.

Substituent effects have also been reported[28] for the iodinolysis of substituted
α-carbethoxybenzylmercuric bromides by iodine in the presence of cadmium
iodide, using solvent toluene plus 1.5 vol. % methanol. Details are given in Table 6.
It may be seen from Tables 5 and 6 that substituent effects in the α-carbethoxybenzyl compounds are rather different from those in the simple benzylmercuric
chlorides. The mechanism of iodinolysis of the α-carbethoxybenzylmercuric
bromides was suggested[28] to be in the boundary region between S_E1 and S_E2.

2.4 BROMINOLYSIS OF BENZYLMERCURIC CHLORIDE BY Br$_2$/Br$^-$

Reaction (15)

$$C_6H_5CH_2HgCl + Br_2 \xrightarrow{Br^-} C_6H_5CH_2Br + HgClBr \tag{15}$$

has been studied[29] using a twenty-fold excess of bromide ion, as ammonium
bromide, and a number of polar solvents. Values of rate coefficients and activation
parameters are collected in Table 7, the reaction being described as second-order
overall, first-order in benzylmercuric chloride and first-order in bromine (*i.e.*
Br$_3^-$)[29]. No conjecture as to mechanism was made, but it seems evident that
terms analogous to 2b and/or 3 in equation (6), p. 138, may make up the actual
rate equation.

TABLE 7

SECOND-ORDER RATE COEFFICIENTS (l.mole^{-1}.sec^{-1}) AND ACTIVATION PARAMETERS
FOR THE BROMINOLYSIS OF BENZYLMERCURIC CHLORIDE BY BROMINE IN THE
PRESENCE OF BROMIDE ION[29]

Solvent	E_a (kcal.mole^{-1})	log A	$\Delta S^{\ddagger}$ (cal.deg^{-1}.mole^{-1})	k_2^{obs} 288 °K	296 °K	298 °K	303 °K
Aq. dioxan[a]	13.2	10.8	−9	6.6			
DMF	15.3	11.4	−6	0.79	1.1	1.7	2.7
Methanol	11.4 (11.5)	9.6	−14.5	10.3	17.6	20.1	27.8

[a] 70 % Dioxan–30 % water; k_2^{obs} = 5.0 (284.5 °K) and 9.8 (293 °K).

2.5 HALOGENOLYSIS OF BENZYLMERCURIC CHLORIDE BY HALOGENS IN SOLVENT CARBON TETRACHLORIDE

In the absence of halide ions, reaction (16) (X = Br, I)

$$C_6H_5CH_2HgCl + X_2 \xrightarrow{CCl_4} C_6H_5CH_2X + HgXCl \tag{16}$$

usually proceeds by a free-radical mechanism. Under constant illumination the kinetic form is first-order in halogen and zero-order in benzylmercuric chloride[30, 31].

When reaction (16) (X = Br) is carried out[30] in the presence of small quantities (0.25–0.50 vol. %) of various aliphatic alcohols, aliphatic ethers, or water, the kinetic form becomes that of the second-order overall*. Furthermore, the reaction rate does not depend on the illimination. It was suggested[30, 32] that in the presence of the above additives, reaction (16) (X = Br) is an electrophilic bimolecular substitution with a complex of the type Br–Br : ORR' as the electrophile.

2.6 BROMINOLYSIS OF sec.-BUTYLMERCURIC BROMIDE BY BROMINE IN SOLVENT CARBON TETRACHLORIDE

The above brominolysis was shown to follow overall first-order kinetics (first-order in bromine and zero-order in sec.-butylmercuric bromide) under constant illumination, and presumably proceeds by some free-radical mechanism[3].

Addition of a small quantity of methanol to the carbon tetrachloride results in the rate of brominolysis becoming independent of the degree of illumination,

---

* Kinetic runs were set up[30, 32] with equal initial concentrations of reactants (benzylmercuric chloride and bromine) and hence only the overall kinetic order can be established.

and a change in the kinetic form to overall second-order*. It was suggested[3] that a bimolecular reaction between *sec.*-butylmercuric bromide and bromine (complexed in some way to methanol) now took place.

2.7 IODINOLYSIS OF DIALKYLMERCURYS BY IODINE

Razuvaev and Savitskii[33] report that the action of iodine on a number of dialkyl- and diarylmercurys in solvent carbon tetrachloride follows second-order kinetics. No rate coefficients were given, but values of E_a and log A were recorded**. Kinetics were presumably carried out under normal laboratory condtions, but Lord and Pritchard[34] report that the iodinolysis of dimethylmercury by iodine in a number of solvents was subject to catalysis by light.

Lord and Pritchard[34] found that when the iodinolysis of dimethylmercury was carried out with rigorous exclusion of light, the reaction was first-order in dimethylmercury and first-order in iodine. Activation energies and rate coefficients for iodinolysis of dimethylmercury in a number of solvents were determined. For solvent carbon tetrachloride, the second-order rate coefficient at 28 °C was found to be 0.073 l.mole^{-1}.min^{-1} and E_a = 7.7 kcal.mole^{-1}. The corresponding values of Razuvaev and Savitskii[33] are k_2 = 0.11 l.mole^{-1}.min^{-1} and E_a = 9.5 kcal. mole^{-1}***.

3. Halogenolysis of tetraalkyltins

Nearly all of the kinetic work in this field is due to Gielen and Nasielski and to Tagliavini, who have investigated the brominolysis and iodinolysis of tetraalkyltins using a variety of solvents. Since the kinetic methods and the results depend largely on the particular solvent, the following discussion is drawn up on the basis of the solvents employed.

---

* The kinetic form was claimed[3] to be first-order in each reactant, but the only kinetic data cited consist of a single run in which the initial concentrations of reactants are 5×10^{-3} M (*sec.*-BuHgBr) and 4.68×10^{-3} M (bromine). Under these conditions it is difficult to deduce anything other than the overall kinetic order.

** E_a was given in kcal.mole^{-1} and A in l.mole^{-1}.min^{-1}. Note that log A is referred to as log k_0, both in the original Russian paper and in Chemical Abstracts[33].

*** The value of 0.11 l.mole^{-1}.min^{-1} for k at 28 °C has been calculated from the given values of E_a = 9.5 and log A = 5.96.

3.1 IODINOLYSIS OF TETRAALKYLTINS BY I_2/I^- IN SOLVENT METHANOL

Reaction (17)

$$R_4Sn + I_2 \overset{I^-}{\rightleftharpoons} RI + R_3SnI \tag{17}$$

was studied by Gielen and Nasielski[13] who used a large excess of iodide ion in order to convert essentially all of the iodine into I_3^-. Under these conditions, the reaction follows overall second-order kinetics with the observed rate equation expressed as

$$v = k_2^{obs} \cdot [R_4Sn][I_3^-] \tag{18}$$

Following the analysis detailed in Section 2.2 (p. 138), the observed second-order rate coefficient, k_2^{obs}, may be assumed to consist of the three terms.

$$k_2^{obs} = \frac{k_2}{K[I^-]} + k_2^b + \frac{k_3}{K} \tag{19}$$

In Table 8 are given the results of experiments in which the concentration of iodide ion was varied. It may be seen that the product $k_2^{obs} \cdot [I^-]$ is constant. Furthermore, a plot[13] of k_2^{obs} against $1/[I^-]$ yields a straight line of slope 5.33×10^{-4} sec^{-1} (that is k_2/K) and of intercept $(-5 \pm 6) \times 10^{-5}$ l.mole^{-1}.sec^{-1} (that is $k_2^b + k_3/K$). Hence within experimental error $(k_2^b + k_3/K) = 0$ and only the term corresponding to $k_2/K[I^-]$ in equation (19) contributes to k_2^{obs}. This term is denoted as term 2a in equation (6)*, p. 138, and is

$$v = k_2 \cdot [R_4Sn][I_2] \tag{20}$$

Thus reaction (17) takes place by a simple second-order reaction between tetraalkyltin and iodine. The second-order rate coefficient for reaction (17), k_2, is given by the expression

TABLE 8

DEPENDENCE OF THE OBSERVED SECOND-ORDER RATE COEFFICIENT, k_2^{obs}, FOR
REACTION (17) (R = Me) ON IODIDE ION CONCENTRATION[13]

Solvent methanol, ionic strength 0.51 M, temp. 20 °C

$[I^-]$ (mole.l^{-1})	$1/[I^-]$	k_2^{obs} (l.mole^{-1}.sec^{-1})	$k_{obs}^2 \cdot [I^-]$
1.10×10^{-1}	9.09	4.79×10^{-3}	5.27×10^{-4}
7.00×10^{-2}	14.29	7.73×10^{-3}	5.41×10^{-4}
6.00×10^{-2}	16.67	8.62×10^{-3}	5.12×10^{-4}
3.40×10^{-2}	28.57	15.20×10^{-3}	5.32×10^{-4}

* The coefficient k_2 in equations (19) and (20) corresponds to the coefficient k_2^a in equation (6); for simplicity the superscript "a" has been omitted.

TABLE 9

CALCULATION OF THE SECOND-ORDER RATE COEFFICIENT, k_2, FOR REACTION (17)
$(R = Me)$[13]

Solvent methanol, temp. 20 °C

k_2^{obs} $(l.mole^{-1}.sec^{-1})$	K $(l.mole^{-1})$	$[I^-]$ $(mole.l^{-1})$	k_2 $(l.mole^{-1}.sec^{-1})$	μ (M)
3.33×10^{-2}	2.04×10^4	1.00×10^{-2}	6.8	0.01
3.68×10^{-2}	2.04×10^4	1.00×10^{-2}	7.5	0.1
3.55×10^{-3}	2.04×10^4	1.00×10^{-1}	7.3	0.1
4.20×10^{-2}	2.04×10^4	1.00×10^{-2}	8.6	0.21
1.52×10^{-2}	2.1×10^{4a}	3.50×10^{-2}	11.2	0.51

[a] Value assumed by the author; quoted values[13] are 1.89×10^4 ($\mu = 10^{-4}$) and 2.04×10^4 ($\mu = 0.01$–0.20).

$$k_2 = k_2^{obs} \cdot K[I^-] \tag{21}$$

Examples of the determination of values of k_2 are given in Table 9; the required values of K (the formation constant for the I_3^- ion) were separately determined.

Gielen and Nasielski[13] also studied the influence of temperature on the value of k_2^{obs} and calculated that $E_a^{obs} = 14.4$ kcal.mole^{-1}. From equation (21) it follows that

$$E_a(k_2) = E_a(k_2^{obs}) + E_a(K) \tag{22}$$

and since $E_a(K)$ was separately found to be -8 kcal.mole^{-1}, the activation energy for the reaction between tetramethyltin and iodine in solvent methanol[13] is thus 6.4 kcal.mole^{-1}.

Values of k_2 for the iodinolysis of a number of tetraalkyltins are given in Table 10.

The mechanism of iodinolysis was suggested[13] to be a simple bimolecular substitution of the tetraalkyltin by iodine, proceeding by mechanism S_E2(open) through a transition state such as (VIII). Evidence for an open, polar transition state of this type is provided[13] by the marked increase in the value of k_2 with increase in the ionic strength of the solvent medium (Table 9).

(VIII)   (IX)

The salt effect on the value of k_2 (Table 9) can be analysed in terms of the equation used by Abraham and Behbahany (see p. 85). For methanol at 20 °C,

TABLE 10

SECOND-ORDER RATE COEFFICIENTS ($l.mole^{-1}.sec^{-1}$) FOR THE IODINOLYSIS OF
TETRAALKYLTINS BY IODINE IN SOLVENT METHANOL AT 20 °C[13]

R_4Sn	k_2	μ (M)
Me_4Sn	6.8	0.01
Et_4Sn	0.8	0.025
Pr^n_4Sn	0.1	0.005
Bu^n_4Sn	0.04	0.005
Pr^i_4Sn	0.004[a]	
$(Allyl_4Sn$	$\simeq 10^7$)	

[a] From data given in ref. 35.

$J = 1.030$, and from the results in Table 9 the value of $[\log_{10}(k/k_0)]/I$ as $I \rightarrow 0$ is about 0.65. It may then be calculated that $Z^2d = 0.63$, where Z is the charge separation in the transition state and d is the distance between the charges $+Z$ and $-Z$ (the treatment assumes a point dipole model for the transition state). The value of 0.63 for Z^2d is appreciably less than that found for the transition state in the substitution of tetraethyltin by mercuric chloride in solvent methanol, where $Z^2d = 2.21$. It is not easy to assign a value to the distance d in the transition state. If this distance is regarded as that between the tin atom and the leaving iodine atom, then a reasonable value would be about 6 A, leading to a value of 0.32 for Z. It appears as though the transition state, (VIII) or (IX), carries a substantial charge separation (in solvent methanol) and that the assignment[13] of mechanism S_E2(open) to iodinolyses of tetraalkyltins in solvent methanol is perfectly reasonable.

In later work[35], the iodinolysis of a number of tetraalkyltins of type $RSnMe_3$ and $RSnEt_3$ (R = Me, Et, Pr^n, Pr^i, Bu^n, and Bu^t) was investigated. Second-order rate coefficients were calculated by the method shown in Table 9. Product analyses of the resulting mixture of alkyl iodides were carried out, and it was then possible to determine the rate coefficients for the cleavage of particular alkyl groups from tin[35]. Details are given in Tables 11 and 12, the relevant statistical corrections being applied to the rate coefficients where necessary. It may be noted that the rate coefficients for cleavage of the symmetrical tetraalkyltins calculated from Table 11 are slightly different from those given in Table 10 (k_2 for $Me_4Sn = 6.8$ (Table 10) and 7.08 (Table 11); k_2 for $Et_4Sn = 0.8$ (Table 10) and 0.88 (Table 11)).

Faleschini and Tagliavini[36] have also determined the second-order rate coefficients for iodinolysis of a number of unsymmetrical tetraalkyltins, but at 35.3 °C rather than at 20°. From their recorded rate coefficient[36] for cleavage of $Me-SnEt_3$, viz. 7.08 $l.mole^{-1}.sec^{-1}$, it appears that under the conditions used by Faleschini

TABLE 11

STATISTICALLY CORRECTED SECOND-ORDER RATE COEFFICIENTS $(l.mole^{-1}.sec^{-1})$ FOR THE CLEAVAGE OF ALKYL–TIN BONDS BY IODINE IN SOLVENT METHANOL AT 20 °C AND IONIC STRENGTH 0.025 M [35]

	k_2		k_2
Me–SnMe$_3$	1.77	Me–SnEt$_3$	3.58
Et–SnMe$_3$	0.256	Et–SnEt$_3$	0.22
Pr^n–SnMe$_3$	0.056	Pr^n–SnEt$_3$	0.065
Bu^n–SnMe$_3$	0.132	Bu^n–SnEt$_3$	0.060
Pr^i–SnMe$_3$	0.01	Pr^i–SnEt$_3$	0.004
Bu^t–SnMe$_3$	0.00		
Me–SnBu$^n_3$	1.42[a]	Me–SnMe$_3$	1.77
Et–SnBu$^n_3$	0.58?[a]	Et–SnEt$_3$	0.22
Pr^n–SnBu$^n_3$	0.16?[a]	Pr^n–SnPr$^n_3$	0.025
Bu^n–SnBu$^n_3$	0.01	Bu^n–SnBu$^n_3$	0.010
		Pr^i–SnPr$^i_3$	0.001

[a] Calculated from values given in ref. 36 (see text).

TABLE 12

STATISTICALLY CORRECTED SECOND-ORDER RATE COEFFICIENTS $(l.mole^{-1}.sec^{-1})$ FOR THE CLEAVAGE OF METHYL–TIN AND ETHYL–TIN BONDS BY IODINE IN SOLVENT METHANOL AT 20 °C AND IONIC STRENGTH 0.025 M [35]

Me–SnMe$_3$	1.77		Me–SnMe$_3$	1.77
Me–SnMe$_2$Et	2.34		Me–SnEt$_3$	3.58
Me–SnMe$_2$Pr^n	1.95		Me–SnPr$^n_3$	1.68[a]
Me–SnMe$_2$Bu^n	1.95		Me–SnBu$^n_3$	1.42[a]
Me–SnMe$_2$Pr^i	1.51		Me–SnPr$^i_3$	0.17[a]
Me–SnMe$_2$Bu^t	0.01			
		Et–SnMe$_3$	0.256	
		Et–SnEt$_3$	0.22	
		Et–SnEt$_2$Pr^n	0.245	
		Et–SnEt$_2$Bu^n	0.247	
		Et–SnEt$_2$Pr^i	0.255	

[a] Calculated from values given in ref. 36 (see text).

and Tagliavini the cleavage of Me–SnEt$_3$ proceeds some 1.98 times as rapidly as under the conditions used by Gielen and Nasielski. In order to compare the various cleavages more exactly, I have divided the rate coefficients reported by Faleschini and Tagliavini by 1.98. These "adjusted" rate coefficients are given also in Tables 11 and 12. It should be noted that the rate coefficients for the cleavage of given alkyl–tin bonds reported by Faleschini and Tagliavini[36] depend on the observation that only the smaller alkyl group in any given unsymmetrical compound was split off as RBr. Whilst this would be expected for the compounds MeSnPr$^n_3$, MeSnBu$^n_3$, and MeSnPr$^i_3$ (Tables 11 and 12), it seems rather unlikely that this would be so for the compounds $EtSnBu^n_3$ and $Pr^nSnBu^n_3$ (Table 11). In the light of recent criticism[37] of the product analyses of Faleschini and Tagliavini, the values given in Table 11 for the cleavage $Et-SnBu^n_3$ and $Pr^n-SnBu^n_3$ must be viewed with some caution.

Gielen and Nasielski[13] originally thought that the sequence of reactivity shown in Table 10 was due to steric interaction between the incoming iodine molecule and the leaving group SnR_3 in the transition state (VIII). If this were so, then for the cleavage of the methyl–tin bond in a series of compounds with different leaving groups, the larger the leaving group the lower should be the rate of cleavage. Reference to Table 12 shows that this prediction is not upheld; the rate of cleavage of the Me–Sn bond is almost the same in the compounds $Me-SnMe_3$, $Me-SnMe_2Pr^i$, and $Me-SnBu^n_3$. Only when a *tert.*-butyl group or three isopropyl groups are introduced into the leaving group does the rate of Me–Sn cleavage decrease. This was recognised by Gielen and co-workers[35] who then advanced the proposition that steric inhibition of solvation and steric inhibition to the approach of the iodine molecule are important factors giving rise to the various sequences of reactivity in Tables 11 and 12. They wrote a mechanism for the iodinolysis which involves coordination of the nucleophilic solvent methanol (Nu) to the tetraalkyltin, reaction (23), in a step prior to the rate-determining substitution (24):

$$Nu + R_4Sn \rightleftharpoons \overset{+}{Nu}-\overset{-}{Sn}R_4 \tag{23}$$

$$\overset{+}{Nu}-\overset{-}{Sn}R_4 + I_2 \rightarrow Nu \cdot R_3SnI + RI \tag{24}$$

Although no coordination complexes of tetraalkyltins with nucleophiles (*i.e.* Lewis bases) have been reported, it is known that trialkyltin chlorides can form coordination complexes with various Lewis bases. Bolles and Drago[38] have shown that trimethyltin chloride forms 1 : 1 complexes with bases such as acetone, acetonitrile, dimethylsulphoxide, and pyridine. All of these complexes are formed exothermally (in inert solvents such as CCl_4), with values of ΔH^0_{299} ranging from -4.8 to -8.2 kcal.mole^{-1} for the reaction

$$Me_3SnCl + Nu \rightarrow Me_3SnCl \cdot Nu \tag{25}$$

Recent thermochemical measurements by Abraham *et al.*[39] have shown, however, that a solute such as tetraethyltin dissolves in nucleophilic solvents endothermally in comparison with dissolution in an inert solvent (see Table 13). In view of these measurements, it is difficult to visualise any interaction between tetraethyltin and nucleophilic solvents that could correspond to the situation represented by equation (23).

The sequences of reactivity shown in Tables 10–12 are dominated by the effect of the alkyl group undergoing substitution rather than by the effect of the alkyl groups attached to the leaving tin atom (compare, for example, the sequences

TABLE 13

HEATS OF SOLUTION OF TETRAETHYLTIN (kcal.mole^{-1}) AT 298 °K

Solvent	ΔH_s^0
Water	4.4
Acetonitrile	2.7
Methanol	1.9
tert.-Butanol	1.5[a]
CCl$_4$	0.4

[a] At 303 °K.

shown in Table 11 with those shown in Table 12). Without giving the detailed analysis again, the general argument of Abraham and Spalding[40] on steric effects in the substitution of tetraalkyltins by mercuric iodide (see p. 70) may also be applied to the cleavage of alkyl–tin bonds by halogens. This argument indicates that in a transition state such as (VIII) the main steric interactions are those between the alkyl group undergoing substitution and (i) the electrophile (I$_2$) and (ii) the set of atoms SnC$_3^\alpha$. This argument would also apply if transition state (IX) were in force (cf. ref. 65).

The iodinolysis of tetraallyltin (see Table 10) is dealt with in Chapter 10, Section 4.1, p. 206.

3.2 IODINOLYSIS OF TETRAALKYLTINS BY IODINE IN SOLVENT ACETIC ACID

Tetraalkyltins react with iodine in solvent acetic acid, in the absence of added iodide ions, according to the bimolecular substitution

$$R_4Sn + I_2 \xrightarrow{\text{CH}_3\text{CO}_2\text{H}} RI + R_3SnI \tag{26}$$

TABLE 14

SECOND-ORDER RATE COEFFICIENTS (l.mole^{-1}.sec^{-1}) FOR THE IODINOLYSIS OF TETRAALKYLTINS BY IODINE IN SOLVENT ACETIC ACID AT 20 °C[41]

R_4Sn	k_2
Me$_4$Sn	0.22 (0.244, ref. 35)
Et$_4$Sn	0.082 (0.080, ref. 35)
Pr$^n_4$Sn	0.0094
Bu$^n_4$Sn	0.0078
Pr$^i_4$Sn	$\simeq$0.0001

TABLE 15

STATISTICALLY CORRECTED SECOND-ORDER RATE COEFFICIENTS ($l.mole^{-1}.sec^{-1}$) FOR
THE CLEAVAGE OF ALKYL–TIN BONDS BY IODINE IN SOLVENT ACETIC ACID AT 20 °C[35]

Me–SnMe$_3$	0.0610	Me–SnMe$_3$	0.0610
Et–SnMe$_3$	0.00950	Et–SnEt$_3$	0.0200
Pr^n–SnMe$_3$	0.00166	Pr^n–SnPr$^n_3$	0.0024[a]
Bu^n–SnMe$_3$	0.00317	Bu^n–SnBu$^n_3$	0.0020[a]
Pr^i–SnMe$_3$	0.00046	Pr^i–SnPr$^i_3$	0.00002[a]
Bu^t–SnMe$_3$	$\simeq$0.00005		

Me–SnMe$_3$	0.0610
Me–SnMe$_2$Et	0.0872
Me–SnMe$_2$Pr^n	0.0635
Me–SnMe$_2$Bu^n	0.0635
Me–SnMe$_2$Pr^i	0.058
Me–SnMe$_2$Bu^t	$\simeq$0.005

[a] Taken from ref. 41.

The second-order rate coefficients given by Gielen and Nasielski[41] for reaction
(26) are listed in Table 14. A number of unsymmetrical tetraalkyltins were also
studied[35], and rate coefficients for the cleavage of given alkyl–tin bonds calculated
from the overall rate coefficients for reaction (26) together with product analyses
of the resulting mixture of alkyl iodides. Results are given in Table 15, and it may
be seen that, in general, they are similar to those obtained using solvent methanol.
A transition state of the S_E2(open) type, (VIII) was suggested by Gielen and
Nasielski for substitutions according to (26).

3.3 IODINOLYSIS OF TETRAALKYLTINS BY IODINE IN SOLVENT DIMETHYLSULPHOXIDE (DMSO)

As for the substitution of tetraalkyltins by iodine in solvent acetic acid, the rate
equation

$$v = k_2 \cdot [R_4Sn][I_2] \tag{27}$$

is also followed when DMSO is employed as the solvent. Recorded values[42] for
the second-order rate coefficient, k_2, are 0.48 $l.mole^{-1}.sec^{-1}$ and 0.021 $l.mole^{-1}.$
sec^{-1} for the substitution of tetramethyltin and tetraethyltin, respectively, by
iodine in DMSO at 20.0 °C.

3.4 BROMINOLYSIS OF TETRAALKYLTINS BY Br_2/Br^- IN SOLVENT DIMETHYLFORMAMIDE (DMF)

Under conditions such that essentially all of the bromine is converted into the species Br_3^-, the reaction

$$R_4Sn + Br_2 \xrightarrow{Br^-} RBr + R_3SnBr \tag{28}$$

is simple second-order, first-order in tetraalkyltin and first-order in Br_3^-, when DMF is used as the solvent, *viz.*

$$v = k_2^{obs} \cdot [R_4Sn][Br_3^-] \tag{29}$$

Experiments revealed that the observed rate law (29) corresponded to the rate law

$$v = k_2 \cdot [R_4Sn][Br_2] \tag{30}$$

where

$$k_2 = k_2^{obs} \cdot K[Br^-] \tag{31}$$

and K is the equilibrium constant for formation of Br_3^-, *viz.*

$$Br_2 + Br^- \rightleftharpoons Br_3^- \tag{32}$$

The kinetic situation is thus analogous to that in the iodination of tetraalkyltins by iodine in presence of iodide ion, discussed in Section 3.1, p. 146. With $[Br^-] = 7.5 \times 10^{-3}$ M, values of k_2^{obs} at 20 °C were found[41] to be 138×10^{-4} (Me$_4$Sn), 64×10^{-4} (Et$_4$Sn), and 8.5×10^{-4} (Pr$^n{}_4$Sn), in l.mole^{-1}.sec^{-1}. The value of K for reaction (32) in solvent DMF is not known at 20°, but at 25° it is[17] 2×10^6 l.mole^{-1} (see Table 2, p. 139). Since the equilibrium constants for triiodide ion and tribromide ion formation usually increase with decreasing temperature, we may estimate K to be about 2.3×10^6 l.mole^{-1} at 20°. The resulting approximate values for k_2, together with the accurate relative values of k_2, are thus

R_4Sn	Me$_4$Sn	Et$_4$Sn	Pr$^n{}_4$Sn	
$k_2(l.mole^{-1}.sec^{-1})$	240	110	14.5	(33)
Relative k_2	100	46	6.1	

It was shown[41] that reaction (28) (R = Me) was accelerated when the ionic strength of the medium was increased, and hence that the mechanism of the reaction, in solvent DMF, is probably that of S_E2(open).

3.5 BROMINOLYSIS OF TETRAALKYLTINS AND TRIALKYLTIN BROMIDES
BY BROMINE, AND BROMINOLYSIS OF TETRAALKYLTINS BY Br_2/Br^-, IN
SOLVENT ACETIC ACID

In the absence of bromide ions, the reaction

$$R_4Sn + Br_2 \rightarrow RBr + R_3SnBr \tag{34}$$

in solvent acetic acid follows[41] the simple rate law

$$v = k_2 \cdot [R_4Sn][Br_2]$$

Second-order rate coefficients for reaction (34), and also for the reaction

$$R_3SnBr + Br_2 \rightarrow RBr + R_2SnBr_2 \tag{35}$$

are given in Table 16. Values of k_2^{obs} were, in addition, determined for brominolyses
in the presence of bromide ion. Details are given in Table 17. The ratio $k_2(Pr^n_4Sn)/$
$k_2(Bu^n_4Sn)$ is 1.15 compared with the ratio $k_2^{obs}(Pr^n_4Sn)/k_2^{obs}(Bu^n_4Sn)$ of 1.16.
Hence the results in Tables 16 and 17 may be combined[41] to yield the following
values for k_2, (l.mole^{-1}.sec^{-1}) at 20 °C.

$$Me_4Sn\ (9.6),\ Et_4Sn\ (8.1),\ Pr^n_4Sn\ (1.15),\ Bu^n_4Sn\ (1.00),\ Pr^i_4Sn\ (0.25) \tag{36}$$

The values of k_2 and k_2^{obs} given in Tables 16 and 17 for brominolysis of tetra-
n-propyltin and tetra-n-butyltin enable a value for K, the equilibrium constant for
reaction (32), to be estimated using equation (31). Under the conditions given in
Table 17, K is found to be 82 l.mole^{-1}, in fair agreement with the reported[22]
value of 52 l.mole^{-1}. at 25 °C (see Table 2, p. 139).

Gielen and Nasielski[41] suggested that the reactivity sequence for reaction
(35), shown in Table 16, was due to an increased importance of inductive effects
compared with the steric sequence (36) found for reaction (34).

The brominolysis of a number of unsymmetrical tetraalkyltins was later investigated[35], and rate coefficients for cleavage of given alkyl–tin bonds determined
by the methods described previously. In this later work[35], a value of $k_2^{obs} = 1.42$

TABLE 16

SECOND-ORDER RATE COEFFICIENTS (l.mole^{-1}.sec^{-1}) FOR BROMINOLYSIS OF TETRAAL-
KYLTINS AND TRIALKYLTIN BROMIDES BY BROMINE IN SOLVENT ACETIC ACID AT
20 °C[41]

Compound	k_2	Compound	$10^4 k_2$
Pr^n_4Sn	1.15	Me_3SnBr	7.37
Bu^n_4Sn	1.00	Et_3SnBr	10.8
Pr^i_4Sn	0.25	Pr^n_3SnBr	1.29

TABLE 17

OBSERVED SECOND-ORDER RATE COEFFICIENTS, k_2^{obs}, FOR BROMINOLYSIS OF TETRAAL-
KYLTINS BY BROMINE IN SOLVENT ACETIC ACID AT 20 °C IN THE PRESENCE OF 0.1 M
LITHIUM BROMIDE[41]

R_4Sn	k_2^{obs} $(l.mole^{-1}.sec^{-1})$
Me_4Sn	1.16
Et_4Sn	0.98
$Pr^n{}_4Sn$	0.141
$Bu^n{}_4Sn$	0.121

TABLE 18

STATISTICALLY CORRECTED SECOND-ORDER RATE COEFFICIENTS $(l.mole^{-1}.sec^{-1})$ FOR
THE CLEAVAGE OF ALKYL–TIN BONDS BY BROMINE IN SOLVENT ACETIC ACID AT 20 °C[35]

$Me–SnMe_3$	2.92[a]	$Me–SnMe_3$	2.92[b]
$Et–SnMe_3$	1.21[a]	$Et–SnEt_3$	2.48[b]
$Pr^n–SnMe_3$	0.36[a]	$Pr^n–SnEt_3$	0.35[b]
$Bu^n–SnMe_3$	0.55[a]	$Bu^n–SnBu^n{}_3$	0.30[b]
$Pr^i–SnMe_3$	0.03[a]	$Pr^i–SnPr^i{}_3$	0.07[b]
		$Me–SnMe_3$	2.92[a]
		$Me–SnMe_2Et$	4.33[a]
		$Me–SnMe_2Pr^n$	3.46[a]
		$Me–SnMe_2Bu^n$	3.40[a]
		$Me–SnMe_2Pr^i$	3.20[a]

[a] These values[35] correspond to $k_2(Me_4Sn) = 11.7$ $l.mole^{-1}.sec^{-1}$.
[b] Taken from sequence (36), p. 154, after multiplication by the factor 1.214 so as to bring the
rate coefficients onto a scale such that $k_2(Me_4Sn) = 11.7$ $l.mole^{-1}.sec^{-1}$.

$l.mole^{-1}.sec^{-1}$ was found for the substitution of tetramethyltin in the presence
of 0.1 M lithium bromide (cf. the value of 1.16 $l.mole^{-1}.sec^{-1}$ obtained earlier[41]).
If K is taken to be 82 $l.mole^{-1}$, the corresponding value of $k_2(Me_4Sn)$ is 11.7
$l.mole^{-1}.sec^{-1}$. The data given in Table 18 are all compatible with this latter
figure. The general results shown in Table 18 are quite similar to those for the
iodinolyses of tetraalkyltins in solvent acetic acid (Table 15, p. 152), in terms of
relative rates.

3.6 IODINOLYSIS OF TETRAALKYLTINS BY IODINE IN SOLVENT
CHLOROBENZENE

Reaction (37)

$$R_4Sn + I_2 \xrightarrow{PhCl} RI + R_3SnI \qquad\qquad (37)$$

was studied[43] using an excess of tetraalkyltin over iodine. Under these conditions, the reaction was found to follow a kinetic law first-order in iodine. The order with respect to tetraalkyltin could not be determined, but it was assumed that this also is unity and that the overall order of reaction (37) is two. All reactions were followed in the absence of light and it was ascertained that the observed rate coefficients were not affected by either addition of glass wool or of triphenylamine. Reaction (37) was thus felt to be a simple bimolecular electrophilic substitution; rate coefficients* were given[43] as follows for reactions at 20 °C.

R_4Sn	Me_4Sn	Et_4Sn	Pr^n_4Sn	Bu^n_4Sn	Pr^i_4Sn	
10^4k_2	1.67	10	1.30	1	9.4	(38)

Gielen and Nasielski[43] suggested that reaction (37) proceeded through a cyclic transition state of the S_E2(cyclic) type, for which we may write (X) as a possible structure. In view of later work[35] on the related brominolysis of tetraalkyltins in solvent chlorobenzene, it would seem possible that reaction (37) might proceed by mechanism S_E2(co-ord) (cf. reactions (41) and (42), p. 158).

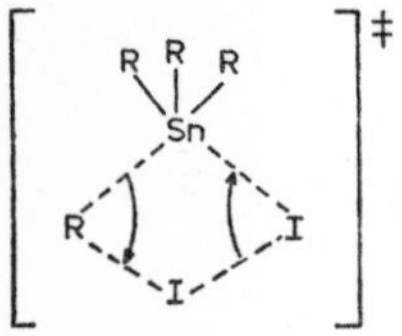

(X)

3.7 BROMINOLYSIS OF TETRAALKYLTINS BY BROMINE IN SOLVENT CHLOROBENZENE

A study by Gielen and Nasielski[43] showed that the reaction

$$R_4Sn + Br_2 \xrightarrow{PhCl} RBr + R_3SnBr \tag{39}$$

is of the second-order overall, first-order in bromine and hence, by deduction, first-order in tetraalkyltin. The rate of reaction (39) is not affected by the introduction of glass wool into the reaction vessel, and it was concluded that the process is an electrophilic substitution. It may be noted that Gielen and Nasielski[43] conducted their experiments in the absence of light in order to preclude incursion of photochemical mechanisms. Rate coefficients reported for reaction (39) are given in Table 19.

Later work[35] suggested that reaction (39) did not follow simple second-order kinetics, but that a two-term rate law was obeyed, viz.

---

* Units of the rate coefficients were not given but are almost certain to be l.mole^{-1}.sec^{-1}.

TABLE 19

SECOND-ORDER RATE COEFFICIENTS $(l.mole^{-1}.sec^{-1})$ FOR THE BROMINOLYSIS OF TETRAALKYLTINS BY BROMINE IN SOLVENT CHLOROBENZENE AT 20 °C[43]

R_4Sn	k_2^{obs}
Me$_4$Sn	0.116 (0.112)[a]
Me$_4$Sn	0.132[b]
Me$_4$Sn	0.205[c]
Et$_4$Sn	1.42 (1.40)[a]
Pr$^n_4$Sn	0.54
Bu$^n_4$Sn	0.51
Octyl$^n_4$Sn	0.62
Dodecyl$^n_4$Sn	0.49
Pr$^i_4$Sn	1.61 (0.35)[a]

[a] Values in parentheses are later values from ref. 35 and are to be preferred.
[b] In the presence of 0.357 M acetic acid.
[c] In the presence of 0.435 M acetic acid.

$$v = k_2^{obs}[R_4Sn][Br_2] + k_3^{obs}[R_4Sn][Br_2]^2 \qquad (40)$$

Values of k_2^{obs} and k_3^{obs} are given in Table 20, and it may be seen that, whereas only the second-order term in equation (40) is of importance for substitution of tetramethyltin and tetraethyltin, yet the third-order term is of as equal importance as the second-order term for the substitution of tetraisopropyltin. The value of k_2^{obs} obtained in the latter case (0.35) is naturally rather different to the original value of 1.61 $l.mole^{-1}.sec^{-1}$ given in Table 19. However, the values of k_2^{obs} given in Table 19 for the n-tetraalkyls are quite valid.

The brominolysis of a number of unsymmetrical tetraalkyltins was also studied[35] and second-order rate coefficients for the cleavage of given alkyl–tin bonds determined by combination of rate studies with product analyses. Details are given in Table 21 and it should be noted that for all of the unsymmetrical tetraalkyls listed in this table only the second-order term in equation (40) is of importance[35]. It is convenient now to discuss the two terms in equation (40) separately.

TABLE 20

RATE COEFFICIENTS FOR THE BROMINOLYSIS OF TETRAALKYLTINS BY BROMINE IN CHLOROBENZENE AT 20 °C[35]

R_4Sn	k_2^{obs} $(l.mole^{-1}.sec^{-1})$	k_3^{obs} $(l^2.mole^{-2}.sec^{-1})$	v_3/v_2[a]
Me$_4$Sn	0.112	10	0.09
Et$_4$Sn	1.40	70	0.07
Pr$^i_4$Sn	0.35	600	1.7

[a] The ratio of the third-order term to the second-order term when $[Br_2] = 10^{-3} M$.

TABLE 21

STATISTICALLY CORRECTED SECOND-ORDER RATE COEFFICIENTS $(l.mole^{-1}.sec^{-1})$ FOR THE CLEAVAGE OF ALKYL–TIN BONDS BY BROMINE IN SOLVENT CHLOROBENZENE AT 20 °C[35]

Me–SnMe$_3$	0.0285	Me–SnMe$_3$	0.0285
Et–SnMe$_3$	0.0208	Et–SnEt$_3$	0.350[a]
Pr^n–SnMe$_3$	0.0090	Pr^n–SnPr$^n_3$	0.135[b]
Bu^n–SnMe$_3$	0.0090	Bu^n–SnBu$^n_3$	0.127[b]
Pr^i–SnMe$_3$	0.0206	Pr^i–SnPr$^i_3$	0.088[a]
Bu^t–SnMe$_3$	$\simeq$0.0550		
		Me–SnMe$_3$	0.0285
		Me–SnMe$_2$Et	0.0550
		Me–SnMe$_2$Pr^n	0.0600
		Me–SnMe$_2$Bu^n	0.0617
		Me–SnMe$_2$Pr^i	0.0562
		Me–SnMe$_2$Bu^t	$\simeq$0.04

[a] From Table 20.
[b] From Table 19.

3.7.1 The second-order term in equation (40)

Gielen and co-workers[35] suggested that the second-order term in equation (40) corresponds to a mechanism in which a five-co-ordinate tin complex (XI) is formed as an intermediate. It seems evident that (XI) can be present in but very

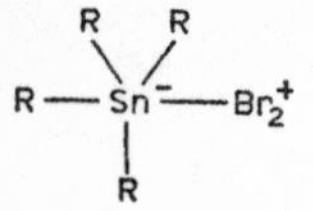

(XI)

low concentration at any given time, since tetraalkyltins do not normally function as Lewis acids. Furthermore, NMR studies have indicated[44] that bromine does not complex to any extent with trimethyltin bromide, a much stronger Lewis acid than the tetraalkyltins, in solvent chlorobenzene. A mechanism incorporating (XI) as an intermediate is given in (41) and (42), and corresponds to mechanism S_E2-(co-ord), viz.

$$R_4Sn + Br_2 \underset{k_1'}{\overset{k_2}{\rightleftharpoons}} R_4Sn–Br_2 \tag{41}$$

$$R_4Sn–Br_2 \overset{k_1}{\rightarrow} RBr + R_3SnBr \tag{42}$$

Application of the steady-state treatment to reactions (41) and (42) leads to an equation for the observed second-order rate coefficient completely analogous to that given in Chapter 3, Section 1.4 (p. 13), viz.

$$k_2^{obs} = k_1 \cdot k_2/(k_1 + k_1') \tag{43}$$

It is now clear why substituent effects on k_2^{obs} are so complicated, (see Table 21), since any given alkyl substituent may affect the values of no less than three rate coefficients. If, however, the relative rates of alkyl–tin cleavages in an unsymmetrical tetraalkyltin are considered, any variations in k_2 and k_1' are eliminated and only the relative values of k_1 are significant, e.g.

$$
\text{Me}_3\text{EtSn–Br}_2^{\overset{-}{}\overset{+}{}} \quad
\begin{array}{l}
\xrightarrow{\; k_1(\text{Et}) \;} \text{EtBr} + \text{Me}_3\text{SnBr} \\[4pt]
\xrightarrow[\; 3k_1(\text{Me}) \;]{} \text{MeBr} + \text{EtMe}_2\text{SnBr}
\end{array}
\tag{44}
$$

The term $3\,k_1(\text{Me})$ arises because there are three methyl groups available in the intermediate $\text{Me}_3\text{EtSn–Br}_2^{\,-\;+}$ but only one ethyl group, and the observed concentration of MeBr thus has to be divided by the statistical factor of 3 in order to calculate $k_1(\text{Me})$.

The necessary data for the determination of the ratios $k_1(\text{Me})/k_1(\text{R})$ for compounds Me_3SnR are included in Table 21, and other data are available[35] for the determination of the ratios $k_1(\text{Et})/k_1(\text{R})$ for compounds Et_3SnR. Results are given in Table 22. In a recent paper[37], product distributions from brominolyses of EtSnBu^n_3 and $\text{Pr}^n\text{SnBu}^n_3$ are reported, and if a value of $k_1(\text{Me})/k_1(\text{Bu}^n)$ of 100/13 is assumed for the hypothetical case of MeSnBu^n_3, a third sequence may be constructed (see Table 22). It is essential to note that the values given in Table 22 are relative values, and that the actual value of, say, $k_1(\text{Et})$ will not be the same in the three series.

Reverting to equation (43), some progress may be made into substituent effects if it is assumed that $k_1 \ll k_1'$ (as is quite possible), for then equation (43) reduces to

$$
k_2^{obs} = K_{(41)} \cdot k_1
\tag{45}
$$

where $K_{(41)}$ is the equilibrium constant for formation of the intermediate (XI). Values of the equilibrium constant for the reaction

TABLE 22

RELATIVE VALUES OF k_1 (REACTIONS (42) AND (44)) FOR THE TRANSFER OF ALKYL GROUPS FROM TIN TO BROMINE IN SOLVENT CHLOROBENZENE AT 20 °C

Series	R					
	Me	Et	Pr^n	Bu^n	Pr^i	Bu^t
Me$_3$SnR[35]	100	38	15	14	37	140
Et$_3$SnR[35]	100	46	17	13	29	
Bu$^n_3$SnR[37]	(100)	46	15	13		

$$R_3SnBr + I^- \rightleftharpoons R_3SnBrI^- \tag{46}$$

have been obtained at 20 °C using solvent acetone, and lead[45] to the sequence

$$
\begin{array}{lcccc}
K_{(46)}(l.mole^{-1}) & 91 & 168 & 123 & 71 \\
R_3SnBr & Me_3SnBr & Et_3SnBr & Bu_3{}^nSnBr & Pr_3{}^iSnBr
\end{array} \tag{47}
$$

We might, therefore, use these values of $K_{(46)}$ to provide an estimate of substituent effects on the values of $K_{(41)}$, whilst using the data in Tables 21 and 22 to deduce substituent effects on the values of k_1 and on the overall rate coefficient, k_2^{obs}. At the same time, it is convenient also to refer to Table 24, p. 164, which contains data on the related brominolysis of tetraalkyltins in the non-polar solvent carbon tetrachloride. The total substituent effects may be broken down into three sections.

(a) *Substituent effects on the value of* $K_{(41)}$

Two effects seem to be in force here, a polar effect in the sense Me < Et < Pr^i < Bu^t and a steric effect in the sense Me > Et > Pr^n > Bu^n > Pr^i > Bu^t. The total combination leads to the results shown in sequence (47) for the trialkyltin bromides. It may be noted that the polar effect is quite the reverse of the expected inhibition of nucleophilic coordination to tin by electron releasing groups attached to the tin atom. It appears, from sequence (47), that variations in $K_{(41)}$ are likely to make comparatively small contributions to the overall variation in k_2^{obs}.

(b) *The effect of the moving group on the value of* k_1

The transition state for reaction (42), to which k_1 refers, may be written as

(XII)

From the data given in Table 22, it is clear that the effect of the moving group in (XII) on k_1 is in the order Me > Et > Pr^n ~ Bu^n < Pr^i < Bu^t. Inspection of Table 24, p. 164, suggests that for the related brominolyses in solvent carbon tetrachloride the order is Me < Et < Pr^n < Pr^i. Thus the moving group effect seems to be a combination of two sequences, *viz.*: Me > Et > Pr^n ~ Bu^n > Pr^i > Bu^t and Me < Et < Pr^n ~ Bu^n < Pr^i < Bu^t. Gielen and co-workers[35] suggested that larger groups moved more rapidly because of steric decompression in the five coordinate intermediate (XI), and another factor could be that of the electron-releasing power of the moving group aiding the 1,3-shift to an electron-deficient bromine atom in (XII).

(c) The effect of the leaving SnR$_3$ group on the value of k_1

A close examination of the values of k_2^{obs} (*i.e.* $k_1 K_{(41)}$) given in Table 21 in the light of the reported sequence (47) suggests that the leaving group effect for brominolysis in chlorobenzene is Me$_3$Sn $\ll$ Et$_3$Sn $\sim$ Pr$^n{}_3$Sn $\sim$ Bu$^n{}_3$Sn $>$ Pr$^i{}_3$Sn. For brominolysis in carbon tetrachloride (Table 24, p. 164) the corresponding sequence seems to be Me$_3$Sn $\ll$ Et$_3$Sn $<$ Pr$^i{}_3$Sn.

There are thus several combinations of sequences leading to the overall substituent effect on the rate coefficient k_2^{obs}. The more outstanding sequences may be summarised as follows.

In the non-polar solvent carbon tetrachloride, the effect of the alkyl group undergoing substitution on the overall rate coefficient is in the order (of increasing values of the rate coefficient)

$$\text{Me} < \text{Et} \sim \text{Pr}^n < \text{Pr}^i (< \text{Bu}^t) \tag{48}$$

The order is moderated in the slightly polar solvent chlorobenzene by a counteracting sequence in the order

$$\text{Me} > \text{Et} > \text{Pr}^n > \text{Pr}^i > \text{Bu}^t \tag{49}$$

The effect of the leaving R$_3$Sn group on the overall rate coefficient, in solvent carbon tetrachloride, is in the sense

$$\text{Me}_3\text{Sn} \ll n\text{-alkyl}_3\text{Sn} < \text{Pr}^i{}_3\text{Sn} \tag{50}$$

Again, a moderating effect in the slightly polar solvent chlorobenzene takes place in the sense

$$\text{Me}_3\text{Sn} > n\text{-alkyl}_3\text{Sn} \gg \text{Pr}^i{}_3\text{Sn} \tag{51}$$

Thus for the symmetrical tetraalkyls undergoing brominolysis in carbon tetrachloride, combination of (48) and (50) leads to the observed order of reactivity Me$_4$Sn $\ll$ n-alkyl$_4$Sn $\ll$ Pr$^i{}_4$Sn (see Table 24, p. 164). The moderating effects in (49) and (51) lead to the observed order Me$_4$Sn $\ll$ n-alkyl$_4$Sn $>$ Pr$^i{}_4$Sn when chlorobenzene is the solvent (see Table 21, p. 158).

3.7.2 The third-order term in equation (40)

The third-order term in equation (40), p. 157, was suggested to correspond to a mechanism in which the intermediate (XI) is now attacked by a second bromine molecule, *viz.*

$$\text{R}_4\text{Sn} + \text{Br}_2 \rightleftharpoons \overset{-}{\text{R}}_4\overset{+}{\text{Sn}}\text{-Br}_2 \tag{52}$$

$$\overset{-}{\text{R}}_4\overset{+}{\text{Sn}}\text{-Br}_2 + \text{Br}_2 \rightarrow \text{RBr} + \text{R}_3\text{SnBr} + \text{Br}_2 \tag{53}$$

A possible transition state for reaction (53) is (XIII), which implies that the substitution takes place with inversion of configuration at the tin atom, in contrast to the second-order mechanism which proceeds through (XII) with retention of configuration at the tin atom. It is interesting that Eaborn and Steward[46] have observed that brominolysis of optically active p-methoxyphenyl-methyl-1-naphthylphenylsilane in benzene and in carbon tetrachloride does indeed proceed with inversion of configuration at silicon; a transition state (XIV) was considered[46] possible.

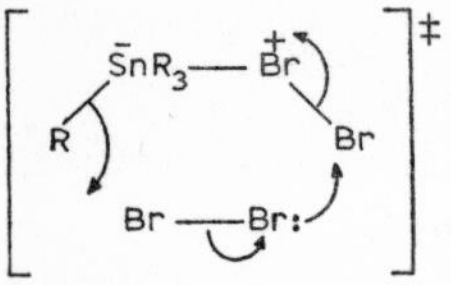

(XIII) 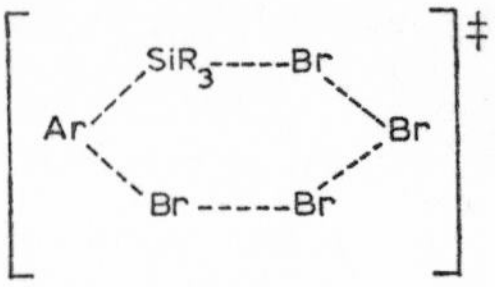   (XIV)

In a final ingenious set of experiments, it was shown[35] that addition of trimethyltin bromide resulted in an increase in the values of k_2^{obs} but in a marked decrease in the values of k_3^{obs} (see Table 23). These results are explicable if a complex $Me_3SnBr \cdot Br_2$ is formed (again in but low concentration), since such a complex would be expected to provide a very active source of electrophilic bromine in the second-order mechanism, through a transition state such as (XV).

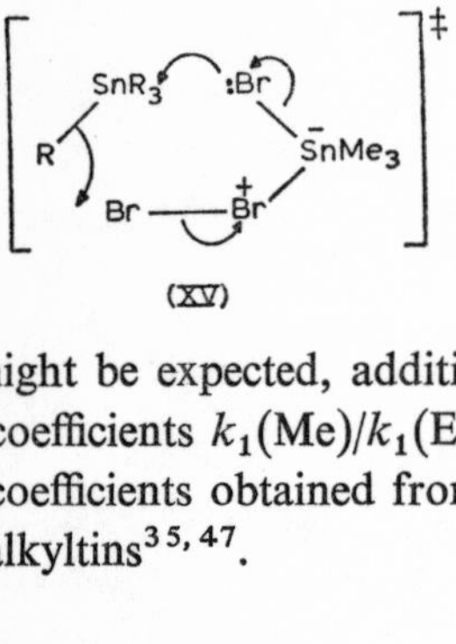

(XV)

As might be expected, addition of trimethyltin bromide affects the ratio of the rate coefficients $k_1(Me)/k_1(Et)$ in process (44) and also affects similar ratios of rate coefficients obtained from the brominolysis of various other unsymmetrical tetraalkyltins[35, 47].

TABLE 23

VARIATION OF THE RATE COEFFICIENTS k_2^{obs} AND k_3^{obs} WITH THE CONCENTRATION OF ADDED TRIMETHYLTIN BROMIDE IN THE BROMINOLYSIS OF TETRAALKYLTINS BY BROMINE IN CHLOROBENZENE AT 20 °C[35]

Substrate	C_0 (M)	$[Me_3SnBr]$ (10^2 mole.l^{-1})	k_2^{obs} (l.mole^{-1}.sec^{-1})	k_3^{obs} (l$_2$.mole^{-2}.sec^{-1})
Me$_4$Sn	3.18×10^{-2}	0.00	0.112	10
Me$_4$Sn	3.18×10^{-2}	1.82	0.138	<5
Me$_4$Sn	3.18×10^{-2}	3.64	0.149	0
Me$_4$Sn	3.18×10^{-2}	7.28	0.175	0
Pr$^i_4$Sn	0.20×10^{-2}	0.00	0.35	600
Pr$^i_4$Sn	0.20×10^{-2}	4.29	0.58	316
Pr$^i_4$Sn	0.20×10^{-2}	8.58	0.87	158

Not only trialkyltin halides, but also tetraalkyltins themselves act in this way as "perturbers"*. Evidence for this effect was obtained[47] by determination of the relative reactivities of alkyl groups in unsymmetrical tetraalkyltins. The following data are typical.

Substrate	Perturber	$k_2^{obs}(Me)/k_2^{obs}(Et)$
Me_3SnEt	None	2.70
Me_3SnEt	$2\% Bu^n_4Sn$	2.31
Me_3SnEt	$4\% Bu^n_4Sn$	1.90

An "autoperturbation" effect was also observed in the case of brominolysis of tetraisopropyltin, the value of k_2^{obs} increasing, and that of k_3^{obs} decreasing, with increasing initial concentration of tetraisopropyltin. No such effect was observed for the substrates tetramethyltin and tetraethyltin, however[47].

It was suggested that the perturbing agent itself formed a complex with bromine, and that this complex acts as a modified bromine molecule towards another tetraalkyltin molecule. If the perturbing agent is denoted as R'_4Sn, and the substrate as R_4Sn, then the relevant reactions are

$$R'_4Sn + Br_2 \rightleftharpoons R'_4Sn \cdot Br_2 \qquad (54)$$

$$R'_4Sn \cdot Br_2 + R_4Sn \rightarrow RBr + R_3SnBr + R'_4Sn \qquad (55)$$

For the autoperturbation effect found for tetraisopropyltin, (54) and (55) still apply, but with $R = R' = Pr^i$.

It is clear that these bromodemetallations, proceeding by the S_E mechanism in the non-polar solvent chlorobenzene, are extremely complex and that a variety of transition states (e.g. (XII), (XIII), and (XV)) are possible. The main function of the more complicated transition states seems to be that of providing a route for bromodemetallation such that collapse of the transition state yields uncharged molecules and not ionic species. In addition, the transition states also serve to distribute any charge separation over a molecular aggregate.

3.8 BROMINOLYSIS OF TETRAALKYLTINS BY BROMINE IN SOLVENT CARBON TETRACHLORIDE

Reaction (56)

$$R_4Sn + Br_2 \xrightarrow{CCl_4} RBr + R_3SnBr \qquad (56)$$

---

* Gielen and co-workers[47] use the term "perturbateur".

has been shown by Gielen and Nasielski[43] to follow overall second-order kinetics, first-order in bromine and hence, by deduction, also first-order in tetraalkyltin. Recorded second-order rate coefficients at 20 °C are given in Table 24. From the observed kinetic form, it appears that reaction (56) follows some type of mechanism S_E2, and Gielen and Nasielski[43] suggested that a transition state such as (XVI) was possible, corresponding to mechanism S_E2(cyclic). From their later results[35] on brominolysis in solvent chlorobenzene, it would appear very likely, however, that reaction (56) proceeds by mechanism S_E2(co-ord), following the course indicated by equations (41) and (42), p. 158.

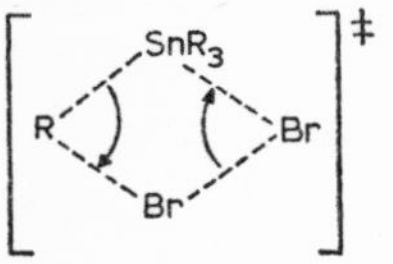

(XVI)

The brominolysis of a number of unsymmetrical tetraalkyltins was later studied by Faleschini and Tagliavini[36] using solvent carbon tetrachloride, but at 35.3 °C. Their recorded second-order rate coefficients for reaction (56) are also given in

TABLE 24

SECOND-ORDER RATE COEFFICIENTS (l.mole^{-1}.sec^{-1}) FOR BROMINOLYSIS OF TETRA-ALKYLTINS BY BROMINE IN SOLVENT CARBON TETRACHLORIDE

Compound	$10^2 k_2^a$ (20 °C)	Compound	$10^2 k_2^b$ (35.5 °C)
Me_4Sn	0.018	Me_4Sn	$\simeq 0.05^c$
Et_4Sn	1.70	$MeSnEt_3$	1.24
Pr^n_4Sn	0.80	$MeSnPr^n_3$	0.70
Bu^n_4Sn	0.97	$MeSnBu^n_3$	0.80
$Octyl^n_4Sn$	1.25	$MeSnPr^i_3$	9.90
$Dodecyl^n_4Sn$	1.44		
Pr^i_4Sn	15.5	$MeSnBu^n_3$	0.80
		$EtSnBu^n_3$	2.47
		$Pr^nSnBu^n_3$	1.83
		$Et–SnBu^n_3$	0.74^d
		$Bu^n–SnEtBu^n_2$	0.58^d
		$Pr^n–SnBu^n_3$	1.03^d
		$Bu^n–SnPrBu^n_2$	0.27^d

[a] Ref. 43.
[b] Ref. 36.
[c] Value estimated by the author from that at 20 °C.
[d] Calculated by the author using the overall rate coefficients given, together with data on product distribution[37] to obtain these values of rate coefficients for the cleavage of the indicated alkyl–tin bond.

Table 24. Faleschini and Tagliavini[36] stated that for all of the unsymmetrical tetraalkyls they studied, only the smaller alkyl group in each tetraalkyl was split off as RBr. Such selectivity is rather unexpected in view of the poor selectivity observed for brominolysis in solvent chlorobenzene. Indeed, Gielen and co-workers[37] have shown that in the brominolysis of $EtSnBu^n_3$ in carbon tetra-chloride at 35 °C, both the ethyl and the n-butyl groups are split off as alkyl bromides, with a ratio $EtBr/Bu^nBr$ of 0.43. Similarly, the ratio Pr^nBr/Bu^nBr was 1.3 in brominolysis of $Pr^nSnBu^n_3$. In view of this work, the rate coefficients recorded by Faleschini and Tagliavini are valid only as overall rate coefficients and (until further work is done) cannot be used as rate coefficients for the cleavage of particular alkyl–tin bonds. Only for the two sets of compounds examined by Gielen and co-workers[37] can the rate coefficient data be used to deduce these alkyl–tin rate coefficients, and values calculated by the author are given in Table 24.

As suggested earlier, the results in Table 24 indicate that the effect of the moving group (the alkyl group undergoing substitution) is to increase the value of the observed rate coefficient in the sense $Me < Et < Pr^n < Pr^i$, whilst the leaving group increases these values in the sense $Me_3Sn < n\text{-alkyl}_3Sn < Pr^i_3Sn$, see equations (48) and (50), p. 161.

3.9 IODINOLYSIS OF TETRA-n-BUTYLTIN BY IODINE IN SOME NON-POLAR SOLVENTS

Jungers et al.[48] have determined the fourth-order rate coefficients, k_4, in the rate expression

$$v = k_4 \cdot [Bu^n_4Sn][I_2]^3 \qquad (57)$$

obtained for the iodinolysis

$$Bu^n_4Sn + I_2 = Bu^nI + Bu^n_3SnI \qquad (58)$$

Values of these fourth-order rate coefficients were reported for iodinolyses carried

TABLE 25

FOURTH-ORDER RATE COEFFICIENTS ($l^3.mole^{-3}.min^{-1}$) FOR THE IODINOLYSIS OF TETRA-n-BUTYLTIN AT 20 °C[48]

Solvent	k_4	Solvent	k_4
n-Butyl bromide	37.5	Toluene	2.00
α-Methylnaphthalene	20.0	Ethylbenzene	1.25
Carbon tetrachloride	7.5	p-Xylene	0.40
Cyclohexane	3.00	m-Xylene	0.38
Benzene	2.50	Mesitylene	0.13

out in various non-polar solvents and numerous binary and tertiary mixtures of non-polar solvents[48]. Some of the values are given in Table 25. In addition, the (accelerating) effect of a number of polar solutes on the value of k_4 was also studied.

4. Halogenolysis of tetraalkylleads

4.1 IODINOLYSIS OF TETRAALKYLLEADS BY I_2/I^- IN POLAR SOLVENTS

Following some earlier preliminary studies[49], Riccoboni et al.[50] studied the iodinolysis of tetramethyl-, tetraethyl-, and tetra-n-propyllead by iodine in the presence of a large excess of iodide ions, using as solvents acetonitrile, methanol, ethanol, and n-propanol. The effect of varying the iodide ion concentration was studied, using tetramethyllead, for all four solvents. It was found[50] that the plots of k_2^{obs} against $1/[I^-]$ gave good straight lines of zero intercept. Hence by the procedure of Gielen and Nasielski[13] (Section 2.2, p. 138) it may be deduced that

TABLE 26

CALCULATION OF SECOND-ORDER RATE COEFFICIENTS (l.mole^{-1}.sec^{-1}) FOR THE IODINOLYSIS OF TETRAMETHYLLEAD BY IODINE

$$k_2 = k_2^{obs} K[I^-]$$

Solvent	Temp. (°C)	μ	k_2^{obs}	$[I^-]$ (mole.l^{-1})	K (l.mole^{-1})	k_2
MeCN[50]	25	0.057	0.168	0.057	2.52×10^{7a}	2.41×10^5
MeOH[50]	25	0.057	10.80	0.057	1.67×10^{4b}	1.03×10^4
EtOH[50]	25	0.057	1.48	0.057	3.32×10^{4c}	2.80×10^3
Pr^{n}OH[50]	25	0.057	1.14	0.057	2.48×10^{4c}	1.61×10^3
MeCOMe[51]	35	0.1	4.36×10^{-4i}		10^{8d}	4.4×10^4
MeCN[42]	20	0.1	9.0×10^{-3i}		3×10^{7e}	2.7×10^5
MeOH[42]	20	0.1	5.1	0.10	2.04×10^{4f}	1.04×10^4
DMSO[42]	20	0.1	0.38	0.10	2.5×10^{5g}	9.5×10^3
MeCO$_2$H[42]	20	0.1	1.16	0.10	1.7×10^{5h}	2×10^4

[a] Ref. 18, value given at 25 °C and $\mu = 0.1$.
[b] From the quoted value[13] of 2.04×10^4 at 20 °C and $\mu = 0.1$, together with the earlier figures[52] for the temperature variation of K.
[c] Ref. 50.
[d] Ref. 20 gives a value of 2×10^8 at 25 °C and $\mu = 0.1$ (see text).
[e] From the value (2.52×10^7) given[18] at 25 °C and $\mu = 0.1$ and an estimate of the temperature variation of K.
[f] Ref. 13.
[g] Ref. 53 gives a value of 2.5×10^5 at 22 °C.
[h] Ref. 59.
[i] Value for $k_2^{obs}[I^-]$.

reaction (59) is a bimolecular reaction between tetraalkyllead and iodine, *viz.*

$$R_4Pb + I_2 \xrightarrow{I^-} RI + R_3PbI \tag{59}$$

and that the actual rate coefficient (k_2) for reaction (59) is related to the observed second-order rate coefficient (k_2^{obs}) by the expression

$$k_2 = k_2^{obs} \cdot K \cdot [I^-] \tag{60}$$

where K is the equilibrium constant for formation of I_3^- under the particular experimental conditions, and $[I^-]$ is the concentration of iodide ion used in a given kinetic experiment. An evaluation of k_2 thus requires a knowledge of the value of K under the experimental conditions; details of the calculation of values of k_2 are given in Table 26, for the substitution of tetramethyllead. Relative values of k_2 for a series of tetraalkylleads undergoing substitution under the same experimental conditions do not depend on K, and such relative rate coefficients are given in Table 27. Gielen and Nasielski[42] also investigated the kinetics of reaction (59) using the solvents acetonitrile, dimethylsulphoxide (DMSO), and methanol. For substitutions in the solvents DMSO and methanol it was found[42] that plots of k_2^{obs} against $1/[I^-]$ were straight lines of zero intercept (or what is equivalent, that values of $k_2^{obs} \cdot [I^-]$ were independent of iodide ion concentration). Details of the calculations of k_2 are given in Table 26, and relative rates of substitution are given in Table 27.

When acetonitrile was used as the solvent, Gielen and Nasielski[42] found that the values of $k_2^{obs} \cdot [I^-]$ were not constant; details are given in Table 28 for the case of the substitution of tetramethyllead. It was deduced[42] that the relation between $k_2^{obs} \cdot [I^-]$ and $[I^-]$ followed the expression

$$k_2^{obs} \cdot [I^-] = 0.0090 + 0.029[I^-] \tag{61}$$

TABLE 27

RELATIVE SECOND-ORDER RATE COEFFICIENTS FOR THE IODINOLYSIS OF SOME TETRAALKYLLEADS BY IODINE

Solvent	Temp. (°C)	Me_4Pb	Et_4Pb	Pr^n_4Pb	Ref.
MeCN	25	100	38.6	8.3	50
MeOH	25	100	38.9	13.4	50
EtOH	25	100	47.5	7.2	50
Pr^{n}OH	25	100	45.3	6.2	50
MeCOMe	35	100	37.6		51
MeCN	20	100	55.5		42
MeOH	20	100	35.3		42
DMSO	20	100	23.6		42
MeCO$_2$H	20	100	56.6		42

TABLE 28

THE EFFECT OF IODIDE ION CONCENTRATION ON THE RATE OF IODINOLYSIS OF TETRA-
METHYLLEAD IN SOLVENT ACETONITRILE

$[I^-]$ (mole.l^{-1})	k_2^{obs} (l.mole^{-1}.sec^{-1})	$k_2^{obs}[I^-]$ (sec^{-1})		
		Determined	Calc. from eqn. (61)	Calc. from eqn. (62)
Temp. 20 °C, ionic strength 0.1 M[42]				
0.0103	0.89	0.0092	0.0093	
0.0201	0.50	0.0101	0.0096	
0.0505	0.22	0.0111	0.0105	
0.1000	0.116	0.0116	0.0119	
Temp. 25 °C, ionic strength 0.194 M[51]				
0.029	0.47	0.0136		0.0136
0.043	0.33	0.0141		0.0138
0.097	0.15	0.0145		0.0147

They suggested that the term in $[I^-]$ in equation (61)* corresponded to a mechanism in which the iodide ion acted as a nucleophilic catalyst. These results of Gielen and Nasielski have been disputed by Pilloni and Tagliavini[51] who declare that in the iodinolysis of tetramethyllead in solvent acetonitrile "... the product $k_2^{obs}\cdot[I^-]$ is fairly constant within the limit of experimental error". Their results are given in Table 28. It may be seen that the values of $k_2^{obs}\cdot[I^-]$ are not quite constant, and can be fitted to an expression of the same type as equation (61), viz.**

$$k_2^{obs}\cdot[I^-] = 0.0132+0.015[I^-] \qquad (62)$$

Table 28 also contains values of $k_2^{obs}\cdot[I^-]$ calculated from equations (61) and (62). The variation of $k_2^{obs}\cdot[I^-]$ with $[I^-]$ is certainly much smaller for the results of Pilloni and Tagliavini than it is for the results of Gielen and Nasielski. How much of this variation is due to experimental error and how much is a genuine variation would seem to the present author to be a moot point. The results of Pilloni and Tagliavini[51] suggest that only a very small percentage of the total reaction proceeds by a mechanism involving nucleophilic assistance by the iodide ion (in the opinion of Pilloni and Tagliavini there is no nucleophilic assistance at all), whilst the results of Gielen and Nasielski[42] indicate that part of the total reaction does involve nucleophilic assistance. For example, at an iodide ion concentration of 0.02 M, the proportion of reaction proceeding by this mechanism is 6.4%

* This term arises from that part of the overall reaction following the rate law $v = k_3\cdot[R_4Pb][I_2][I^-]+k_2^b\cdot[R_4Pb][I_3^-]$, see Section 2.2, p. 138.
** Note that equations (61) and (62) are not exactly comparable, since they refer to kinetic experiments carried out under different conditions of temperature and ionic strength.

(on equation 61) or 2.3 % (on equation 62). The relative rate coefficients reported by Gielen and Nasielski are given in Table 26.

Pilloni and Tagliavini[51] also reported rate studies using acetone as solvent. They obtained a value of 4.36×10^{-4} sec^{-1} for the product $k_2^{obs} \cdot [I^-]$, using tetramethyllead as the substrate, at 35 °C and ionic strength 0.1 M. They suggested that the value of K under these conditions was probably around 10^6 l.mole^{-1} and hence $k_2 \simeq 4.36 \times 10^2$ l.mole^{-1}.sec^{-1}. A reported value[20] for K in solvent acetone and ionic strength 0.1 M is, however, no less than 2×10^8 l.mole^{-1} at 25 °C, and the value at 35 °C is not likely to be under 10^8 l.mole^{-1}. On this basis, $k_2 \simeq 4.4 \times 10^4$ l.mole^{-1}.sec^{-1} (see Table 26). Table 27 gives the relative rate coefficients obtained using acetone solvent[51].

It may be seen from Table 26 that the rate of substitution of tetramethyllead by iodine increases along the solvent series

$$\text{Pr}^n\text{OH} < \text{EtOH} < \text{DMSO} < \text{MeOH} < \text{MeCOMe} < \text{MeCN} \qquad (63)$$

Solvent effects on the rates of halogenodemetallations are discussed in general in Section 6, p. 173.

The addition of sodium perchlorate considerably increases the value of k_2^{obs} for the iodinolysis of tetramethyllead in solvents methanol, ethanol, and acetonitrile[50]. Since values of K do not vary very much with ionic strength (at least with solvent methanol[13]), it seems fairly clear that k_2 itself must increase with increase in ionic strength. Kinetic salt effects were not studied with the other solvents listed in Table 26, but it seems reasonable to suggest that in all these cases the substitution proceeds by mechanism S_E2(open) through a transition state such as (XVII) or (XVIII).

(XVII)   (XVIII)

The sequences of reactivity shown in Table 27 are all examples of a "steric sequence". Pilloni and Tagliavini[51] suggest that these sequences are due to the larger alkyl groups inhibiting attack of the solvent on the tetraalkyllead, in a pre-rate-determining coordination. On the analysis of Abraham and Spalding[40], however, these sequences could be held to arise from transition-state interactions between the moving group (R) and the incoming electrophilic (I_2), and between the moving group (R) and the leaving group (PbR$_3$). The rather smaller interactions (Table 27) involved when the leaving group is PbR$_3$ than when the leaving group is SnR$_3$ (see Table 7, Chapter 11, Section 5.1, p. 226) may then be due to the longer R - - - Pb bond compared with the R - - - Sn bond*, since if the moving group R is further away from the leaving group MR_3 interactions between R and MR_3 will naturally be reduced.

4.2 IODINOLYSIS OF TETRAALKYLLEADS BY IODINE IN SOLVENTS BENZENE AND CARBON TETRACHLORIDE

Pilloni and Tagliavini[51] also studied the kinetics of iodinolysis of tetraalkylleads, using non-polar solvents in the absence of iodide ion. Second-order rate coefficients for the reaction

$$R_4Pb+I_2 = RI+R_3PbI \qquad (64)$$

are given in Table 29. The effect of the addition of polar compounds to solvent carbon tetrachloride on the rate coefficients for reaction (64) (R = Me) was also studied. It was shown that addition of acetonitrile, methanol, ethanol, and n-propanol resulted in an increase in the rate coefficient, and that this increase was proportional to the square of the concentration of the additive. The rate coefficient was increased some ten-fold on the addition of about 0.07 M of the alcohols or of about 0.03 M acetonitrile.

It was suggested[51] that these powerful accelerations of rate are due to ". . . the importance of the solvent molecules around the metal centre". Whether this is thought to be an effect on the initial state or on the transition state (or both) is not clear.

The order of reactivity of the three tetraalkylleads studied in this work is remarkably similar to that of the corresponding tetraalkyltins, as may be seen from the data assembled in Table 30. It seems possible, therefore, that similar mechanisms obtain for the halogenolysis of tetraalkylleads and tetraalkyltins and that reaction (64), in non-polar solvents, proceeds by mechanism S_E2(cyclic) or mechanism S_E2(co-ord).

TABLE 29

SECOND-ORDER RATE COEFFICIENTS ($l.mole^{-1}.sec^{-1}$) FOR THE IODINOLYSIS OF TETRAAL-KYLLEADS BY IODINE[51]

Solvent	Me_4Pb	Et_4Pb	Pr^n_4Pb
Benzene, 35 °C	1.9	22.8	7.1
Carbon tetrachloride, 31 °C	0.0355	1.08	0.43

---

* Bond lengths are Sn–C = 2.18 A in Me_4Sn (ref. 54) and Pb–C = 2.30 A in Me_4Pb (ref. 55).

TABLE 30

RELATIVE SECOND-ORDER RATE COEFFICIENTS FOR THE HALOGENOLYSIS OF TETRAAL-
KYLTINS AND TETRAALKYLLEADS BY HALOGENS

Halogen/solvent	Temp. (°C)	M	Me_4M	Et_4M	Pr^n_4M
I_2/Benzene[a]	35	Pb	100	1200	370
Br_2/PhCl[b]	20	Sn	100	1230	470
I_2/CCl$_4$[a]	31	Pb	100	3040	1200
Br_2/CCl$_4$[c]	20	Sn	100	9500	4400

[a] Ref. 51, see Table 29 (p. 170).
[b] Ref. 35, see Table 19 (p. 157) for values of k_2^{obs}.
[c] Ref. 43, see Table 24 (p. 164).

4.3 THE BROMINOLYSIS OF TETRAALKYLLEADS BY Br_2/Br^- IN SOLVENT METHANOL

The kinetics of the reaction

$$R_4Pb + Br_2 \xrightarrow{Br^-} RBr + R_3PbBr \qquad (65)$$

were studied under conditions such that the concentration of bromine was maintained constant in any given experiment[56]. The kinetic order with respect to the
tetraalkyllead was found to be unity, and from experiments in which the bromine
concentration was varied from run to run the order with respect to bromine was
also found to be unity. The second-order rate coefficients in the rate equation

$$v = k_2 \cdot [R_4Pb][Br_2] \qquad (66)$$

were found to be 11.0×10^4 l.mole^{-1}.sec^{-1} (R = Me) and 9.4×10^4 l.mole^{-1},
sec^{-1} (R = Et) at 25 °C and at an ionic strength of 0.2 M. At lower ionic strengths,
the rate coefficient is appreciably less, viz. 7.9×10^4 l.mole^{-1}.sec^{-1} (R = Et) at
25 °C and ionic strength 0.05 M.

It was concluded[56] that reaction (65) proceeds by mechanism S_E2(open), for
which the transition state (XIX) was written.

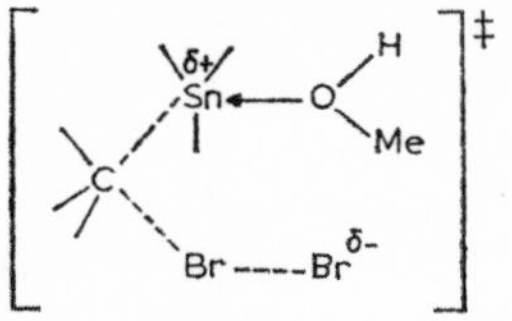

(XIX)

5. Iodine bromide cleavage of tetraalkyltins

Redl et al.[57] have shown that in the reaction of tetraalkyltins with IBr, using solvents methanol and cyclohexane, the products formed are given by

$$R_4Sn + IBr = RI + R_3SnBr \tag{67}$$

Kinetic studies showed that rate equation

$$v = k_2[R_4Sn][X_2] + k_3[R_4Sn][X_2]^2 \tag{68}$$

was followed; values of k_2 and k_3 are given in Table 31, together with corre-

TABLE 31

RATE COEFFICIENTS FOR REACTION OF TETRAALKYLTINS WITH IODINE BROMIDE AND BROMINE

Reagent	Solvent	Temp. (°C)	Me_4Sn		Pr_4Sn	
			k_2 ($l.mole^{-1}.sec^{-1}$)	k_3 ($l^2.mole^{-2}.sec^{-1}$)	k_2 ($l.mole^{-1}.sec^{-1}$)	k_3 ($l^2.mole^{-2}.sec^{-1}$)
Br_2[a]	CCl_4	20	1.84×10^{-4}		8×10^{-3}	
IBr	CCl_4	25	1.7×10^{-3}	2.2	$<5 \times 10^{-4}$	10.1
Br_2	Cyclohexane	25	No reaction	No reaction	2.2×10^{-3}	$\simeq 0.0$
IBr	Cyclohexane	25	1.1×10^{-3}	2.5×10^{-1}	No reaction	No reaction

[a] Ref. 43.

sponding values for brominolysis. The values of k_2 vary so greatly with the tetraalkyltin and with the electrophilic reagent, that Redl et al.[57] suggested that the mechanism for the second-order path could alter, not only from Br_2 to IBr, but also with change in the tetraalkyltin.

For the third-order path, two possible mechanisms, (69) and (70), were put forward, viz.

$$R_4Sn + IBr \overset{K}{\rightleftharpoons} Complex$$

$$Complex + IBr \overset{k'_2}{\rightarrow} Products \tag{69}$$

$$2\,IBr \overset{K}{\rightleftharpoons} Intermediate$$

$$Intermediate + R_4Sn \overset{k'_2}{\rightarrow} Products \tag{70}$$

In either case the observed third-order rate coefficient is given by $k_3 = Kk'_2$.

6. Solvent effects on the iodinolysis of tetraalkyltins and tetraalkylleads

Second-order rate coefficients for reaction (71) (M = Sn, Pb)

$$R_4M + I_2 \rightarrow RI + R_3MI \tag{71}$$

have been obtained, using a variety of solvents. The available data are collected in Table 32 for the case of R = Me and Et. It is evident that polar solvents, especially dipolar aprotic solvents, considerably acelerate reaction (71). A separation of such solvent influences into initial-state effects and transition-state effects may be accomplished through the equation

$$\Delta G_t^0(\text{Tr}) = \Delta G_t^0(R_4M) + \Delta G_t^0(I_2) + \delta \Delta G^{\ddagger} \tag{72}$$

where $\Delta G_t^0(Y)$ refers to the standard free energy of transfer of a species Y from the reference solvent 1 (methanol, in the present case) to any other solvent 2. The transition state in reaction (71) is denoted as Tr, and the term $\delta \Delta G^{\ddagger}$ is given by $\delta \Delta G^{\ddagger} = \Delta G_2^{\ddagger} - \Delta G_1^{\ddagger}$. Details of the calculations are in Tables 33 and 34. It should be noted that the standard state adopted is not the same as the one used to calculate the original rate coefficients, and hence that the rate coefficients (l.mole^{-1}.sec^{-1}) in Table 32 have been converted to units of mole-fraction^{-1}. sec^{-1} before the calculation of values of $\delta \Delta G^{\ddagger}$.

From Tables 33 and 34 it may be seen that the rate increases associated with polar solvents (*i.e.* the decrease in $\delta \Delta G^{\ddagger}$) are due both to a reduction in $\Delta G_t^0(\text{Tr})$ and to an increase in $\Delta G_t^0(\text{Reactants})$ on transfer from a non-polar solvent to a polar solvent. Solvent effects on the reactants in equation (71) are considerable, in many cases being as large as solvent effects on the transition state.

TABLE 32

SECOND-ORDER RATE COEFFICIENTS (l.mole^{-1}.sec^{-1}) FOR THE IODINOLYSIS OF TETRAAL-KYLTINS AND TETRAALKYLLEADS

Solvent	Me_4Sn^a	Et_4Sn^a	Me_4Pb^b	Et_4Pb^b
Acetonitrile			2.41×10^5	9.3×10^4
Acetone			3.8×10^4	1.4×10^4
Methanol	6.70^c	0.83^c	1.03×10^4	4.0×10^3
Ethanol			2.8×10^3	1.33×10^3
Dimethylsulphoxide	0.48	0.021	1.0×10^4	2.4×10^3
n-Propanol			1.61×10^3	7.3×10^2
Acetic acid	0.244	0.080	3.0×10^4	1.7×10^4
Chlorobenzene	1.67×10^{-4}	1.0×10^{-3}		
Benzene			1.1	13.2
Carbon tetrachloride			2.6×10^{-2}	0.78

[a] At 20 °C from refs. 35, 42, and 43.
[b] At 25 °C from Tables 26 (p. 166) and 29 (p. 170).
[c] Values corresponding to $\mu = 0$.

TABLE 33

FREE ENERGIES OF TRANSFER, ON THE MOLE FRACTION SCALE, FROM METHANOL TO OTHER SOLVENTS OF TETRAALKYLTINS, IODINE, AND THE TETRAALKYLTIN/IODINE TRANSITION STATES (kcal.mole^{-1}) AT 298 °K[61]

Solvent	Me_4Sn+I_2			
	$\delta\Delta G^{\ddagger}$	$\Delta G_t^0(Me_4Sn)$	$\Delta G_t^0(I_2)$	$\Delta G_t^0(Tr)$
Methanol	0	0	0	0
Acetic acid	1.9	−0.2	0.9	2.6
Chlorobenzene	6.8	−1.6	−0.1	5.1

Solvent	Et_4Sn+I_2			
	$\delta\Delta G^{\ddagger}$	$\Delta G_t^0(Et_4Sn)$	$\Delta G_t^0(I_2)$	$\Delta G_t^0(Tr)$
Methanol	0	0	0	0
Acetic acid	1.6	−0.1	0.9	2.4
Chlorobenzene	4.5	−2.2	−0.1	2.2

TABLE 34

FREE ENERGIES OF TRANSFER, ON THE MOLE FRACTION SCALE, FROM METHANOL TO OTHER SOLVENTS OF TETRAALKYLLEADS, IODINE, AND THE TETRAALKYLLEAD/IODINE TRANSITION STATES (kcal.mole^{-1}) AT 298 °K[61]

Solvent	Me_4Pb+I_2			
	$\delta\Delta G^{\ddagger}$	$\Delta G_t^0(Me_4Pb)$	$\Delta G_t^0(I_2)$	$\Delta G_t^0(Tr)$
Acetonitrile	−1.7	0	0.4	−1.3
Acetone	−0.4	−0.8	1.2	0.0
Methanol	0	0	0	0
Ethanol	1.0	−0.5	−0.3	0.2
n-Propanol	1.5	−0.8	−0.4	0.3
Acetic acid	−0.8	−0.2	0.9	−0.1
Benzene	5.9	−1.7	−0.4	3.7
Carbon tetrachloride	8.1	−2.0	0.5	6.5

Solvent	Et_4Pb+I_2			
	$\delta\Delta G^{\ddagger}$	$\Delta G_t^0(Et_4Pb)$	$\Delta G_t^0(I_2)$	$\Delta G_t^0(Tr)$
Acetonitrile	−1.7	0.2	0.4	−1.1
Acetone	−0.4	−0.9	1.2	−0.1
Methanol	0	0	0	0
Ethanol	0.9	−0.8	−0.3	−0.2
n-Propanol	1.4	−1.1	−0.4	−0.1
Acetic acid	−1.0	−0.1	0.9	−0.2
Benzene	3.9	−2.4	−0.4	1.1
Carbon tetrachloride	5.6	−2.6	0.5	3.5

Information as to the nature of the transition state in reaction (71) may be obtained through a comparison of $\Delta G_t^0(\text{Tr})$ values with those for other transition states and those for various model solutes. Data are available[62] on solvent effects on the transition state in the Menschutkin reaction

$$\text{Me}_3\text{N} + p\text{-NO}_2\text{C}_6\text{H}_4\text{CH}_2\text{Cl} \rightarrow p\text{-NO}_2\text{C}_6\text{H}_4\text{CH}_2\overset{+}{\text{N}}\text{Me}_3 + \text{Cl}^- \tag{73}$$

and also on the [Et$_4$NI] ion pair. These are summarised in Table 35; the $[\text{Et}_4\text{Pb}/\text{I}_2]^{\ddagger}$ and $[\text{Me}_3\text{N}/\text{RCl}]^{\ddagger}$ transition states* tend to follow the same pattern in ΔG_t^0 values, but values of $\Delta G_t^0(\text{Et}_4\text{NI})$ are quite different from those for the transition states. Abraham[62] has already concluded that the transition state in reaction (73) carries only a comparatively low separation of charge, perhaps of the order of 0.3–0.4 units and from the data in Table 35 it appears that the $[\text{Et}_4\text{Pb}/\text{I}_2]^{\ddagger}$ transition state also carries a similar separation of charge.

TABLE 35

COMPARISON OF FREE ENERGIES OF TRANSFER, ON THE MOLE FRACTION SCALE, FROM METHANOL OF TRANSITION STATES AND ION PAIRS (kcal.mole^{-1}) AT 298 °K[61, 62]

	ΔG_t^0		
Solvent	$[Me_3N/RCl]^{\ddagger a}$	$[Et_4Pb/I_2]^{\ddagger}$	$[Et_4NI]$
Acetonitrile	−1.6	−1.1	1.2
Acetone	−0.9	−0.1	1.9
Methanol	0	0	0
Ethanol	0.5	−0.2	1.1
n-Propanol	0.6[b]	−0.1	1.4
Benzene	1.0	1.1	6.4
Carbon tetrachloride	4.0[b]	3.5	10.5

[a] $R = p\text{-NO}_2\text{C}_6\text{H}_4\text{CH}_2$.
[b] Approximate values.

REFERENCES

1 F. R. JENSEN AND L. H. GALE, *J. Am. Chem. Soc.*, 81 (1959) 1261; 82 (1960) 148.
2 F. R. JENSEN, L. D. WHIPPLE, D. K. WEDEGAERTNER AND J. A. LANDGREBE, *J. Am. Chem. Soc.*, 81 (1959) 1262; 82 (1960) 2466.
3 O. A. REUTOV, É. V. UGLOVA, I. P. BELETSKAYA AND T. B. SVETLANOVA, *Izv. Akad. Nauk SSSR, Ser. Khim.*, (1964) 1383; *Bull. Acad. Sci. USSR*, (1964) 1297.
4 O. A. REUTOV, É. V. UGLOVA, I. P. BELETSKAYA AND T. B. SVETLANOVA, *Izv. Akad. Nauk SSSR, Ser. Khim.*, (1968) 1151; *Bull Acad. Sci. USSR*, (1968) 1101.
5 H. M. WALBORSKY, F. J. IMPASTATO AND A. E. YOUNG, *J. Am. Chem. Soc.*, 86 (1964) 3283.
6 D. E. APPLEQUIST AND A. H. PETERSON, *J. Am. Chem. Soc.*, 83 (1961) 862.
7 D. E. APPLEQUIST AND G. N. CHMURNY, *J. Am. Chem. Soc.*, 89 (1967) 875.

* $R = p\text{-NO}_2\text{C}_6\text{H}_4\text{CH}_2$.

8 K. SISIDO, S. KOZIMA AND K. TAKIZAWA, *Tetrahedron Letters*, (1967) 33; K. SISIDO, T. MIYANISI, T. ISIDA AND S. KOZIMA, *J. Organometal. Chem.*, 23 (1970) 117.

9 P. BAEKELMANS, M. GIELEN AND J. NASIELSKI, *Tetrahedron Letters*, (1967) 1149.

10 N. S. ZEFIROV, P. P. KADZYAUSKAS AND YU. K. YUR'EV, *Zh. Obshch. Khim.*, 36 (1966) 1735; *J. Gen. Chem. USSR Engl. Transl.*, 36 (1966) 1731.

11 J. L. KELLER, *Thesis*, U.C.L.A., 1948, quoted by S. WINSTEIN AND T. G. TRAYLOR, ref. 12.

12 S. WINSTEIN AND T. G. TRAYLOR, *J. Am. Chem. Soc.*, 78 (1956) 2597.

13 M. GIELEN AND J. NASIELSKI, *Bull. Soc. Chim. Belges*, 71 (1962) 32.

14 L. I. KATZIN AND E. GEBERT, *J. Am. Chem. Soc.*, 77 (1955) 5814; R. W. RAMETTE AND R. W. SANDFORD, JR., *J. Am. Chem. Soc.*, 87 (1965) 5001.

15 L. RICCOBONI, G. PILLONI, G. PLAZZOGNA AND G. TAGLIAVINI, *J. Electroanal. Chem.*, 11 (1966) 340.

16 J. NASIELSKI, O. BUCHMAN, M. GROSJEAN AND E. HANNECART, *Bull. Soc. Chim. Belges*, 77 (1968) 15.

17 R. L. BENOIT, M. GUAY AND J. DESBARRES, *Can. J. Chem.*, 46 (1968) 1261; A. J. PARKER, *J. Chem. Soc. A*, (1966) 220; R. L. BENOIT, *Inorg. Nucl. Chem. Letters*, 4 (1968) 723.

18 J. DESBARRES, *Bull. Soc. Chim. France*, (1961) 502.

19 R. L. BENOIT AND M. GUAY, *Inorg. Nucl. Chem. Letters*, 4 (1968) 215.

20 I. V. NELSON AND R. T. IWAMOTO, *J. Electroanal. Chem.*, 7 (1964) 218.

21 J. E. DUBOIS AND F. GARNIER, *Bull. Soc. Chim. France*, (1965) 1715.

22 T. W. NAKAGAWA, L. J. ANDREWS AND R. M. KEEFER, *J. Phys. Chem.*, 61 (1957) 1007.

23 I. P. BELETSKAYA, O. A. REUTOV AND T. P. GUR'YANOVA, *Izv. Akad. Nauk SSSR, Otd. Khim. Nauk*, (1961) 1589; *Bull. Acad. Sci. USSR, Div. Chem. Sci.*, (1961) 1483.

24 I. P. BELETSKAYA, O. A. REUTOV AND T. P. GUR'YANOVA, *Izv. Akad. Nauk SSSR, Otd. Khim. Nauk*, (1961) 1997; *Bull. Acad. Sci. USSR, Div. Chem. Sci.*, (1961) 1863.

25 O. R. REUTOV, I. P. BELETSKAYA AND T. P. FETISOVA, *Dokl. Akad. Nauk SSSR*, 166 (1966) 861; *Proc. Acad. Sci. USSR*, 166 (1966) 158.

26 O. A. REUTOV, I. P. BELETSKAYA AND T. P. FETISOVA, *Izv. Akad. Nauk SSSR, Ser. Khim.*, (1967) 990; *Bull. Acad. Sci. USSR*, (1967) 960.

27 O. A. REUTOV, I. P. BELETSKAYA AND T. P. FETISOVA, *Dokl. Akad. Nauk SSSR*, 155 (1964) 1095; *Proc. Acad. Sci. USSR*, 155 (1964) 347.

28 O. A. REUTOV, G. A. ARTAMKINA AND I. P. BELETSKAYA, *Dokl. Akad. Nauk SSSR*, 153 (1963) 588; *Proc. Acad. Sci. USSR*, 153 (1963) 939.

29 O. A. REUTOV, I. P. BELETSKAYA AND T. A. AZIZYAN, *Izv. Akad. Nauk SSSR, Otd. Khim. Nauk*, (1962) 424; *Bull. Acad. Sci. USSR, Div. Chem. Sci.*, (1962) 393.

30 I. P. BELETSKAYA, O. A. REUTOV AND T. A. AZIZYAN, *Izv. Akad. Nauk SSSR, Otd. Khim. Nauk*, (1962) 223; *Bull. Acad. Sci. USSR, Div. Chem. Sci.*, (1962) 204.

31 I. P. BELETSKAYA, O. A. REUTOV AND T. P. GUR'YANOVA, *Izv. Akad. Nauk SSSR, Otd. Khim. Nauk*, (1961) 2178; *Bull Acad. Sci. USSR, Div. Chem. Sci.*, (1961) 2036.

32 I. P. BELETSKAYA, T. A. AZIZYAN AND O. A. REUTOV, *Izv. Akad. Nauk SSSR, Ser. Khim.*, (1963) 1332; *Bull. Acad. Sci. USSR*, (1963) 1208.

33 G. A. RAZUVAEV AND A. V. SAVITSKII, *Dokl. Akad. Nauk SSSR*, 85 (1952) 575; *Chem. Abstr.*, 47 (1953) 9911.

34 A. LORD AND H. O. PRITCHARD, *J. Phys. Chem.*, 70 (1966) 1689.

35 S. BOUÉ, M. GIELEN AND J. NASIELSKI, *J. Organometal. Chem.*, 9 (1967) 443.

36 S. FALESCHINI AND G. TAGLIAVINI, *Gazz. Chim. Ital.*, 97 (1967) 1401.

37 S. BOUÉ, M. GIELEN, J. NASIELSKI, J. AUTIN AND M. LIMBOURG, *J. Organometal. Chem.*, 15 (1968) 267.

38 T. F. BOLLES AND R. S. DRAGO, *J. Am. Chem. Soc.*, 88 (1966) 3921.

39 M. H. ABRAHAM, R. J. IRVING AND G. F. JOHNSTON, *J. Chem. Soc. A*, (1970) 199; M. H. ABRAHAM, unpublished work.

40 M. H. ABRAHAM AND T. R. SPALDING, *J. Chem. Soc. A*, (1969) 399.

41 M. GIELEN AND J. NASIELSKI, *Bull. Soc. Chim. Belges*, 71 (1962) 601.

42 M. GIELEN AND J. NASIELSKI, *J. Organometal. Chem.*, 7 (1967) 273.

43 M. GIELEN AND J. NASIELSKI, *J. Organometal Chem.*, 1 (1963) 173.

44 S. BOUÉ, M. GIELEN AND J. NASIELSKI, *Bull. Soc. Chim. Belges*, 76 (1967) 559.

45 M. Gielen, J. Nasielski and R. Yernaux, *Bull. Soc. Chim. Belges*, 72 (1963) 594.
46 C. Eaborn and O. W. Steward, *J. Chem. Soc.*, (1965) 521.
47 S. Boué, M. Gielen and J. Nasielski, *J. Organometal Chem.*, 9 (1967) 481.
48 J. C. Jungers, L. Sajus, I. de Aguirre and D. Decroocq, *Rev. Inst. Franc. Pétrole Ann. Combust. Liquides*, 20 (1965) 513; 21 (1966) 285. G. Martino and J. C. Jungers, *Bull. Soc. Chim. France*, (1970) 3392.
49 L. Riccoboni and L. Oleari, *Ric. Sci. Rend., Sez. A*, 3 (1963) 1031; L. Riccoboni, G. Pilloni, G. Plazzogna and C. Bernardin, *Ric. Sci. Rend., Sez. A*, 3 (1963) 1231.
50 L. Riccoboni, G. Pilloni, G. Plazzogna and G. Tagliavini, *J. Electroanal. Chem.*, 11 (1966) 340.
51 G. Pilloni and G. Tagliavini, *J. Organometal. Chem.*, 11 (1968) 557.
52 O. Buchman, M. Grosjean and J. Nasielski, *Helv. Chim. Acta*, 47 (1964) 1679.
53 F. W. Hiller and J. H. Krueger, *Inorg. Chem.*, 6 (1967) 528.
54 L. O. Brockway and H. O. Jenkins, *J. Am. Chem. Soc.*, 58 (1936) 2036.
55 C. Wong and V. Schomaker, *J. Chem. Phys.*, 28 (1958) 1007.
56 M. Gielen, J. Nasielski, J. E. Dubois and P. Fresnet, *Bull. Soc. Chim. Belges*, 73 (1964) 293.
57 G. Redl, B. Altner, D. Anker and M. Minot, *Inorg. Nuclear Chem. Letters*, 5 (1969) 861.
58 C. Barraqúe, J. Vedel and B. Tremillon, *Anal. Chim. Acta*, 46 (1969) 263.
59 G. Durand and B. Tremillon, *Anal. Chim. Acta*, 49 (1970) 135.
60 R. W. Johnson and R. G. Pearson, *Chem. Commun.*, (1970) 986; R. G. Pearson and W. R. Muir, *J. Am. Chem. Soc.*, 92 (1970) 5519.
61 M. H. Abraham, unpublished work.
62 M. H. Abraham, *Tetrahedron Letters*, (1970) 5233; *J. Chem. Soc. B*, (1971) 299.
63 F. R. Jensen, V. Madan and D. H. Buchanan, *J. Am. Chem. Soc.*, 93 (1971) 5283.
64 G. M. Whitesides and D. J. Boschetto, *J. Am. Chem. Soc.*, 93 (1971) 1529.
65 F. R. Jensen and D. D. Davis, *J. Am. Chem. Soc.*, 93 (1971) 4048.

Chapter 9

Other Electrophilic Substitutions
of Organometallic Compounds

In this chapter a number of electrophilic substitutions that have been the subject of kinetic studies are reviewed. Many of these substitutions have been suggested to proceed via mechanisms denoted, in the terminology used in the present work, as S_E2(cyclic) and S_E2(co-ord). Additionally, a short account of the stereochemical course of the carbonation reaction has been included, even though no kinetic studies have been reported.

1. The stereochemical course of carbonation of alkyllithium and alkylmagnesium compounds

Carbonation of optically active 2-octyllithium[1], 2-butyllithium[2], and 1-methyl-2,2-diphenylcyclopropyllithium[3] proceeds with retention of optical activity ($> 20 \%$, $> 83 \%$, 100% respectively) and retention of configuration. Full retention of configuration was also observed[4] for the carbonation of *cis*- and *trans*-2-methylcyclopropyllithium and of *exo*- and *endo*-norbornyllithium. When the optically active 1-methyl-2,2-diphenylcyclopropyllithium was allowed to react with magnesium bromide and the resulting product carbonated, the corresponding cyclopropylcarboxylic acid was obtained fully active. It was thus deduced[5] that 1-methyl-2,2-diphenylcyclopropylmagnesium bromide underwent carbonation again with retention of optical activity and configuration. *Endo*-norbornyl-magnesium bromide is also carbonated with retention of (geometric) configuration[6].

No kinetic studies of reactions (1) and (2) have been reported, but transition states of the S_E2(cyclic) type are often written for these reactions.

$$RLi + CO_2 = RCO_2Li \tag{1}$$

$$RMgX + CO_2 = RCO_2MgX \tag{2}$$

2. Addition reactions of organomagnesium compounds

A wealth of papers has been published on the kinetics and mechanism of the addition of organomagnesium compounds to ketones, nitriles, etc. It is not proposed to discuss these papers, especially since reviews by Ashby[7] and Wakefield[8] are available. Recent work on this topic includes papers by Smith and Billet[9], Holm and Blankholm[10], and Ashby et al.[11]. For the addition of dimethylmagnesium to 4-methylmercaptoacetophenone in solvent ether, Smith and Billet[9] conclude that a very fast reversible formation of a complex ($K = 6$ l.mole^{-1}) is followed by a first-order collapse of the complex, viz.

$$Me_2Mg + Ketone \overset{K}{\rightleftharpoons} Complex \rightarrow Product \tag{3}$$

Ashby et al.[7, 11], however, suggest that the addition of methylmagnesium bromide to benzophenone, again in solvent ether, proceeds by a termolecular mechanism, written as

$$MeMgBr + Ketone \rightleftharpoons Complex$$
$$Complex + MeMgBr \rightarrow Product \tag{4}$$

3. The reaction between dialkylzincs and benzaldehyde

The reaction of diethylzinc with benzaldehyde has been reported[12] to be first-order in diethylzinc and first-order in benzaldehyde. At the temperature of refluxing ether (presumably 35 °C) the second-order rate coefficient was found to be 1.0×10^{-2} l.mole^{-1}.min^{-1} in solvent ether. Addition of small quantities (~ 1 mole % of the initial concentration of diethylzinc) of zinc or magnesium halides considerably increased the value of the rate coefficient.

Di-n-propylzinc reacts with benzaldehyde in ether at 25 °C to give both addition products and reduction products. The second-order rate coefficient for total reaction is 2.1×10^{-3} l.mole^{-1}.min^{-1}, but is increased in value on addition of tetra-n-butylammonium halides. Addition of these halides also considerably increases the ratio of addition to reduction[30].

4. Alkylation of triarylbromomethane by alkylmercuric halides

Reutov and co-workers[13] have shown that the action of triphenylbromomethane on α-carbethoxybenzylmercuric bromide proceeds according to equation (5) (Ar = Ph)

$$Ar_3CBr + PhCH(CO_2Et)HgBr \xrightarrow{DCE} \underset{\underset{CAr_3}{|}}{PhCHCO_2Et} + HgBr_2 \tag{5}$$

in solvent 1,2-dichloroethane (DCE). Kinetic studies on reaction (5) were complicated, because the mercuric bromide formed complexed with triphenylbromomethane, and the kinetics were therefore followed in presence of a 100-fold excess of mercuric bromide[14]. Under these conditions, the electrophilic reagent is a complex of the triarylbromomethane and mercuric bromide. The reaction

$$(p\text{-Me-}C_6H_4)_3CBr\cdot HgBr_2 + p\text{-Y-}C_6H_4CH(CO_2Et)HgBr$$

$$\xrightarrow{\text{DCE}} p\text{-Y-}C_6H_4CHCO_2Et + 2\ HgBr_2 \qquad (6)$$
$$\underset{C(C_6H_4\text{-}p\text{-Me})_3}{|}$$

was found to obey a second-order kinetic law; rate coefficients are given in Table 1.

TABLE 1

SECOND-ORDER RATE COEFFICIENTS ($l.mole^{-1}.sec^{-1}$) FOR THE SUBSTITUTION OF p-Y-$C_6H_4CH(CO_2Et)HgBr$ BY THE TRI-p-TOLYLBROMOMETHANE/MERCURIC BROMIDE COMPLEX, REACTION (6), IN SOLVENT 1,2-DICHLOROETHANE AT 50 °C[14]

p-Y	k_2	p-Y	k_2
Pr^i	25.7	Cl	2.9
Et	21.8	Br	2.2
Bu^t	10.3	I	1.9
H	7.7	NO_2	1.4
F	5.1		

The transition state for reaction (6) was supposed[14] to be of an "open" type (I), and on this basis the electrophile is thus $Ar_3\overset{\delta+}{C} - - - \overset{\delta-}{HgBr_3}$. As can be seen from the data in Table 1, electron-donating substituents in the organomercury substrate accelerate reaction (6) and electron-withdrawing substituents retard reaction (6). It may therefore be presumed that formation of the new C–C bond is more important than the breaking of the old C–HgBr bond, in transition state (I).

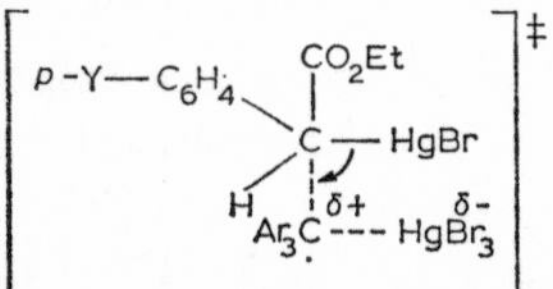

(I)

Some studies on reaction (5) (Ar = p-tolyl) were also carried out in the absence of added mercuric bromide[14]. It was found that the two reactants rapidly formed a complex, and that this complex then decomposed in accordance with a first-order rate law. The unimolecular decomposition may thus be written as

$$
\begin{array}{c}
p\text{-Y}-C_6H_4 \quad CO_2Et \\
\diagdown \; | \\
C\text{----}HgBr \\
H \diagup \; | \\
Ar_3C\text{----}Br^+
\end{array}
\longrightarrow
\left[
\begin{array}{c}
p\text{-Y}-C_6H_4 \quad CO_2Et \\
\diagdown \; | \\
C\text{---}HgBr \\
H \diagup \; | \\
Ar_3C\text{---}Br^+
\end{array}
\right]^{\ddagger}
\longrightarrow \text{products} \qquad (7)
$$

$$\text{(II)} \hspace{5cm} \text{(III)}$$

It was also observed, qualitatively, that the effect of p-substituents in the compound p-Y–C_6H_4CH(CO_2Et)HgBr was in the order of increasing rate of reaction Y = alkyl < H < halogen < NO_2. This is completely the reverse of the order shown in Table 1 and is the order expected for reactions following mechanism S_E1. Reutov and co-workers[14] suggested that in (III) the dominant feature is the breaking of the old C–HgBr bond, and that this bond is to some extent broken before the electrophilic attack of the Ar_3C group occurs.

For different triarylbromomethanes, the reactivity order[14] was found to be $(p\text{-Me-}C_6H_4)_3CBr < Ph_3CBr \gg (p\text{-NO}_2\text{-}C_6H_4)_3CBr$. Thus not only is the breaking of the old C–HgBr bond important (perhaps aided in the sense p-Me > p-H > p-NO$_2$) but also the formation of the new C–C bond must play some part (this bond formation would be aided in the sense p-Me < p-H < p-NO$_2$). Combination of these effects results in the compound Ph_3CBr being the most effective electrophile of the three studied.

5. Cleavages of alkylboronic acids

5.1 THE ACTION OF HYDROGEN PEROXIDE ON ALKYLBORONIC ACIDS

The alkyl–boron bond is readily cleaved by hydrogen peroxide (these reactions are normally carried out in presence of alkali) according to

$$R-B\langle + HO_2^- = ROB\langle + HO^- \qquad (8)$$

and it is generally accepted (see ref. 15) that reaction (8) involves retention of configuration in the moving alkyl group, R. A kinetic study[16] of the cleavage of alkylboronic acids by hydrogen peroxide showed that the reactions followed clean second-order kinetics, first-order in hydrogen peroxide and first-order in boronic acid, both for reactions carried out in aqueous perchloric acid and in presence of an alkaline buffer. For the alkali reactions, a mechanism involving an initial reversible coordination was suggested, viz.

$$RB(OH)_2 + {}^-OOH \underset{}{\overset{K}{\rightleftharpoons}} [RB(OH)_2OOH]^- \qquad (9)$$

$$
\begin{array}{c}
R \\
| \\
HO-B-O-OH \\
| \\
OH
\end{array}
\xrightarrow{k_1}
(HO)_2B-OR + HO^- \qquad (10)
$$

TABLE 2

RATE COEFFICIENTS FOR THE REACTIONS OF BORONIC ACIDS WITH HYDROGEN
PEROXIDE IN SOLVENT WATER AT 25 °C[16]

R	$10^{11}K_a$	$10^3 k_2^{obs\,a}$ $(l.mole^{-1}.sec^{-1})$	$k_1(rel.)$	$10^3 k'_2{}^{obs\,b}$ $(l.mole^{-1}.sec^{-1})$	$k'_1(rel.)$
Me	2.52	0.127	1	0.24	1
Bu^n	1.83	4.80	52	7.24	42
Bu^s	2.5	23.3	185		
Bu^t	4.32	71.8	330	94	229
1-bicyclo-heptyl	3.05	110	680		
Ph	138	16	2.3	31.9	2.4
Vinyl	32	6.8	4.2		
Benzyl	72	87.5	24		

ᵃ k_2^{obs} refers to reactions run in 0.158 M sodium acetate plus 0.042 M acetic acid.
ᵇ $k'_2{}^{obs}$ refers to reactions run in 3.77 M perchloric acid.

Traylor and co-workers[16] referred to this mechanism as S_E2, but in the present
terminology it is denoted as S_E2(co-ord). The observed second-order rate coefficient is given by (*cf.* Chapter 3, Section 1.4, p. 13).

$$k_2^{obs} = Kk_1 \qquad (11)$$

In order to obtain an estimate of the relative values of K, for different boronic
acids, the acidity of these boronic acids was also measured. In Table 2 are given
values of K_a see (12), and also the relative values of k_1 calculated from the relation
$k_1(rel.) = k_2^{obs}/K_a$.

$$RB(OH)_2 + H_2O \overset{K_a}{\rightleftharpoons} RB(OH)_3^- + H^+ \qquad (12)$$

It is of interest to see that K_a (and hence, by analogy, K) does not vary appreciably
when R is a saturated alkyl group. Similar behaviour has been observed in the
complexation of iodide ion with trialkyltin bromides (see sequence (47), p. 160).
The rate coefficients $k_1(rel.)$ refer to reaction (10), for which reaction the transition
state (IV) has been suggested. ·

(IV)

In (IV) the group R migrates to an electron-deficient centre, and it might, therefore,
be expected that electron release by the group R would aid reaction. The sequence
in $k_1(rel.)$ of the Me < Bu^n < Bu^s < Bu^t could thus be due to the inductive ($+I$)
effect of the alkyl groups. If there were any steric strain in the four-coordinate

boron complex, migration of R would result in release of this strain. Traylor and co-workers[16] suggested that such relief of steric strain was not an important factor in determining the sequence of k_1(rel.), however.

For reactions run under acidic conditions, the observed second-order rate coefficient, $k_2'^{obs}$, may similarly be related to the relative first-order coefficients for complex decomposition by the equation $k'(\text{rel.}) = k_2'^{obs}/K_a$. In Table 2 are also given values of $k_2'^{obs}$ and $k'(\text{rel.})$; they vary in a similar way to the values of k_2^{obs} and $k(\text{rel.})$.

5.2 THE OXIDATION OF ALKYLBORONIC ACIDS BY CHROMIC ACID IN AQUEOUS SOLUTION

Ware and Traylor[17] have shown that the reaction of chromic acid with *tert.*-butylboronic acid, *viz.*

$$CrO_3 + t\text{-BuB(OH)}_2 \xrightarrow{H_2O} t\text{-BuOH} + H_3BO_3 + Cr^{3+} \tag{13}$$

follows overall second-order kinetics, first-order in Cr^{VI} and first-order in the boronic acid. From a study of the dependence of the observed second-order rate coefficient, k_2^{obs}, on pH it was deduced that the only effective oxidants in (13) were $HCrO_4^-$ and $H_3CrO_4^+$. For the former reactant, (14) and (15) were written to describe the mechanism of oxidation, *viz.*

$$HCrO_4^- + RB(OH)_2 \rightleftharpoons \left[RB(OH)_2OCrO_3H\right]^- \tag{14}$$

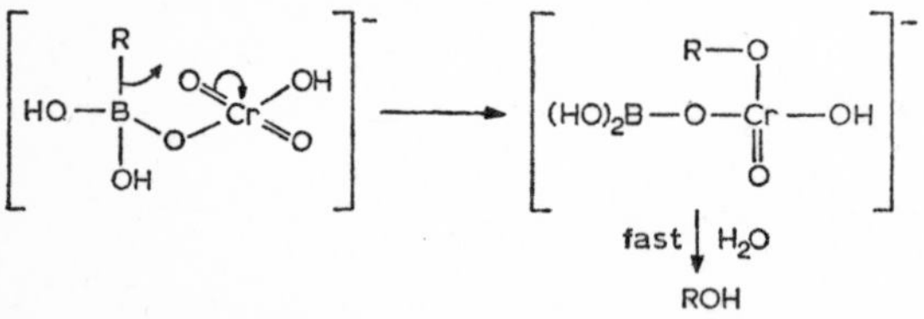

$$\tag{15}$$

In the present terminology the mechanism corresponds to S_E2(co-ord). Values of k_2^{obs} (l.mole^{-1}.sec^{-1}) at 30 °C were found for reactions (14) and (15) to be

R	Me	Et	Bu^t
k_2^{obs}	2.4×10^{-7}	6.6×10^{-4}	7.5×10^{-2}

$$\tag{16}$$

It is of some interest that the cleavage of alkylboronic acids by HOO^-, $HOOH_2^+$ and $HCrO_4^-$, all proceeding by mechanism S_E2(co-ord), yield the same sequence for the effect of the alkyl group on the observed second-order rate coefficient, *viz.* Me < *n*-alkyl < *sec.*-alkyl < *tert.*-alkyl. This sequence may therefore be regarded as that typical of mechanism S_E2(co-ord), insofar as the effect of the alkyl group undergoing substitution is considered.

6. Addition of trialkylaluminiums to nitriles, esters, and ketones

6.1 ADDITION OF TRIMETHYL- AND TRIETHYLALUMINIUM TO BENZONITRILE IN SOLVENT XYLENE, AND TO ESTERS AND KETONES

Pasynkiewicz and Kuran[18] showed that when benzonitrile was allowed to react with trimethylaluminium in a 1 : 1 molar ratio* in solvent xylene at 120 °C, a complex (V) was formed rapidly and irreversibly, *viz.*

$$Al_2Me_6 + 2\ PhCN \overset{fast}{=\!=\!=} 2\ PhCN \cdot AlMe_3 \quad (V) \tag{17}$$

The complex then rearranges intramolecularly in accord with a first-order rate law, with a value of 5.46×10^{-4} min^{-1} for the first-order rate coefficient for the reaction

$$PhCN \cdot AlMe_3 \overset{slow}{\longrightarrow} PhMeC : NAlMe_2 \tag{18}$$

Rearrangement of the complex (V) was illustrated[18] as

$$\tag{19}$$

The relative reactivity of various aluminium compounds towards benzonitrile[19] and benzyl cyanide[20] was in the order of increasing reactivity $(MeAlCl_2)_2 < (Me_2AlCl)_2 < (Me_3Al)_2$. This order was suggested[18] to arise through the electron-withdrawing power of the chlorine atoms inhibiting the 1,3-shift of the methyl group in the intramolecular rearrangement. Addition of triethylaluminium to a series of esters $p\text{-Y-}C_6H_4CO_2Et$ (1 : 1 molar ratio of triethylaluminium to ester, and in solvent xylene at 150°), yielding after hydrolysis $p\text{-Y-}C_6H_4COEt$, took place in the order of increasing facility Y = MeO < Me < H < Cl < CN. It was suggested[21] that a complex is again formed and that electron-attracting groups, Y, aid the 1,3-shift of the ethyl group in the rearrangement (VI).

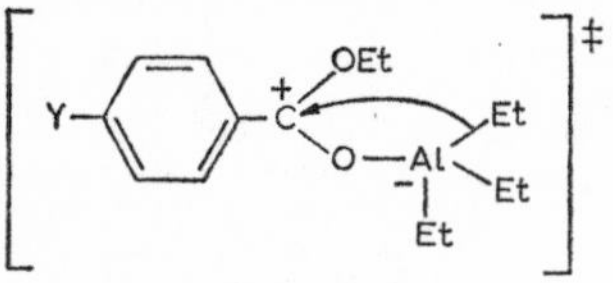

(VI)

Various preparative studies have confirmed the existence of 1 : 1 complexes between benzonitrile and trimethyl- and triethylaluminium[22, 23], and have shown

---

* Throughout Section 6 molar ratios are given on the basis of the formula R_3Al, even though the trialkylaluminium may actually be present entirely, or largely, as the dimeric species R_6Al_2.

that, whereas these complexes (such as (V)) are stable at room temperature, they rearrange on heating in accordance with equation (18)*.

When a molar ratio of benzonitrile to trimethylaluminium of 1 : 2 is used, the addition mechanism changes. Pasynkiewicz and Kuran[24] showed that under these conditions, using solvent xylene at 90–100 °C, the reaction becomes second-order and they suggested that the complex (V) is now attacked by another molecule of trimethylaluminium, viz.

$$Al_2Me_6 + PhCN \overset{fast}{\rightleftharpoons} PhCN \cdot AlMe_3 + AlMe_3 \tag{20}$$

$$PhCN \cdot AlMe_3 + AlMe_3 \xrightarrow{slow} PhMeC : NAlMe_2 + AlMe_3 \tag{21}$$

The transition state for reaction (21) was pictured[24] as (VII).

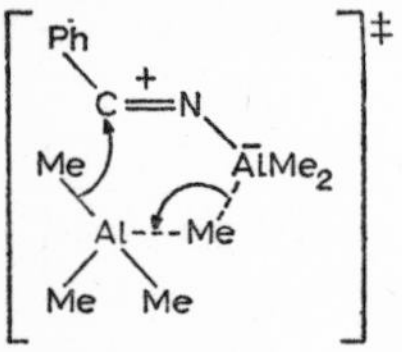

(VII)

6.2 ADDITION OF TRIMETHYLALUMINIUM TO BENZOPHENONE

The addition of trimethylaluminium to benzophenone in solvent benzene at 25 °C has been studied by Ashby et al.[25]. They find that for reactions using a 1 : 1 molar ratio of trimethylaluminium to benzophenone, a complex (VIII) is formed rapidly. This complex then decomposes in a rate-controlling step which is first-order (at least, over the first 50–75 % reaction). The mechanism was described[25] by equations (22)–(25).

$$Al_2Me_6 \rightleftharpoons 2\, Me_3Al \tag{22}$$

$$Ph_2CO + AlMe_3 \rightleftharpoons Ph_2CO \cdot AlMe_3 \qquad (VIII) \tag{23}$$

$$Ph_2CO \cdot AlMe_3 \xrightarrow{slow} Ph_2C(Me)OAlMe_2 \tag{24}$$

$$Ph_2C(Me)OAlMe_2 + AlMe_3 \rightleftharpoons Ph_2C(OMe)OAlMe_2 \cdot AlMe_3 \tag{25}$$

Equilibria (22)**, (23) and (25) were formulated as being set up rapidly, and the equilibrium (23) presumably lies very much in favour of the complex (VIII).

---

* The product formulated as PhMeC : NAlMe₂ is actually dimeric in benzene[23].

** In solvent benzene, trimethylaluminium is largely dimeric and the equilibrium in equation (22) must lie very much to the left hand side.

Decomposition of (VIII) follows a scheme related to that shown in (19), the transition state being (IX).

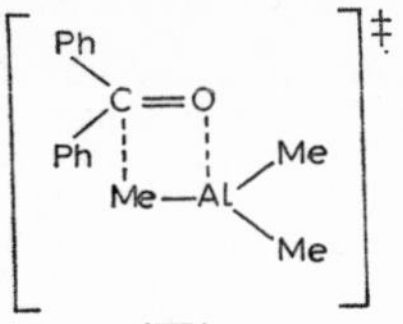

(IX)

When a molar ratio of trimethylaluminium to benzophenone of 2 : 1 was used, again with solvent benzene at 25 °C, a change of mechanism was observed[25] and the reaction formulated as

$$Al_2Me_6 \rightleftharpoons 2\ Me_2Al \qquad (26)$$

$$Ph_2CO + Me_3Al \rightleftharpoons Ph_2CO \cdot AlMe_3 \qquad (VIII) \qquad (27)$$

$$Ph_2CO \cdot AlMe_3 + Me_3Al \rightarrow Ph_2C(Me)OAlMe_2 + Me_3Al \qquad (28)$$

$$Ph_2C(Me)OAlMe_2 + Me_3Al \rightleftharpoons Ph_2C(Me)OAlMe_2 \cdot AlMe_3 \qquad (29)$$

Now (VIII), instead of rearranging intramolecularly, is attacked by another molecule of trimethylaluminium monomer (step (28)) through a transition state formulated[25] as (X).

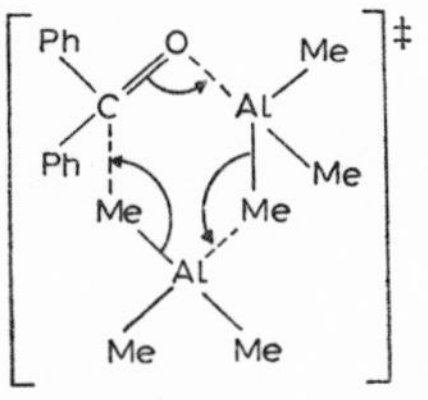

(X)

Although more detailed, the mechanism given by (26)–(29) is close to that shown in reactions (20) and (21). The transition states (VII) and (X) are also clearly related to each other.

In contrast to the complicated mechanisms proposed for the addition of trimethylaluminium to benzophenone in solvent benzene, addition in solvent ether is straightforward[26]. Trimethylaluminium probably exists as the monomeric complex $Me_3Al \cdot OEt_2$ in ether, and addition was found to obey the second-order rate equation, $dP/dt = k_2^{obs} \cdot [Me_3Al][Ph_2CO]$, where P denotes the product of addition. The mechanism of addition was described[26] by the reactions

$$Ph_2CO + Me_3Al \cdot OEt_2 \underset{k'_1}{\overset{k_2}{\rightleftharpoons}} Ph_2CO \cdot AlMe_3 + Et_2O \qquad (VIII) \qquad (30)$$

$$Ph_2CO \cdot AlMe_3 \overset{k_1}{\rightarrow} Ph_2C(Me)OAlMe_2 \qquad (31)$$

A complex (VIII) is formed, but only in very low concentration at any given time, and then rearranges through a transition state such as (IX). Application of the steady-state treatment yields the expression

$$\frac{dP}{dt} = \frac{k_1 k_2}{k_1 + k_1'} [Me_3Al][Ph_2CO] \tag{32}$$

Hence

$$\frac{dP}{dt} = k_2^{obs}[Me_3Al][Ph_2CO] \tag{33}$$

A value of 0.148 l.mole^{-1}·h^{-1} was obtained[26] for k_2^{obs} at 25 °C.

The mechanism described by (30) and (31), yielding the rate expressions (32) and (33), may be denoted as S_E2(co-ord). The more complicated mechanisms advanced for the other trialkylaluminium addition reactions dealt with in Sections 6.1 and 6.2 do not fall into the category of simple S_E2(co-ord) reactions. Neither does any reaction in which the intermediate complex is formed rapidly and irreversibly, and the complex then slowly decomposes to the products.

Ashby and Yu[31] have also investigated the reduction of benzophenone (to benzhydrol) by triisobutylaluminium in solvent ether. They suggest that a complex between the ketone and the trialkyl is formed rapidly and reversibly; in a subsequent slow step the complex rearranges intramolecularly to yield $Ph_2CHO·AlBu^i_2$ and isobutene.

7. Disproportionation of tetraalkylsilanes catalysed by aluminium bromide

The disproportionation (34) (R = Et)

$$2\ RSiMe_3 \xrightarrow{Al_2Br_6} Me_4Si + R_2SiMe_2 \tag{34}$$

was shown by Russell[27] to follow the kinetic law

$$-d[EtSiMe_3]/dt = k[Al_2Br_6][EtSiMe_3]^{1.5} \tag{35}$$

when solvents benzene and cyclohexane were used. Rate coefficients for reaction (34) (R = Et) in solvent benzene were recorded as follows (with units of k as $l^{1.5}$.mole$^{-1.5}$.sec^{-1})

$10^6 k$	22.7	61.8	138.5	
Temp. (°C)	40	60	80	(36)

Calculated activation parameters were $E_a = 9.5$ kcal.mole^{-1} and $\Delta S^{\ddagger} = -48.6$ cal.deg^{-1}.mole^{-1}. Reaction (34) (R = Et) is much slower in solvent cyclohexane, with $10^6 k = 2.96 \, l^{1.5}.mole^{-1.5}.sec^{-1}$ at 80 °C.

A complicated reaction sequence was proposed by Russell[27] in order to account for the rate expression (35). The elementary reaction in which disproportionation took place was thought to be reaction (37) (R = Et) through a transition state (XI) (R = Et)[27].

$$RSiMe_3 \text{ --- } AlBr_3 + RSiMe_3 \rightarrow Me_4Si + R_2SiMe_2 + AlBr_3 \tag{37}$$

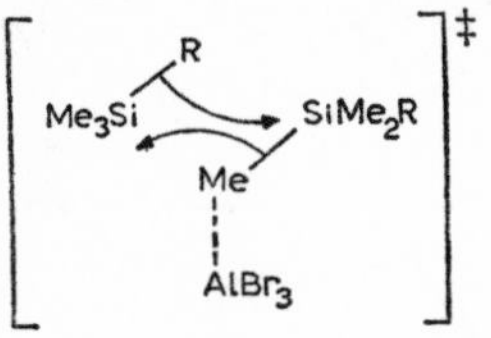

(XI)

Russell and Nagpal[28] later measured the relative rates of reaction (34) at 40 °C, using solvent benzene, and reported the following order of reactivity with respect to the group R.

R	allyl	Et	Pr^n	Pr^i	Bu^n	Bu^i	cyclo-Pe	cyclo-Hex	
Relative rate	71	1	0.61	0.0048	0.15	0.025	0.059	0.0021	(38)

Transition state (XI) involves electrophilic attack at the group R. However, the sequence of relative rates given in (38) may not necessarily correspond to the relative rates of the appropriate elementary reaction (37), since the rate coefficient in equation (35) is a composite coefficient and includes various rate coefficients and equilibrium constants of steps prior to reaction (37).

8. The oxidation of tetraalkyltins by chromium trioxide in solvent acetic acid

Tetra-*n*-butyltin is oxidised by chromium trioxide (1 : 1 molar ratio) to yield tri-*n*-butyltin acetate and compounds derived from the *n*-butyl group cleaved from the tin atom (mainly *n*-butyraldehyde and butyric acid)[29]. In the initial stages of the oxidation, the reaction follows the simple rate expressions $v = k_2[Bu^n_4Sn][CrO_3]$. Values of the second-order rate coefficient for the oxidation of a number of tetraalkyltins by CrO_3 in solvent acetic acid at 20 °C were reported[29] to be as follows

R_4Sn	Me_4Sn	Et_4Sn	Pr^n_4Sn	Bu^n_4Sn	Pr^i_4Sn	
$k_2(l.mole^{-1}.sec^{-1})$	0.04	5.4	3.6	2.8	8.1	(39)

A cyclic transition state was suggested[29] and the oxidation may be represented as

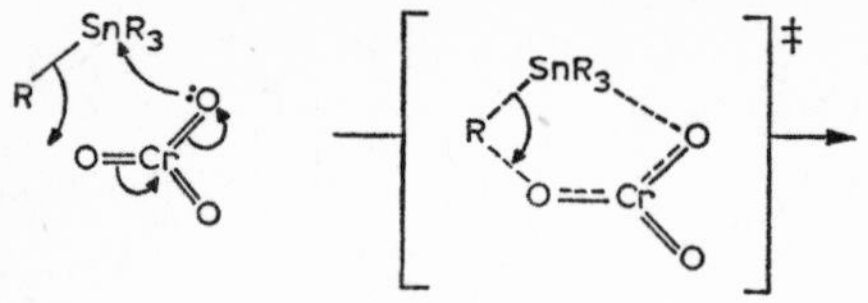

$$(40)$$

In the present terminology, such a mechanism could be denoted as S_E2(cyclic).

9. The displacement of cobalt(III) and chromium(III) from pyridiomethylcobaltate ions and -chromium(III) ions by nitrosating agents

Bartlett and Johnson[32] have shown that the pyridiomethylcobalt(III) ions and the pyridiomethylchromium(III) ions, (XII) and (XIII),

$$[\overset{+}{H}NC_5H_4CH_2Co(CN)_5]^{3-} \qquad \text{(XII)}$$

$$[\overset{+}{H}NC_5H_4CH_2Cr(H_2O)_5]^{2+} \qquad \text{(XIII)}$$

react with nitrous acid in aqueous acidic solution to yield the oxime $\overset{+}{H}NC_5H_4CH=$ NOH and the oxime hydrolysis product $\overset{+}{H}NC_5H_4CHO$. The displacement in aqueous perchloric acid was found to follow the rate equation

TABLE 3

RATE COEFFICIENTS FOR THE DISPLACEMENT OF COBALT(III) AND CHROMIUM(III) FROM Y-PYRIDIOMETHYLPENTACYANOCOBALTATE IONS (XII) AND Y-PYRIDIOMETHYL-PENTAAQUOCHROMIUM IONS (XIII) IN AQUEOUS ACID AT 24.8 °C[32]

		Electrophile		
		$H_2NO_2^+$ k_3 [a]	$HgCl_2$ k_2 [b]	$TlCl_3$ k_2 [b]
Y	Substrate			
2	$[\overset{+}{H}NC_5H_4CH_2Co(CN)_5]^{3-}$	0.38	0.57	3.3
3	$[\overset{+}{H}NC_5H_4CH_2Co(CN)_5]^{3-}$	24	$\simeq 5.0$	$\simeq 150$
4	$[\overset{+}{H}NC_5H_4CH_2Co(CN)_5]^{3-}$	16	8.5	58
4	$[\overset{+}{H}NC_5H_4CH_2Cr(H_2O)_5]^{2+}$	1.3	3.2	6.1

[a] Values of k_3 in $l^2.mole^{-2}.sec^{-1}$.
[b] Values of k_2 in $l.mole^{-1}.sec^{-1}$ from ref. 33.

$$v = k_3[\text{substrate}][\text{HNO}_2][\text{H}^+] \tag{41}$$

indicating that the active nitrosating species under these conditions is $\text{H}_2\text{NO}_2{}^+$. Values of k_3 are given in Table 3, together with the second-order rate coefficients for displacement by the electrophiles HgCl_2 and TlCl_3, also in aqueous solution. It was suggested[32] that the displacement of cobalt(III) and of chromium(III) by $\text{H}_2\text{NO}_2{}^+$ proceeded by the bimolecular S_E2(open) mechanism, as shown in equation (42) below for the cobalt complex. Support for this mechanism is provided by a comparison of the relative rates of substitution of the 2-, 3-, and 4-isomers of (XII) by $\text{H}_2\text{NO}_2{}^+$, HgCl_2, and TlCl_3. The latter two electrophiles have already[33] been shown to displace cobalt(III) and chromium(III) from (XII) and (XIII) by the S_E2(open) mechanism.

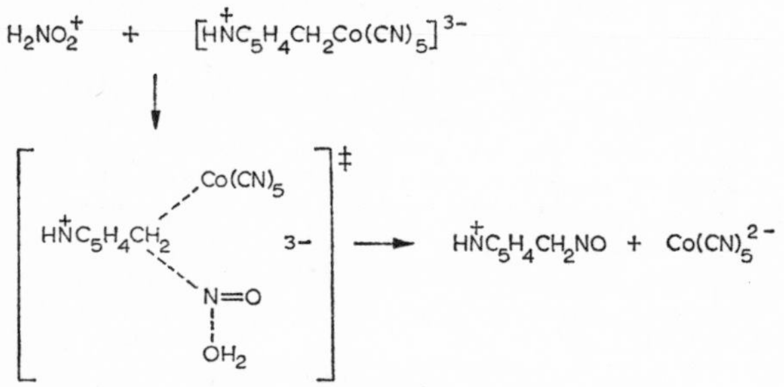

$$\tag{42}$$

$$\tag{43}$$

In the presence of chloride ion, displacement of chromium(III) from (XIII) was accelerated (at constant ionic strength), and the rate equation was now found to be given by the expression

$$v = k_3[\text{substrate}][\text{HNO}_2][\text{H}^+] + k_4[\text{Cl}^-][\text{substrate}][\text{HNO}_2][\text{H}^+] \tag{44}$$

The fourth-order term corresponds to reaction of NOCl with the substrate since, from equilibria (45) and (46),

$$\text{HNO}_2 + \text{H}_3\text{O}^+ \rightleftharpoons \text{H}_2\text{NO}_2{}^+ + \text{H}_2\text{O} \tag{45}$$

$$\text{H}_2\text{NO}_2{}^+ + \text{Cl}^- \rightleftharpoons \text{NOCl} + \text{H}_2\text{O} \tag{46}$$

it follows that $[\text{HNO}_2][\text{H}^+] \propto [\text{H}_2\text{NO}_2{}^+]$ and that $[\text{H}_2\text{NO}_2{}^+][\text{Cl}^-] \propto [\text{NOCl}]$. The value of k_4 in equation (44) was found to be 3.4 $l^3.\text{mole}^{-3}.\text{sec}^{-1}$ for the substrate 4-pyridiomethylpentaaquochromium(III) ion. Bartlett and Johnson[32] again described the mechanism as S_E2(open), and wrote for attack by NOCl the scheme

$$\text{NOCl} \quad + \quad \left[\overset{+}{\text{H}}\text{NC}_5\text{H}_4\text{CH}_2\text{Cr}(\text{H}_2\text{O})_5\right]^{2+}$$

$$\text{HNC}_5^-\text{H}_4\text{CH}_2\text{NO} + \text{Cr}(\text{H}_2\text{O})_5^{3+} + \text{Cl}^-$$

(47)

Again, reaction (43) occurs subsequent to the bimolecular electrophilic displacement (47).

REFERENCES

1 R. L. LETSINGER, *J. Am. Chem. Soc.*, 72 (1950) 4842.
2 D. Y. CURTIS AND W. J. KOEHL, JR., *J. Am. Chem. Soc.*, 84 (1962) 1967.
3 H. M. WALBORSKY, F. J. IMPASTATO AND A. E. YOUNG, *J. Am. Chem. Soc.*, 86 (1964) 3283.
4 D. E. APPLEQUIST AND A. H. PETERSON, JR., *J. Am. Chem. Soc.*, 83 (1961) 862; D. E. APPLEQUIST AND G. N. CHMURNY, *J. Am. Chem. Soc.*, 89 (1967) 875.
5 H. M. WALBORSKY AND A. E. YOUNG, *J. Am. Chem. Soc.*, 86 (1964) 3288.
6 F. R. JENSEN AND K. L. NAKAMAYE, *J. Am. Chem. Soc.*, 88 (1966) 3437.
7 E. C. ASHBY, *Quart. Rev. (London)*, 21 (1967) 259.
8 B. J. WAKEFIELD, *Organometal. Chem. Rev.*, 1 (1966) 131.
9 S. G. SMITH AND J. BILLET, *J. Am. Chem. Soc.*, 89 (1967) 6948; 90 (1968) 4108. See also *Tetra hedron Letters*, (1969) 4467.
10 T. HOLM AND I. BLANKHOLM, *Acta Chem. Scand.*, 22 (1968) 708; T. HOLM, *Acta Chem. Scand.*, 23 (1969) 579.
11 E. C. ASHBY, R. B. DUKE AND H. M. NEUMANN, *J. Am. Chem. Soc.*, 89 (1967) 1964; but see E. C. ASHBY, F. W. WALKER AND H. M. NEUMANN, *Chem. Commun.*, (1970) 330.
12 B. MARX, *Compt. Rend. Sér. C*, 266 (1968) 1646.
13 I. P. BELETSKAYA, O. A. MAKSIMENKO AND O. A. REUTOV, *Zh. Org. Khim.*, 2 (1966) 1129; *J. Org. Chem. USSR*, 2 (1966) 1124.
14 I. P. BELETSKAYA, O. A. MAKSIMENKO, V. B. VOL'EVA AND O. A. REUTOV, *Zh. Org. Khim.*, 2 (1966) 1132; *J. Org. Chem. USSR*, 2 (1966) 1127; I. P. BELETSKAYA, O. A. MAKSIMENKO AND O. A. REUTOV, *Izv. Akad. Nauk SSSR, Ser. Khim.*, (1966) 662; *Bull. Acad. Sci. USSR*, (1966) 627; *Dokl. Akad. Nauk SSSR*, 168 (1966) 333; *Proc. Acad. Sci. USSR*, 168 (1966) 473.
15 A. G. DAVIES AND B. P. ROBERTS, *J. Chem. Soc. C*, (1968) 1474.
16 H. C. MINATO, J. C. WARE AND T. G. TRAYLOR, *J. Am. Chem. Soc.*, 85 (1963) 3024.
17 J. C. WARE AND T. G. TRAYLOR, *J. Am. Chem. Soc.*, 85 (1963) 3026.
18 S. PASYNKIEWICZ AND W. KURAN, *Roczniki Chem.*, 39 (1965) 979.
19 S. PASYNKIEWICZ, W. KURAN AND E. SOSZYŃSKA, *Roczniki Chem.*, 38 (1964) 1285.
20 S. PASYNKIEWICZ, K. STAROWIEYSKI AND Z. RZEPKOWSKA, *J. Organometal. Chem.*, 10 (1967) 527.
21 S. PASYNKIEWICZ, L. KOZERSKI AND B. GRABOWSKI, *J. Organometal. Chem.*, 8 (1967) 233.
22 K. STAROWIEYSKI, S. PASYNKIEWICZ AND M. BOLESLAWSKI, *J. Organometal Chem.*, 10 (1967) 393.
23 J. E. LLOYD AND K. WADE, *J. Chem. Soc.*, (1965) 2662; J. R. JENNINGS, J. E. LLOYD AND K. WADE, *J. Chem. Soc.*, (1965) 5083.
24 S. PASYNKIEWICZ AND W. KURAN, *Roczniki Chem.*, 39 (1965) 1199.
25 E. C. ASHBY, J. LAEMMLE AND H. M. NEUMANN, *J. Am. Chem. Soc.*, 90 (1968) 5179.
26 E. C. ASHBY AND J. T. LAEMMLE, *J. Org. Chem.*, 33 (1968) 3398.
27 G. A. RUSSELL, *J. Am. Chem. Soc.*, 81 (1959) 4815, 4825, 4831.

28 G. A. RUSSELL AND K. L. NAGPAL, *Tetrahedron Letters*, (1961) 421.
29 C. DEBLANDRE, M. GIELEN AND J. NASIELSKI, *Bull. Soc. Chim. Belges*, 73 (1964) 214.
30 M. CHASTRETTE AND R. AMOUROUX, *Tetrahedron Letters*, (1970) 5165.
31 E. C. ASHBY AND S. H. YU, *J. Org. Chem.*, 35 (1970) 1034.
32 E. H. BARTLETT AND M. D. JOHNSON, *J. Chem. Soc. A*, (1970) 523.
33 E. H. BARTLETT AND M. D. JOHNSON, *J. Chem. Soc. A*, (1970) 517.

Chapter 10

Electrophilic Substitutions of Allyl-metal Compounds

The allyl–metal bond is usually broken very readily by electrophilic reagents and allyl groups are cleaved from mercury, boron, silicon, germanium, tin, and lead much more easily than are the saturated alkyl groups. Kinetic studies, however, have not been very numerous; such studies are reviewed in this chapter.

Direct comparison of the rate of cleavage of an allyl–metal bond with that of an ethyl–metal bond is possible in three particular cases, and it will be shown later that the ratio of the second-order rate coefficients for allyl–metal cleavage to ethyl–metal cleavage are as shown below.

Bond cleaved	Electrophile	Solvent	$k_2(allyl)/k_2(ethyl)$	Page no.
R–SnEt$_3$	HgCl$_2$	Acetonitrile	10^5	198
R–HgI	H$_3$O$^+$	Water	3×10^6	201
R–SnR$_3$	I$_2$	Methanol/acetone	4×10^7	206

Thus for three characteristic electrophilic substitutions — a metal-for-metal exchange, an acidolysis, and an iodinolysis (iododestannation) — the allyl–metal bond is cleaved some 10^6–10^7 times as rapidly as is the ethyl–metal bond. The ratio $k_2(allyl)/k_2(n\text{-propyl})$ will correspondingly be about 10^7–10^8.

In the cleavage of allyl–metal compounds, the available mechanisms of substitution are increased by the inclusion of mechanisms involving rearrangement, S'_E, as detailed in Chapter 3, Section 2, p. 16.

1. Base catalysed S$_E$1 cleavage of 3-phenallylsilicon and -tin compounds

The 3-phenallyl group is cleaved from silicon and tin under the influence of aqueous alcoholic alkali according to reactions (1) (R = 3-phenallyl, R' = alkyl, M = Si or Sn) and (2), viz.[1].

$$\text{Base}^- + \text{R}'_3\text{MR} \xrightarrow{\text{slow}} \text{Base–MR}'_3 + \text{R}^- \tag{1}$$

$$\text{R}^- + \text{Solvent} \xrightarrow{\text{fast}} \text{RH} + \text{Base}^- \tag{2}$$

References p. 210

TABLE 1

SECOND-ORDER RATE COEFFICIENTS ($l.mole^{-1}.sec^{-1}$) FOR THE BASE-CATALYSED
CLEAVAGE OF 3-PHENALLYLSILICON AND –TIN COMPOUNDS IN SOLVENT AQUEOUS
ALCOHOLS[1]

Compound	Solvent	Temp. (°C)	$10^4 k_2$	E_a (kcal.mole^{-1})	log A (l.mole^{-1}. sec^{-1})
Me$_3$SiCH$_2$CH–CHPh	"60%" EtOH	49.2	2.36		
Me$_3$SiCH$_2$CH–CHPh	"60%" EtOH	40	0.97	19.2	9.2
Me$_3$SiCH$_2$CH–CHPh	"60%" MeOH	40	1.43		
Et$_3$SnCH$_2$CH–CHPh	"60%" MeOH	40	46.0	18.8	10.0
Bu$^n_3$SnCH$_2$CH–CHPh	"60%" MeOH	40	32		

Since the base is regenerated in the fast step (2), each kinetic run is first-order in substrate. The first-order rate coefficients may be converted into second-order rate coefficients through the equation

$$k_2 = k_1^{obs}/[\text{Base}^-] \qquad (3)$$

and Table 1 gives values of k_2 obtained by Roberts and El Kaissi[1] in this way. Although sodium hydroxide was the source of the base, it is not possible to specify the nature of the attacking species owing to equilibria such as

$$\text{OH}^- + \text{MeOH} \rightleftharpoons \text{MeO}^- + \text{H}_2\text{O} \qquad (4)$$

On the basis of an isotope effect $k(\text{H}_2\text{O})/k(\text{D}_2\text{O}) = 0.50$ for cleavage of Me$_3$SiCH$_2$CH=CHPh in aqueous methanol, Roberts and El Kaissi[1] suggested that bond making and bond breaking in the transition state (I) are of equal importance.

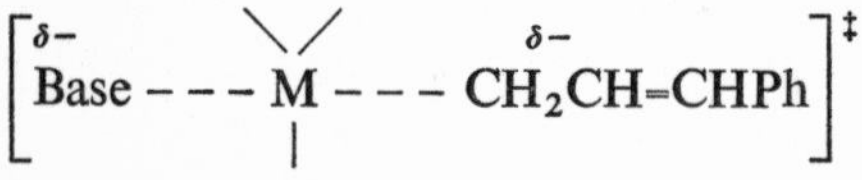

$$\left[\overset{\delta-}{\text{Base}} - - - \overset{\diagdown \diagup}{\underset{|}{\text{M}}} - - - \overset{\delta-}{\text{CH}_2\text{CH=CHPh}} \right]^{\ddagger}$$

(I)

If a carbanion is formed free in solution at any time, then attack by solvent is possible at the original α-carbon atom (to yield β-methylstyrene) or at the original γ-carbon atom (to yield allylbenzene), viz.

$$\overset{-}{\text{CH}_2}\text{–CH=CH–Ph} \leftrightarrow \text{CH}_2=\text{CH–}\overset{-}{\text{CH}}\text{–Ph}$$

$$\downarrow \text{solvent} \qquad\qquad \downarrow \text{solvent}$$

$$\text{CH}_3\text{–CH=CH–Ph} \qquad \text{CH}_2=\text{CH–CH}_2\text{–Ph}$$

The observed hydrocarbon product was β-methylstyrene only[1], thus the cleavages involve no rearrangement and may be designated as S_E1-OR^- (where R = H, Me, or Et depending on the actual basic species involved).

2. Metal-for-metal exchanges of allyl–metal compounds

2.1 SUBSTITUTION OF ALLYLIC DERIVATIVES OF GROUP IVA ELEMENTS BY MERCURIC SALTS

Roberts[2] has studied the substitution of trimethylallylsilicon by mercuric chloride in solvent acetonitrile. The substitution is complex, since there is formed not only the expected trimethylsilicon chloride (50 %) and allylmercuric chloride (50 %) but also an unidentified silicon-containing compound denoted as X. Furthermore, it was also shown that the action of trimethylsilicon chloride on trimethylallylsilicon afforded X and unchanged trimethylsilicon chloride. The overall reaction was formulated as

$$Me_3SiCH_2CH=CH_2 + HgCl_2 \xrightarrow{slow} Me_3SiCl + CH_2=CHCH_2HgCl \qquad (6)$$

$$Me_3SiCH_2CH=CH_2 + Me_3SiCl \xrightarrow{fast} Me_3SiCl + X + etc. \qquad (7)$$

Reaction (6) is itself not simple, because Roberts observed that when solutions of trimethyllallylsilicon and mercuric chloride were mixed, an increase in the ultraviolet absorption occurred over and above that due to any reactants and products. Roberts suggested that a 1 : 1 complex was formed and wrote (6) as*

$$Me_3SiCH_2CH=CH_2 + HgCl_2 \underset{}{\overset{K}{\rightleftharpoons}} \pi\text{-complex} \xrightarrow{k_1} Me_3SiCl + CH_2=CHCH_2HgCl \quad (8)$$

Now provided that the complex is in equilibrium with the reactants, the observed second-order rate coefficients for reaction (6) will be given by

$$k_2^{obs} = Kk_1 \qquad (9)$$

Due to the stoichiometry of the sum of reactions (6) and (7), the actual rate law is given by

$$v = k_2^{obs} \cdot (a-x)(b-2x) \qquad (10)$$

where $a = [HgCl_2]_0$, $b = [Me_3SiCH_2CH=CH_2]_0$, and $x = [Me_3SiCl] = [CH_2=CHCH_2HgCl]$. Rate coefficients calculated on the basis of equation (10) appear to increase in value somewhat during any given kinetic run, however.

Values of k_2^{obs} l.mole^{-1}.sec^{-1} obtained by Roberts at various temperatures, using solvent acetonitrile, were given as follows,

---

* Note that the nomenclature of the various rate coefficients and equilibrium constants used in this Section is not the same as that used by Roberts.

$$0.042(25 \text{ °C}), 0.056(30 \text{ °C}), 0.083(35 \text{ °C}), \text{ and } 0.137(43.7 \text{ °C}) \tag{11}$$

and lead[2] to the activation parameters $E_a = 12.0$ kcal.mole^{-1} and log $A = 7.31$. Relative values of k_2^{obs} for substitution of trimethylallylsilicon by various mercuric salts in acetonitrile were

$$\text{HgI}_2(0.18), \text{ HgBr}_2(0.9), \text{ HgCl}_2(1.0), \text{ and } \text{Hg(OAc)}_2(29), \tag{12}$$

and on this basis Roberts[2] suggested that the substitution involved an open transition state, although no specific transition state was formulated.

In a subsequent paper, the action of mercuric salts on allylgermanium compounds was described. Solvents acetonitrile and ethanol were used, and the simple stoichiometry (13) (R = H, Me) found to obtain[3].

$$\text{Et}_3\text{GeCH}_2\text{CH=CHR} + \text{HgX}_2 = \text{Et}_3\text{GeX} + \text{XHgCH}_2\text{CH=CHR} \tag{13}$$

The observed second-order rate coefficients for reaction (13) are given in Table 2.

The reactivity of the compounds $\text{Et}_3\text{MCH}_2\text{CH=CH}_2$ towards mercuric salts in solvent acetonitrile was reported[3] to be in the sense M = Si(1) < Ge(167) ≪ Sn, with the allyl–tin bond cleaved so rapidly that the rate of cleavage could not be measured. From the data given in Table 2, it thus seems that the second-order rate coefficient for the reaction

$$\text{Et}_3\text{SnCH}_2\text{CH=CH}_2 + \text{HgCl}_2 \xrightarrow{\text{MeCN}} \text{Et}_3\text{SnCl} + \text{ClHgCH}_2\text{CH=CH}_2 \tag{14}$$

must be at least 500 l.mole^{-1}.sec^{-1} at 25 °C. Now Abraham and Hogarth[4] have determined the second-order rate coefficient for the reaction

$$\text{Et}_4\text{Sn} + \text{HgCl}_2 \xrightarrow{\text{MeCN}} \text{Et}_3\text{SnCl} + \text{EtHgCl} \tag{15}$$

to be 0.0187 l.mole^{-1}.sec^{-1} at 25 °C, and thus the cleavage of an Sn–allyl bond by mercuric chloride proceeds at least 10^5 times as rapidly as that of an Sn–ethyl bond under comparable conditions. As implied by equation (13), cleavage of $\text{Et}_3\text{GeCH}_2\text{CH=CHCH}_3$ by mercuric bromide in solvent acetonitrile yields the unrearranged compound $\text{BrHgCH}_2\text{CH=CHCH}_3$ (isolated in 88 % yield). Although this would seem to indicate substitution by mechanism S_E and not by mechanism S_E', it is possible for the rearranged compound to be formed initially and then itself to undergo rearrangement, viz.

$$\text{CH}_2\text{=CH–CH–CH}_3 \rightleftharpoons \text{CH}_2\text{CH=CH–CH}_3 \tag{16}$$
$$\quad\quad\quad\quad |\quad\quad\quad\quad\quad\quad |$$
$$\quad\quad\quad \text{HgBr}\quad\quad\quad\quad \text{HgBr}$$

Roberts[3] suggested that reaction (13) (R = H, Me) might also involve an intermediate complex (cf. mechanism (8)), although he did not venture further details.

If some sort of π-complex is involved, then we might describe the substitutions as in (17) for an S_E process and as in (18) for an S_E' process.

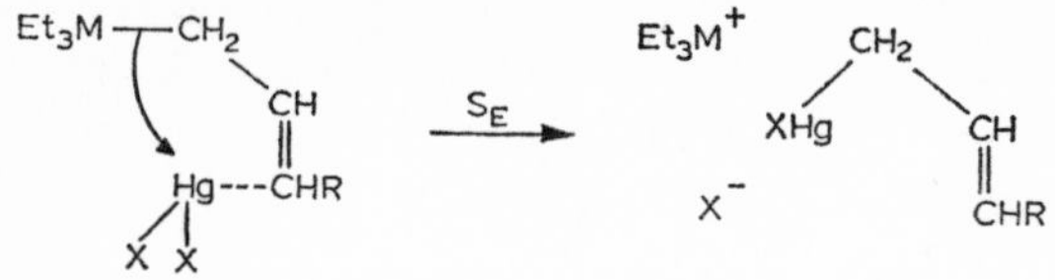

$$\text{(17)}$$

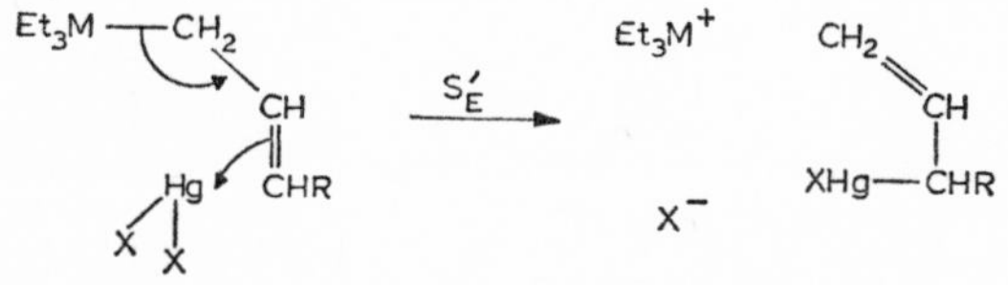

$$\text{(18)}$$

When ethanol was used as the solvent, Roberts[3] observed sequence (12), qualitatively, to represent the reactivity of $Et_3GeCH_2CH=CH_2$ towards various mercuric salts. The unrearranged compound $BrHgCH_2CH=CHCH_3$ was again isolated (in 25 % yield) from the reaction between $Et_3GeCH_2CH=CHCH_3$ and mercuric bromide, but this is not decisive evidence for mechanism S_E (see above). Whether or not π-complexes are intermediates in the substitution (13) (R = H, Me) when ethanol is the solvent is not clear. If so, then reactions (17) and (18) may again be suggested.

Roberts[5] has extended his work to the cleavage of allyl–tin compounds, using solvent ethanol. Following the work of Abraham and Spalding on the cleavage of tetraalkyltins (see p. 70), Roberts now represented the reaction of allyl compounds with mercuric salts as*

$$R_3SnCH_2CH=CHR' + HgX_2 \xrightarrow[\text{slow}]{k_2} R'CH=CHCH_2HgX + R_3SnX$$

$$R_3SnX + HgX_2 \overset{\text{rapid}}{\rightleftharpoons} R_3SnX \cdot HgX_2$$

Second-order rate coefficients for the slow step are given in Table 2, together with activation parameters, where recorded. The order of reactivity towards mercuric iodide in ethanol is $R_3SnCH_2CH=CH_2 > R_3SnCH_2CH=CHMe \simeq R_3SnCH_2CH=CHPh$. Roberts[5] suggests that the sequence indicates that the slow electrophilic step proceeds by mechanism S_E2, since the electron-releasing methyl group should favour attack by mechanism S_E2'. However, in the acidolysis of allyltin compounds (see Table 5, p. 205) the sequence of reactivity is also $R_3SnCH_2CH=CH_2 > R_3SnCH_2CH=CHMe$ and yet the $R_3SnCH_2CH=CHMe$ compound reacts almost entirely by the S_E2' pathway, equation (32b) (R = Me). Because of the difficulty that the allylmercuric halide formed in the cleavage can itself undergo an allylic

---

* It is possible that such a set of reactions was in force for the related cleavages of allyl–silicon and allyl–germanium compounds studied by Roberts[2, 3] with mercuric iodide as the electrophile.

TABLE 2

SECOND-ORDER RATE COEFFICIENTS ($l.mole^{-1}.sec^{-1}$) FOR THE SUBSTITUTION OF ALLYL-GERMANIUM AND ALLYL–TIN COMPOUNDS BY MERCURIC SALTS[3, 5]

Compound	HgX_2	Solvent	Temp. (°C)	k_2	E_a (kcal.mole^{-1})	$\Delta S^{\ddagger}$ (cal.deg^{-1} mole^{-1})
$Et_3GeCH_2CH=CH_2$	$HgBr_2$	MeCN	24.9	39[a]		
$Et_3GeCH_2CH=CH_2$	$HgCl_2$	MeCN	24.9	45		
$Et_3GeCH_2CH=CHMe$	$HgBr_2$	MeCN	24.9	110		
$Et_3GeCH_2CH=CH_2$	$HgBr_2$	EtOH	25	0.035		
$Et_3GeCH_2CH=CHMe$	$HgBr_2$	EtOH	25	0.149		
$Et_3SnCH_2CH=CH_2$	HgI_2	EtOH	30	2900	5.8	−30
$Et_3SnCH_2CH=CHMe$	HgI_2	EtOH	30	15		
$Et_3SnCH_2CH=CHPh$	HgI_2	EtOH	30	3.36		
$Bu^n_3SnCH_2CH=CH_2$	HgI_2	EtOH	30	1020		
$Et_3SnCH_2CH=CHPh$	$HgCl_2$	EtOH	25	178		
$Et_3SnCH_2CH=CHPh$	$HgBr_2$	EtOH	25	13.6		
$Et_3SnCH_2CH=CHPh$	HgI_2	EtOH	25	2.8	8.4	−31
$Ph_3SnCH_2CH=CH_2$	HgI_2	EtOH	25	25[b]	4.0	−41
$Ph_3SnCH_2CH=CHMe$	HgI_2	EtOH	25	0.54		
$Ph_3SnCH_2CH=CHPh$	HgI_2	EtOH	25	0.46		
$Ph_3SnCH_2CH=CH_2$	$HgCl_2$	EtOH	25	1500		
$Ph_3SnCH_2CH=CH_2$	$HgBr_2$	EtOH	25	330		
$Ph_3SnCH_2CH=CH_2$	HgI_2	EtOH	25	25[b]		

[a] $k_2 = 0.70$ in 98 % MeCN–2 % water.
[b] $k_2 = 83$ in 96 % EtOH–4 % water.

rearrangement, it is not possible to deduce rigorously whether the cleavage takes place by the S_E2 or S_E2' mechanism. Roberts[5] concludes, though, that the cleavage of the allylic tin compounds in Table 2 proceeds by the S_E2(open) mechanism, and writes

3. Acidolysis of allyl-metal compounds

3.1 ACIDOLYSIS OF ALLYLMERCURIC IODIDE BY AQUEOUS ACID

Kreevoy et al.[6] showed that when allylmercuric iodide was allowed to react with aqueous acid, in the presence of 1 equivalent of iodide ion per allylmercuric iodide, propene and mercuric iodide were formed according to*

$$CH_2=CHCH_2HgI + H^+ + I^- = CH_2=CHCH_3 + HgI_2 \qquad (19)$$

When trans-$CH_3CH=CHCH_2HgI$ was similarly treated with aqueous acid and iodide ion, there was formed predominantly (95 %) $CH_2=CHCH_2CH_3$, thus demonstrating that some type of S_E' process is operative. A rate-determining proton transfer (to the γ-carbon atom of the allylmercuric iodide) is indicated by a value of 3.1 for the solvent isotope effect, $k(H_2O)/k(D_2O)$. When the added iodide ion concentration was very low ($[I^-]_0 = [\text{allylHgI}]_0 = 1.0-5.0 \times 10^{-5} M$), reaction (19) was found to follow overall second-order kinetics, first-order in allylmercuric iodide and first-order in H^+. Values of the second-order rate coefficient are given in Table 3 and lead[6] to the activation parameters $\Delta H^{\ddagger} = 16.42 \pm 0.05$ kcal.mole^{-1} and $\Delta S^{\ddagger} = -11.9 \pm 0.2$ cal.mole^{-1}.deg^{-1}.

At 110 °C, the second-order rate coefficient for the acidolysis of allylmercuric iodide will be around 7.5 l.mole^{-1}.sec^{-1}, compared with the value[7] of 4.0×10^{-6} l.mole^{-1}.sec^{-1} for the acidolysis of ethylmercuric iodide at 110 °C**. Thus the

TABLE 3

SECOND-ORDER RATE COEFFICIENTS (l.mole^{-1}.sec^{-1}) FOR THE ACIDOLYSIS OF ALLYL-MERCURIC IODIDE BY PERCHLORIC ACID IN SOLVENT 96 % WATER–4 % METHANOL (v/v)[6]

Temp. (°C)	$10^3 k_2$	Temp. (°C)	$10^3 k_2$
0.0	0.95	46.1	102
15.1	5.0	59.8	250
25.0	14.0	64.1	380
35.0	34	74.7	840
39.2	54	110	7500[a]

[a] Extrapolated value calculated by the author using the reported value[6] of E_a. Initial concentration of substrate $\sim 4 \times 10^{-5} M$. Reactions carried out in the presence of added iodide ion $\sim 4 \times 10^{-5} M$.

---

* In the absence of added iodide ion, only 0.5 mole of propene per mole of allylmercuric iodide is formed.
** This is the value obtained[7] using 1 M perchloric acid, whereas the value of k_2 for the allyl–mercury cleavage refers to zero ionic strength. This latter rate coefficient does not depend very much on ionic strength[6], and in 1 M perchloric acid is probably not more than 13 l.mole^{-1}.sec^{-1} at 110 °C, judging from ionic strength effects at lower temperatures. Use of values of 13 l.mole^{-1}.sec^{-1} and 4.0×10^{-6} l.mole^{-1}.sec^{-1} results in a factor of 3×10^6.

allyl–Hg bond is cleaved by acid about 10^6 times as rapidly as is the ethyl–Hg bond.

Reaction (19) was shown[8] to be subject to general acid catalysis, and it was concluded[6, 9] that the rate-determining step is a proton transfer in which substantially only the proton is moving along the reaction co-ordinate, viz.

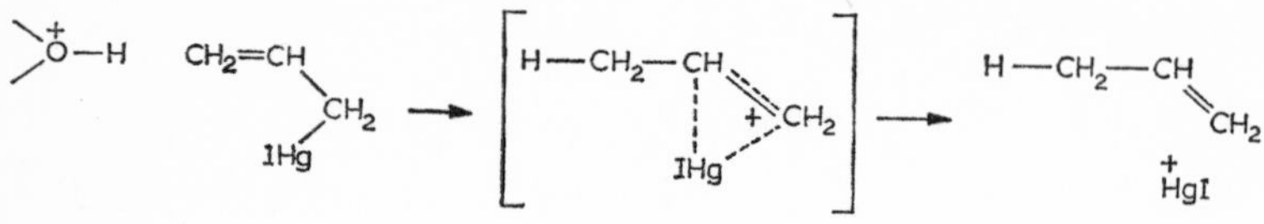

$$HgI^+ + I^- \xrightarrow{\text{fast}} HgI_2 \tag{21}$$

In a subsequent paper[10], details were given of experiments in which the concentration of added iodide was varied from 6×10^{-3} to 25×10^{-3} M. The variation of the observed first-order rate coefficient (reactions were run with a large excess of iodide ion and hydrogen ion over the concentration of allylmercuric iodide) with $[I^-]$ and $[H^+]$ was expressed by the equation

$$k_1 = 1.41 \times 10^{-2}[H^+] + 5.6[H^+][I^-] + 1.8 \times 10^{-2}[H^+][I^-]^2$$
$$+ 3.2 \times 10^{-3}[I^-] + 2.5 \times 10^{-1}[I^-]^2 \tag{22}$$

where the units of the various rate coefficients are appropriate to the various concentration terms. The defining equation for k_1 is, of course,

$$v = k_1 \cdot [CH_2\text{=}CHCH_2HgI] \tag{23}$$

At very low iodide ion concentrations, equation (22) reduces to

$$k_1 = 1.41 \times 10^{-2}[H^+] \tag{24}$$

and combination of equations (23) and (24) shows that the coefficient 1.41×10^{-2} is actually the value of the second-order rate coefficient at 25 °C given in Table 3. The term in $[H^+]$ in equation (22) thus corresponds to the mechanism shown in reaction (20).

The terms in $[H^+][I^-]$ and $[H^+][I^-]^2$ were considered[10] to correspond to mechanisms in which the complexes $CH_2\text{=}CHCH_2HgI_2^-$ and $CH_2\text{=}CHCH_2HgI_3^{2-}$ were subject to attack by hydrogen ion by mechanisms similar to that shown in (20).

Since general acid catalysis of reaction (19) had been observed[8], the terms in $[I^-]$ and $[I^-]^2$, in equation (22) might be due to water-catalysed decompositions of the complexes $CH_2\text{=}CHCH_2HgI_2^-$ and $CH_2\text{=}CHCH_2HgI_3^{2-}$. However, it was felt that the catalytic constants in equation (22) were much too large for this to be the case, and that the above complexes probably decomposed unimolecularly[10]. Taking the singly-charged complex as an example, the full mechanism would be

$$CH_2\text{=}CHCH_2HgI + I^- \rightleftharpoons CH_2\text{=}CHCH_2HgI_2^- \tag{25}$$

$$CH_2\text{=}CHCH_2HgI_2^- \xrightarrow{\text{slow}} CH_2\text{=}CHCH_2^- + HgI_2 \tag{26}$$

$$CH_2\text{=}CHCH_2^- + \text{Solvent} \xrightarrow{\text{fast}} CH_2\text{=}CHCH_3 \tag{27}$$

This mechanism thus corresponds to $S_E1\text{-}I^-$.

At the higher concentrations of added iodide ion used, deviations from the first-order behaviour required by equation (23) were observed. These deviations were ascribed[10] to the incursion of a new mechanism involving the symmetrisation of allylmercuric iodide, catalysed by the iodide ion, viz.

$$2\,CH_2\text{=}CHCH_2HgI \overset{I^-}{\rightleftharpoons} (CH_2\text{=}CHCH_2)_2Hg + HgI_2 \tag{28}$$

The diallylmercury is then itself subject to acidolysis by the hydrogen ion.

3.2 ACIDOLYSIS OF ALLYLMERCURIC CHLORIDE BY HCl IN 90% AQUEOUS DIOXAN

Nesmeyanov et al.[11] have reported the above acidolysis to be a second-order process. Second-order rate coefficients, in $l.mole^{-1}.sec^{-1}$, at various temperatures were given as

$$6.36 \times 10^{-3}\ (20\ °C),\ 1.94 \times 10^{-2}\ (30\ °C),$$
$$6.46 \times 10^{-2}\ (40\ °C),\ \text{and}\ 1.31\ (70\ °C) \tag{29}$$

The figure for 70 °C was an extrapolated one. Activation parameters were calculated[11] to be $E_a = 21.1\ kcal.mole^{-1}$ and $\log A = 13.561$.

No comparable kinetic studies exist on alkylmercuric chlorides as substrates, so that the high reactivity of the allyl compound cannot be assessed quantitatively with respect to the alkyl compounds. It is of interest to note, however, that whereas diethylmercury is cleaved rather slowly by HCl in 90% aqueous dioxan ($k_2 = 1.95 \times 10^{-3}\ l.mole^{-1}.sec^{-1}$ at 50 °C), diallylmercury underwent acidolysis at so high a rate that a value for the rate coefficient could not be obtained[12].

3.3 ACIDOLYSES OF CROTYLMERCURIC BROMIDE AND CINNAMYL-MERCURIC BROMIDE

The acidolysis of trans-crotyl- and trans-cinnamylmercuric salts by a variety of agents has been reported by Winstein and co-workers[13]. Details are given in

TABLE 4

HYDROCARBON PRODUCTS FROM THE ACIDOLYSIS OF CROTYL- AND CINNAMYL-
MERCURIC SALTS[13]

Compound	Reagent	Product
MeCH–CHCH$_2$HgX^a	HCl/ethyl acetate	MeCH$_2$CH–CH$_2$ (> 99 %)
MeCH–CHCH$_2$HgBr	HClO$_4$/acetic acid	MeCH$_2$CH–CH$_2$ (> 99 %)
PhCH–CHCH$_2$HgX^a	HCl/ether	PhCH$_2$CH–CH$_2$ (> 98 %)

^{a} X = Br and OAc.

Table 4 and it may be seen that in all cases these (very rapid) acidolyses lead to rearranged products. No kinetic studies were undertaken, but the mechanisms shown in equations (30) and (31) were suggested[13] for the hydrogen chloride and perchloric acid reactions respectively.

$$\text{(30)}$$

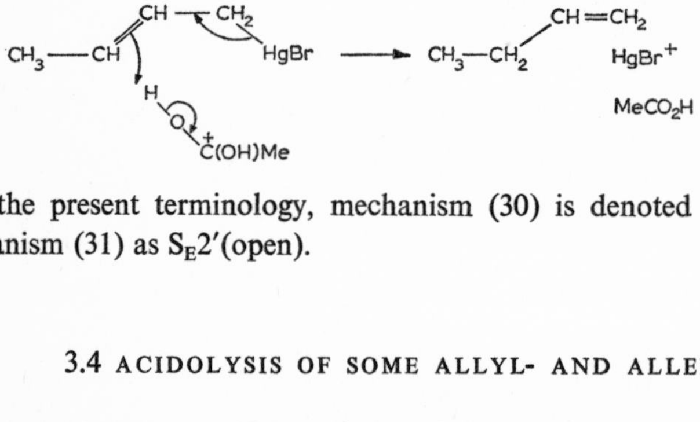

$$\text{(31)}$$

In the present terminology, mechanism (30) is denoted as S_E2'(cyclic) and mechanism (31) as S_E2'(open).

3.4 ACIDOLYSIS OF SOME ALLYL- AND ALLENYLTINS

Trimethylallyltin reacts quite readily with hydrogen chloride in solvent 96 % methanol –4 % water to form trimethyltin chloride and propene. Kinetic studies showed[14] that the acidolysis was first-order in each reactant, and that the second-order rate coefficient for cleavage by hydrogen chloride was identical in value to that for cleavage by perchloric acid. The acidolysis may thus be described by equation (32) (R = H), if we denote the electrophile as H$_2$OMe$^+$.

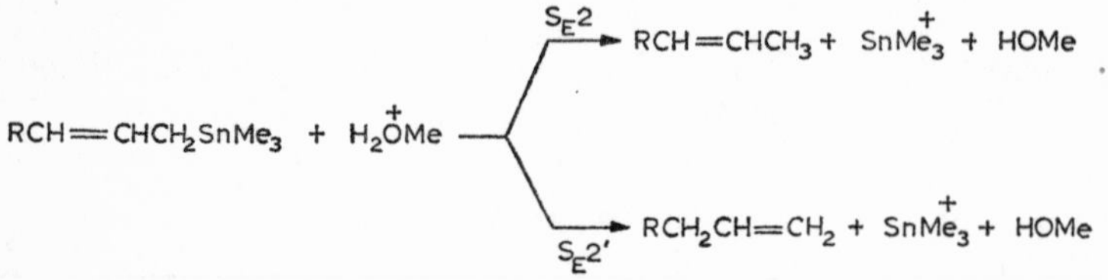

$$\text{(32a)}$$
$$\text{(32b)}$$

Acidolysis of *cis*- and *trans*-trimethylcrotyltin by hydrogen chloride in solvent 96 % methanol –4 % water yielded 98 % 1-butene and only 1–2 % *cis*-2-butene[14]. Hence reaction (32) (R = Me) proceeds almost entirely by the S_E2' pathway (32b) (R = Me). If it is assumed that the acidolysis of the other allytin compounds listed in Table 5 proceeds exclusively by the S_E2' pathway, then the observed rate coefficients for acidolysis, k_2, are also the rate coefficients for the S_E2' pathway, $k_2(S_E2')$. In Table 5 are given the second-order rate coefficients determined by Kuivila and Verdone[14]. These authors suggest that the acidolyses proceed through transition states such as (II) and (III), in which trimethylcrotyltin is shown as a representative substrate. In the present terminology, transition state (II) corresponds to mechanism S_E2'(open), and (III) corresponds to mechanism S_E2'-(cyclic).

TABLE 5

SECOND-ORDER RATE COEFFICIENTS ($l.mole^{-1}.sec^{-1}$) FOR THE ACIDOLYSIS OF ALLYL-AND ALLENYLTINS BY HCl IN SOLVENT METHANOL (96 %) –WATER (4 %) (v/v) AT 25. °C

Compound	k_2	$k_2(S_E2')$	$k_2(S_E2)$	Ref.
$CH_2=C(Me)CH_2SnMe_3$	24.8	24.8[a]		14
$CH_2=CHCH_2SnMe_3$	0.475	0.475[a]		14
cis-$CH_3CH=CHCH_2SnMe_3$	0.0508	0.0508[a]		14
trans-$CH_3CH=CHCH_2SnMe_3$	0.0274	0.0274[a]		14
$CH_2=CHCH_2SnPh_3$	0.00441	0.00441[a]		14
$CH_3CH=CHCH_2SnPh_3$[b]	0.00032	0.0032[a]		14
Me, H — C=C=C — Me, $SnMe_3$	0.519	0.223	0.296	15
Ph, H — C=C=C — Me, $SnMe_3$	0.111	0.0670	0.0441	15
Me, H — C=C=C — Ph, $SnMe_3$	0.125	0.0188	0.1061	15
Me, H — C=C=C — Me, $SnEt_3$	0.299	0.199	0.100	15
Me, H — C=C=C — Me, $SnPh_3$	0.00739	0.00657	0.00082	15

[a] Assuming that acidolysis proceeds only by mechanism S_E2' (see text).
[b] *Cis–trans* ratio unknown.

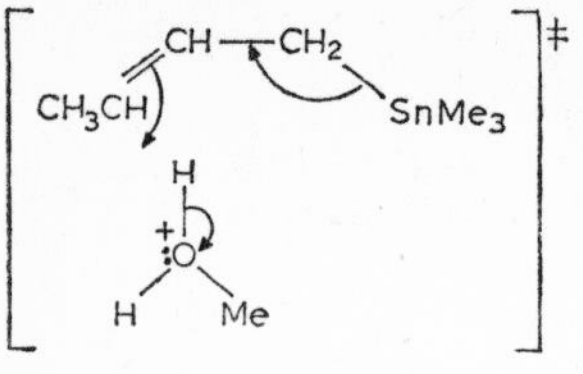

(II) 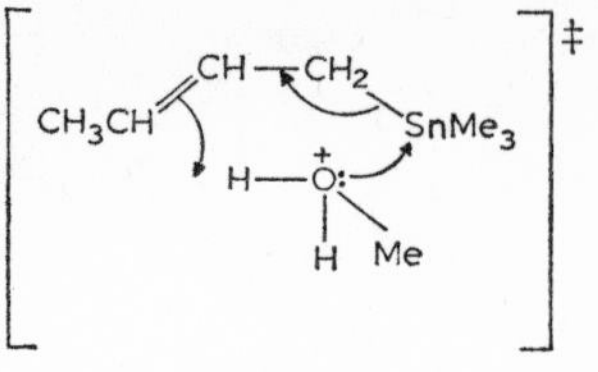   (III)

The acidolysis of allenyltins was later investigated by Kuivila and Cochran[15]. Although these acidolyses formally involve substitution at unsaturated carbon, rather than at saturated carbon, it is of interest to compare them with the allyltin cleavages, and details are also given in Table 5. The overall rate coefficients for acidolyses were partitioned into their S_E2 and S_E2' components by determination of the proportions of allene and acetylene, e.g.

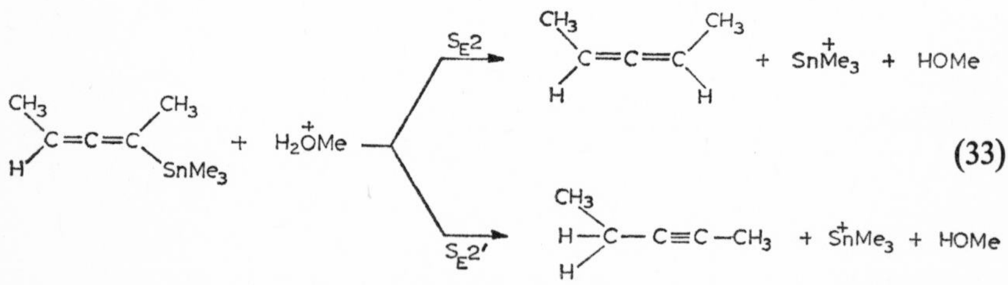

$$(33)$$

In the acidolysis of both allyl- and allenyltin compounds, the value of $k_2(S_E2')$ decreases as the leaving group changes from $-SnMe_3$ to $-SnPh_3$. This decrease is unlikely to be due to a steric effect, since the leaving group is removed from the site of substitution, and implies that $-SnPh_3$ is actually a poorer leaving group than is $-SnMe_3$[15]. It seems, therefore, that phenyl groups do not stabilise the incipient cation $^+SnPh_3$.

The effect of methyl and phenyl substituents in the allyl and allenyl group on the rate of acidolysis is evidently complex.

4. Halogenolysis (halogenodemetallation) of allyl-metal compounds

4.1 IODINOLYSIS OF TETRAALLYLTIN BY I_2/I^- IN SOLVENT ACETONE

The iodinolysis

$$allyl_4Sn + I_2 \xrightarrow{I^-/acetone} allyl-I + allyl_3SnI \tag{34}$$

has been shown to obey the rate equation

$$v = k_2^{obs.}[allyl_4Sn][I_3^-] \tag{35}$$

Gielen and Nasielski[16] suggested (see Chapter 8, Section 2.2, p. 138)* that the obserevd second-order coefficient could be considered to be composed of three terms as follows

$$k_2^{obs} = \frac{k_2}{K[I^-]} + k_2^b + \frac{k_3}{K} \qquad (36)$$

K is the equilibrium constant for I_3^- formation in solvent acetone. Table 6 gives

TABLE 6

THE EFFECT OF IODIDE ION CONCENTRATION ON THE OBSERVED SECOND-ORDER RATE COEFFICIENT, k_2^{obs}, FOR IODINOLYSIS OF TETRAALLYLTIN BY IODINE IN SOLVENT ACETONE AT 20 °C AND IONIC STRENGTH 0.5 M[16]

k_2^{obs} $(l.mole^{-1}.sec^{-1})$	$[I^-]$ $(mole.l^{-1})$	$k_2^{obs} \cdot [I^-]$ (sec^{-1})	$k_2^{obs} \cdot [I^-]$ calc. from. eqn. (37) (sec^{-1})
4.1	0.5	2.05	2.05
6.3	0.25	1.57	1.54
8.85	0.15	1.33	1.34
13.3	0.1	1.33	1.23
22.4	0.05	1.12	1.13

details of the variation of k_2^{obs} with $[I^-]$. This variation may be expressed** as

$$k_2^{obs} \cdot [I^-] = 1.03 + 2.05[I^-] \qquad (37)$$

Thus at an iodide ion concentration of 0.1 M, about 83 % of the total rate coefficient is given by the term $k_2/K[I^-]$ in equation (36); this proportion drops to 50 % at an iodide ion concentration of 0.5 M.

The term $k_2/K[I^-]$ corresponds to a mechanism in which the tetraallyltin is attacked by a molecule of iodine, probably by mechanism S_E2'(open)[16], equation (38) (R = allyl).

$$R_3Sn\!-\!CH_2\!-\!CH\!=\!CH_2 \quad I\!-\!I \longrightarrow R_3Sn^+ + CH_2\!=\!CH\!-\!CH_2I + I^- \qquad (38)$$

The term k_3/K in equation (36) may be equated with the rate law $v''' = k_3[allyl_4Sn][I_2][I^-]$ and hence a mechanism involving nucleophilic assistance by the iodide ion is indicated, whilst the term k_2^b corresponds to the rate law $v'' = k_2^b[allyl_4Sn][I_3^-]$ and hence to a mechanism in which a direct attack of I_3^- on

---

* In Chapter 8, Section 2.2, the coefficient k_2^a corresponds to the coefficient denoted as k_2 in the present section.
** In a later paper, Gielen and Nasielski[17] give the expression $k_2^{obs} = 1.05 + 2[I^-]$, close to that of equation (37) deduced by the author.

tetraallyltin takes place. These latter two mechanisms (equivalent to v'' plus v''') cannot be distinguished from each other.

The ratio of the rates of iodinolysis, tetraallyltin: trimethylparamethoxyphenyltin, was found[16] to be 1.3×10^3 using solvent acetone at 20 °C. Since the reactivity ratio for iodinolysis in solvent methanol at 20 °C of trimethylparamethoxyphenyltin: tetraethyltin is 3.4×10^4, it may be deduced[16] that tetraallyltin is about 4×10^7 times as reactive as tetraethyltin towards iodine.

4.2 IODINOLYSES OF ALLYLTIN COMPOUNDS BY I_2/I^- IN VARIOUS SOLVENTS

Roberts[18] later studied the iodinolysis of several allyltin compounds by I_2/I^- using a number of polar solvents. In the cleavage of $Ph_3SnCH_2CH=CH_2$ using as solvents methanol, ethanol, and dimethylsulphoxide, the term* $(k_2{}^b + k_3/K)$ in equation (36) was found to be negligible, so that $k_2^{obs} = k_2/K[I^-]$. Iodide ion assistance is therefore not important in the above three solvents. In solvents acetone and acetonitrile, however, the term $(k_2{}^b + k_3/K)$ is significant. For example, the variation of k_2^{obs} with $[I^-]$ for iodinolysis of $Ph_3SnCH_2CH=CH_2$ in acetonitrile at 24.9°C is given[18] by (cf. equation (37))

$$k_2^{obs} \cdot [I^-] = 6.60 \times 10^{-2} + 1.90[I^-] \tag{39}$$

Hence at an iodide ion concentration of 10^{-2} M (the usual concentration employed with acetonitrile) the proportion of iodinolysis proceeding by the iodide ion assisted path is 20 %; this proportion increases to 75 % at an iodide ion concentration of 10^{-1} M.

In Table 7 are given details of the calculation of rate coefficients, k_2, for the reaction

$$Ph_3SnCH_2CH=CHR + I_2 \overset{k_2}{\rightleftharpoons} \begin{matrix} RCH=CHCH_2I \\ + \\ RCHICH=CH_2 \end{matrix} + Ph_3SnI \tag{40}$$

from the observed coefficients, through equation (36) by isolation of the term $k_2/K[I^-]$. Roberts[18] gives also a calculation of k_2 for the case of acetone solvent, but it is not clear whether the term $k_2/K[I^-]$ has been separated from the term $(k_2{}^b + k_3/K)$ in this calculation. It should be noted that the values of K in Table 7 are not those used by Roberts; the value of $K = 2.5 \times 10^7$ in solvent acetonitrile at 25 °C now seems to be well authenticated[19, 20] (see p. 139), and two independent sets of workers[20, 21] give $K = 2.5 \times 10^5$ at 22 °C in solvent dimethylsulphoxide.

---

* Note that the various rate coefficients are denoted in this section by different symbols to those used by Roberts[18].

TABLE 7

CALCULATION OF SECOND-ORDER RATE COEFFICIENTS (l.mole^{-1}.sec^{-1}) FOR THE
IODINOLYSIS OF ALLYLTIN COMPOUNDS BY IODINE, REACTION (40), AT 25 °C

Solvent	Substrate	k_2/K^b	K^a	k_2
Acetonitrile	$Ph_3SnCH_2CH=CH_2$	6.60×10^{-2}	2.5×10^7	1.6×10^6
Acetonitrile	$Ph_3SnCH_2CH=CHPh$	3.8×10^{-4}	2.5×10^7	9.5×10^3
Methanol	$Ph_3SnCH_2CH=CH_2$	48	1.99×10^4	9.6×10^5
Methanol	$Ph_3SnCH_2CH=CHPh$	2.19	1.99×10^4	4.4×10^4
Methanol	$Ph_3SnCH_2CH=CHMe$	17.4	1.99×10^4	3.5×10^5
Dimethylsulphoxide	$Ph_3SnCH_2CH=CH_2$	1.80	2.5×10^5	4.5×10^5
Dimethylsulphoxide	$Ph_3SnCH_2CH=CHPh$	0.229	2.5×10^5	5.7×10^4
Dimethylsulphoxide	$Ph_3SnCH_2CH=CHMe$	1.45	2.5×10^5	3.6×10^5

[a] Values of K in l.mole^{-1} from refs. 19–21 (see text).
[b] Values of k_2/K from ref. 18.

The values used by Roberts were 6.3×10^6 and 7.9×10^6 respectively.

Roberts also determined the observed activation parameters for iodinolysis of $Ph_3SnCH_2CH=CHR$ (R = H, Me, and Ph). These values are, of course, composite quantities and are related to the standard thermodynamic changes for the equilibrium

$$I_2 + I^- \overset{K}{\rightleftharpoons} I_3^- \tag{41}$$

as well as to the activation parameters for reaction (40), since, for example, $\Delta H_{obs}^{\ddagger} = \Delta H_{(40)}^{\ddagger} - \Delta H_{(41)}^{\circ}$. Values of $\Delta H_{(41)}^{\circ}$ are known only for the case of solvent methanol, for which[16] $\Delta H_{(41)}^{\circ} = -8.0$ kcal.mole^{-1}. In methanol with 0.1 M sodium iodide, values of $\Delta H_{obs}^{\ddagger}$ and corresponding values of $\Delta H_{(40)}^{\ddagger}$ are (kcal. mole^{-1})

Compound	$\Delta H_{obs}^{\ddagger}$	$\Delta H_{(40)}^{\ddagger}$
$Ph_3SnCH_2CH=CH_2$	4.2	-3.8
$Ph_3SnCH_2CH=CHMe$	6.6	-1.4
$Ph_3SnCH_2CH=CHPh$	5.4	-2.6

In view of the calculated negative values of $\Delta H_{(40)}^{\ddagger}$, Roberts suggested that reaction (40) was itself composed of more than one elementary reaction, and wrote for the unassisted iodinolysis (40) the set of reactions

$$I_2 + \text{Substrate} \overset{K(\pi)}{\rightleftharpoons} \pi\text{-Complex} \tag{42}$$

$$\pi\text{-Complex} \overset{k_1}{\rightarrow} \text{Products} \tag{43}$$

Then since $\Delta H_{(40)}^{\ddagger} = \Delta H_{(42)}^{\circ} + \Delta H_{(43)}^{\ddagger}$, the negative value of $\Delta H_{(40)}^{\ddagger}$ could arise through a very negative value for the standard enthalpy change in reaction (42). Values of $\Delta H_{obs}^{\ddagger}$ in solvents acetone, acetonitrile, and dimethylsulphoxide were all quite small (4–8 kcal.mole^{-1}) so that it is possible that the sequence (42) and (43) obtains for the unassisted iodinolysis in these solvents.

The nature of the iodide ion assisted solvolyses is not clear; Roberts[18] suggests that the assisted iodinolyses proceed with rearrangement (S_E') but that the unassisted reactions take place without allylic rearrangement (S_E).

REFERENCES

1 R. M. G. ROBERTS AND F. EL KAISSI, *J. Organometal. Chem.*, 12 (1968) 79.
2 R. M. G. ROBERTS, *J. Organometal. Chem.*, 12 (1968) 89.
3 R. M. G. ROBERTS, *J. Organometal. Chem.*, 12 (1968) 97.
4 M. H. ABRAHAM AND M. J. HOGARTH, *J. Chem. Soc. A*, (1971) 1474.
5 R. M. G. ROBERTS, *J. Organometal. Chem.*, 18 (1969) 307.
6 M. M. KREEVOY, P. J. STEINWAND AND W. V. KAYSER, *J. Am. Chem. Soc.*, 86 (1964) 5013; 88 (1966) 124.
7 M. M. KREEVOY AND R. L. HANSEN, *J. Am. Chem. Soc.*, 83 (1961) 626.
8 M. M. KREEVOY, T. S. STRAUB, W. V. KAYSER AND J. L. MELQUIST, *J. Am. Chem. Soc.*, 89 (1967) 1201.
9 M. M. KREEVOY, P. STEINWAND AND T. S. STRAUB, *J. Org. Chem.*, 31 (1966) 4291.
10 M. M. KREEVOY, D. J. W. GOON AND R. A. KAYSER, *J. Am. Chem. Soc.*, 88 (1966) 5529.
11 A. N. NESMEYANOV, A. E. BORISOV AND I. S. SAVEL'EVA, *Dokl. Akad. Nauk SSSR*, 172 (1967) 1043; *Proc. Acad. Sci. USSR*, 172 (1967) 155.
12 A. N. NESMEYANOV, A. E. BORISOV AND I. S. SAVEL'EVA, *Dokl. Akad. Nauk SSSR*, 155 (1964) 603; *Proc. Acad. Sci. USSR*, 155 (1964) 280.
13 P. D. SLEEZER, S. WINSTEIN AND W. G. YOUNG, *J. Am. Chem. Soc.*, 85 (1963) 1890.
14 H. G. KUIVILA AND J. A. VERDONE, *Tetrahedron Letters*, (1964) 119.
15 H. G. KUIVILA AND J. C. COCHRAN, *J. Am. Chem. Soc.*, 89 (1967) 7152.
16 M. GIELEN AND J. NASIELSKI, *Bull. Soc. Chim. Belges*, 71 (1962) 32.
17 M. GIELEN AND J. NASIELSKI, *J. Organometal. Chem.*, 7 (1967) 273.
18 R. M. G. ROBERTS, *J. Organometal. Chem.*, 24 (1970) 675.
19 J. DESBARRES, *Bull. Soc. Chim. France*, (1961) 502; J.-C. MARCHON, *Compt. Rend. Ser. C*, 267C (1968) 1123.
20 R. L. BENOIT, *Inorg. Nucl. Chem. Letters*, 4 (1968) 723.
21 F. W. HILLER AND J. H. KRUEGER, *Inorg. Chem.*, 6 (1967) 528.

Chapter 11

Constitutional Effects, Salt Effects, and Solvent Effects in Electrophilic Substitution at Saturated Carbon

There have been only few reports of electrophilic substitutions that follow mechanism S_E1, and hence this chapter is mainly concerned with the various S_E2 mechanisms in which both a molecule of the substrate and a molecule of the electrophile are incorporated into the transition state.

1. Entropies of activation as a criterion of mechanism. Substitution by mechanism S_E2

It has been suggested[1] that the observation of a large negative activation entropy* (e.g. -25 cal.deg^{-1}.mole^{-1}) in reactions between two neutral molecules may be used as evidence that two sites of attachment are involved in the transition state (in other words, the transition state will be of the S_E2(cyclic) type and not of

TABLE 1

ENTROPIES OF ACTIVATION (cal.deg^{-1}.mole^{-1}) FOR SOME BIMOLECULAR ELECTROPHILIC SUBSTITUTIONS AT SATURATED CARBON

Reactants	Solvent	$\Delta S^{\ddagger}$	Transition state	Ref.
MeHgBr+HgBr$_2$	Ethanol	-25^a	Open^b	2
Benzyl HgBr+HgBr$_2$	Quinoline	-26^a	Cyclic	3
Et$_2$Hg, Pr$^n{}_2$Hg+HgI$_2$	Dioxan	-28	Cyclic	4
Pr$^n{}_2$Hg+HgI$_2$	Benzene	-28	Cyclic	5
R$_2$Hg+RHgBr^c	Ethanol	-32^c	Open^d	6
Et$_4$Sn+HgCl$_2$	Methanol	-22	Open^b	8
Et$_4$Sn+HgCl$_2$	tert.-Butanol	-42	Open^b	7
Et$_4$Sn+HgCl$_2$	Acetonitrile	-29	Open^b	9
Et$_4$Sn+HgI$_2$	Methanol	-27	Open^b	8, 9
Et$_4$Sn+HgI$_2$	Acetonitrile	-31	Open^b	9
Me$_4$Sn+I$_2$	Methanol	-35^a	Open^b	10
Me$_3$Sb+SbCl$_3$	DMF	-25	(Cyclic)	1

^{a} Calculated by the author from data given in refs. 2, 3, and 10.
^{b} Rate coefficients for these reactions have been shown to be increased in value on addition of inert salts.
^{c} R = 1,5-dimethylhexyl.
^{d} Positive kinetic salt effects were observed[11] for the exchange (R = Bu^s) in solvent ethanol.

* Values of $\Delta S^{\ddagger}$ quoted in this section all refer to a standard state of 1 mole/litre.

the S_E2(open) type). Table 1 contains data on values of $\Delta S^{\ddagger}$ for several reactions between two neutral molecules; it may be seen that $\Delta S^{\ddagger}$ is generally a large negative number for both reactions proceeding through an open transition state and those proceeding through cyclic transition states. Hence a single observation of a large negative activation entropy cannot be used as a mechanistic criterion, although it is possible that variation of $\Delta S^{\ddagger}$ with, say, solvent composition might prove to be of mechanistic value.

2. Constitutional influences in the electrophilic reagent

2.1 SUBSTITUTION BY MECHANISM S_E2

Many electrophilic reagents are capable not only of electrophilic attack at the carbon atom of the C–M bond, but also of nucleophilic attack at the metal atom of the C–M bond. It is such nucleophilic attack that gives rise to the mechanisms S_E2(cyclic) and S_E2(co-ord). Dessy et al.[12, 13] have several times stressed the importance of internal nucleophilic assistance by the electrophilic reagent in reaction mechanisms of organometallic compounds, and it is the account by Dessy and co-workers[13] of the acidolysis of triethylboron that is probably the classic report in this field.

In solvent diglyme, the rate of acidolysis of triethylboron by carboxylic acids is found to be greater the weaker is the acid, there being a linear correlation between $\log k_2$ and the pK of the carboxylic acid. Thus nucleophilic coordination of the carboxylic acid to the boron atom must be an important feature of the reaction mechanism, and the acidolysis may therefore be denoted as following mechanism S_E2(cyclic) or S_E2(co-ord), perhaps through a transition state such as (I)*.

(I)

It is interesting to note that the rate of acidolysis of diphenylmercury by carboxylic acids in solvent dioxan is greater the stronger is the acid. Thus the predominant feature of this substitution at unsaturated carbon must be electrophilic attack by the acid and not nucleophilic co-ordination to the mercury atom[18].

The acidolysis of diethylcadmium by various nuclear-substituted benzyl alcohols in solvent ether is accelerated by electron-donating substituents in the electrophile and retarded by electron-withdrawing substituents[14]. As in the case of acidolysis of triethylboron, co-ordination of the electrophile to the metal atom

---

* See also Chapter 7, Section 5, p. 122.

must be an important feature of the transition state*.

In these above cases, the electrophilic reagent has been varied in only a minor way, so that the nature of the transition state is probably very similar along the series of reagents (*e.g.* the series of carboxylic acids or the series of substituted benzyl alcohols). If more fundamental changes in the electrophile are introduced, then it is clearly possible for the nature of the transition state to alter along the series of electrophiles.

Ingold and co-workers[15] have suggested that for the substitution of a given substrate by a series of mercuric salts, the rate of substitution should increase with increasing ionicity of the mercuric salt if mechanism S_E2(open) is followed. In other words, the sequence

TABLE 2

RELATIVE REACTIVITIES OF MERCURIC SALTS IN SOME BIMOLECULAR ELECTROPHILIC SUBSTITUTIONS AT SATURATED CARBON

Substrate	Solvent[a]	Relative reactivity of HgX_2 where X is					Ref.
		I	Br	Cl	OAc	NO_3	
Me_2Hg	Methanol	1	6	21			16
Me_2Hg	Dioxan	1	2.3	1.7			16
Bu^s_2Hg	Ethanol		1		75	4×10^3	15
Et_4Sn	Methanol[b]	1*		1.3*	340*		7
Et_4Sn	Acetonitrile[b]	1*		0.5*	2.4		9
RCH_2Cr^{2+} [d]	Water		1**	3.7**	160	140	17
$MeAuPPh_3$	Dioxan	1	0.9	0.5	~30		19
$MeAuPPh_3$	Acetone		1	0.8			19
$EtAuPPh_3$	Dioxan	1	0.9	0.4	>15		19
$RAuPPh_3$ [e]	Dioxan	1	1.0	0.5	5.9	>120	19
Me_3AuPPh_3	Dioxan		1		~50		19

Substrate	Solvent[a]	Relative reactivity of $MeHgX$ where X is					Ref.
		I	Br	Cl	OAc	NO_3	
$MeAuPPh_3$	80 % Aq. dioxan[c]	1	0.7*	0.8	8.8	~1.4×10^4	19
$MeAuPPh_3$	DMSO[c]	1	0.4	0.3	0.3	>1.4×10^3	19
$EtAuPPh_3$	DMSO	1	0.3	0.1	0.1	~3.0×10^3	19

* Positive kinetic salt effects indicate that these reactions proceed by mechanism S_E2(open).
** Product analyses indicate that these reactions proceed by mechanism S_E2(open).
[a] Solvents usually contained small quantities of acetic acid or nitric acid for reactions involving $Hg(OAc)_2$ and $Hg(NO_3)_2$.
[b] Relative reactivity of $HgI_3^- = 0$.
[c] Relative reactivity of $HgBr_3^- = 0$.
[d] R = 4-pyrido$^+$.
[e] R = n-BuC(CN)CO_2Et.

---

* See also Chapter 7, Section 3, p. 111.

$$HgX_3^- < HgI_2 < HgBr_2 < HgCl_2 < Hg(OAc)_2 < Hg(NO_3)_2 \qquad (1a)$$

should be characteristic of mechanism S_E2(open). Data on the relative rates of substitution by mercuric salts are collected in Table 2. It is easy to see that this suggestion of Ingold does not apply to sequences of reactivity of the mercuric halides. In the substitution of tetraethyltin by mercuric chloride and mercuric iodide in solvents methanol and acetonitrile, all four reactions are subject to positive kinetic salt effects and thus probably proceed by mechanism S_E2(open). But the observed sequences of reactivity are $HgI_2 < HgCl_2$ in methanol, and $HgI_2 > HgCl_2$ in acetonitrile.

It therefore remains to see whether the more general sequence (X = halogen)

$$HgX_3^- < HgX_2 < Hg(OAc)_2 < Hg(NO_3)_2 \qquad (1b)$$

is characteristic of substitutions proceeding by mechanism S_E2(open). The more general sequence, however, covers such a wide range of character of electrophiles that it cannot be assumed that the mechanism of substitution remains constant along such a series, or even along any part of such a series. Jensen and Rickborn[56] criticise the use of such sequences of reactivity as (1) on this ground. They point out that mercuric acetate and mercuric nitrate might be able to form cyclic six-centred transition states more easily than a mercuric halide could form a four-centred transition state, viz.

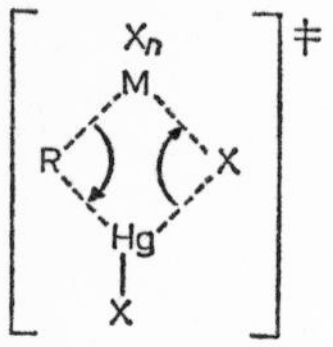

(IIa) 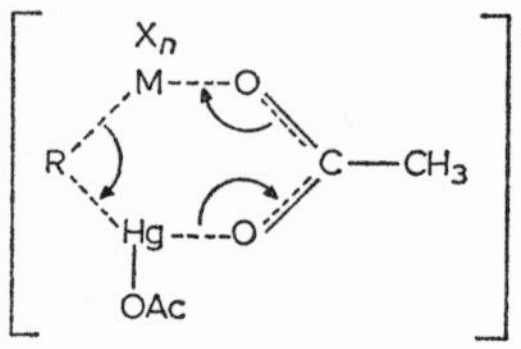   (IIb)

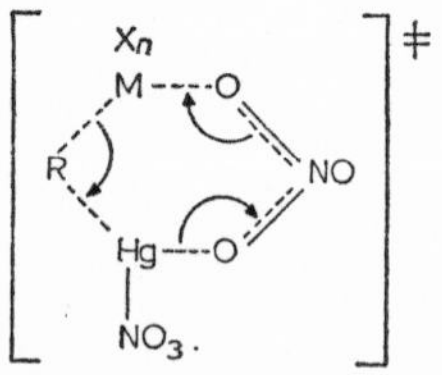

(IIc)

It seems to the author that an observed reactivity sequence $HgX_2 < Hg(OAc)_2$, in the absence of any independent evidence, could therefore be interpreted in a number of different ways.

(i) Both substitutions proceed by mechanism S_E2(open), and the sequence is due to mercuric acetate being a more powerful electrophile than the mercuric halide.

(*ii*) Both substitutions proceed by mechanism S_E2(cyclic), and the sequence is due to mercuric acetate being better able to act as a bridging group in (IIb) than is the mercuric halide in (IIa).

(*iii*) The two substitutions are proceeding by different mechanisms.

An examination of Table 2 reveals that although mercuric acetate and mercuric nitrate have often been used as electrophilic reagents, there are but few instances in which independent evidence as to their mechanism of reaction has been put forward. Positive kinetic salt effects have been observed in the substitution of *sec.*-butylmercuric acetate by mercuric acetate (with lithium nitrate in solvent ethanol)[2], the substitution of di-*sec.*-butylmercury by *sec.*-butylmercuric nitrate (with lithium nitrate in solvent ethanol)[11], and the substitution of tetraethyltin by mercuric acetate (with tetra-*n*-butylammonium perchlorate in methanol)[7]. In the latter case, it was suggested[7] that the observed very large positive kinetic salt effect was possibly due to anion exchange between mercuric acetate and the perchlorate ion.

At the moment, therefore, there is little evidence to suggest that sequence (1b) can be taken as characteristic of mechanism S_E2(open).

2.2 SUBSTITUTION BY MECHANISM S_E1

For a substrate ionising by a simple S_E1 process, it is expected that the rate of ionisation will be independent of the nature of any electrophilic reagent present, provided that the electrophile is present in fairly low concentration.

3. Constitutional influences in the substrate

3.1 THE EFFECT OF *p*-SUBSTITUENTS ON THE ELECTROPHILIC CLEAVAGE OF BENZYL–METAL AND α-CARBETHOXYBENZYL–METAL BONDS

Some of the earliest investigations in this field are due to Eaborn *et al.*[20, 21], who studied the effect of substituents in the benzyl group on the base-catalysed (S_E1-OR$^-$) cleavage of benzyl–silicon and benzyl–tin bonds. Their results for the *p*-substituents used are given in Table 3 (Nos. 1 and 2). The formation of the carbanion, p-X–$C_6H_4CH_2^-$, is strongly aided by electron-withdrawing X groups, Hammett ρ values being[20, 21] 4.88 and 3.23 for Nos. 1 and 2 respectively. There is only one series of compounds, No. 3 (Table 3), for which substituent effects on an uncatalysed S_E1 reaction are known. Once again, electron-withdrawing groups, X, aid formation of the carbanion, p-X–$C_6H_4\overset{-}{C}HCO_2Et$, although the Hammett ρ value is now only $\simeq 1.0$. The substituent effects shown in Nos. 1–3 (Table 3) may be taken as the typical pattern for cleavages following mechanism S_E1.

Reaction No. 4 (Table 3) has been suggested[23] to follow mechanism S_E2(open).

TABLE 3

RELATIVE RATES OF ELECTROPHILIC CLEAVAGE OF THE CARBON–METAL BOND IN SOME p-SUBSTITUTED BENZYL–METAL AND α-CARBETHOXYBENZYL–METAL COMPOUNDS

No.	Reactants[a]	Solvent	p-X									Ref.
			NO_2	I	Br	Cl	F	H	Pr^i	Me	Bu^t	
1	$ArCH_2SiMe_3+HO^-$	H_2O/MeOH	1.8×10^6		19.3	13.9		1		0.19	0.02[b]	20
2	$ArCH_2SnMe_3+HO^-$	H_2O/MeOH	4.0		12.8	7.9	0.55	1		0.21		21
3	$ArCH(CO_2Et)HgBr+HgBr_2$	DMSO		1.3				1			0.7	22
4	$ArCH(CO_2Et)HgBr$ $+p$-tolyl$_3$CBr·$HgBr_2$	$(CH_2Cl)_2$	0.18	0.25		0.38	0.66	1	3.3		1.3	23
5[c]	$ArCH_2HgBr+HgBr_2$	DMSO				0.8	0.8	1	1.2	1.2		24
6	$ArCH_2HgBr+HgBr_2$	Quinoline				0.9	0.8	1	1.6	1.4		25
7[c]	$ArCH_2HgBr+HgBr_2/Br^-$	DMSO				2.4	1.4	1	0.7	1.0		24
8	$2ArCH(CO_2Et)HgBr+2NH_3$	$CHCl_3$	160[d]	6.1	4.9	4.3	1.3	1	0.4	0.3	0.25	26
9	$ArCH(CO_2Et)HgBr+I_2/I^-$	Toluene	62[d]	4.4		3.3	1.4	1	0.5		0.4	27
10	$ArCH_2HgBr+I_2/I^-$	MeOH	Fast			0.95	1.12	1		2.6	11.1[b]	28
11	$ArCH_2HgBr+I_2/I^-$	DMF	Fast		0.76	0.64	0.75	1		2.0	11.1[b]	28

[a] Ar = p-X–C_6H_4.
[b] These values are for X = OMe.
[c] Calculated by the author from the given exchange half-lives[24].
[d] Calculated on the assumption[26,27] that Hammett's equation holds.

In this case reaction is aided by electron-donating substituents ($\rho \simeq -2.0$) as might be expected for attack by an electrophilic reagent at the carbon atom adjacent to the aromatic ring. Reactions Nos. 5 and 6 (Table 3) are simple S_E2 exchanges of mercury-for-mercury and show similar substituent effects to those in reaction No. 4, although on a very reduced scale ($\rho \simeq -0.4$ and -0.8). Reaction No. 6 has been suggested[25] to proceed through a cyclic transition state of the S_E2(cyclic) type (III). A similar transition state is possible in the case of reaction No. 5, but since dimethylsulphoxide is quite a "polar" solvent the transition state could also be formulated as (IV), that is corresponding to mechanism S_E2(open). In any case, the substituent effects shown in Nos. 5 and 6 (Table 3) are consistent with a dominant electrophilic attack of the electrophile.

(III)   (IV)

Reaction No. 7 (Table 3) probably involves[24] (with an excess of bromide ion) the reactants $ArCH_2HgBr_2^-$ and $HgBr_3^-$, and hence substituent effects will reflect the equilibrium constant for formation of the complex $ArCH_2HgBr_2^-$ as well as the rate coefficient for the actual substitution. Similar considerations apply to reaction No. 8 (Table 3), where complexes $ArCH(CO_2Et)HgBr \cdot NH_3$ are formed prior to the electrophilic substitution. Since such complex formation will be aided by electron-withdrawing substituents in the aromatic ring, it is clearly difficult to assess the substituent effects in the actual electrophilic substitution.

The final three listed reactions, Nos. 9–11 (Table 3), are of interest in that in all three cases the electrophilic reagent is probably I_3^- and the substrate is the neutral benzylmercuric bromide*. The series No. 9 is clearly that of electron-attracting substituents aiding reaction ($\rho = 2.3$) and it is thus reasonable to conclude that nucleophilic coordination to mercury in the transition state (V) is more important than electrophilic attack at the saturated carbon atom.

(V)

* An alternative interpretation is that a complex $ArCH_2HgBrI^-$ (or $ArCH(CO_2Et)HgBrI^-$) undergoes substitution by a neutral iodine molecule (see p. 139).

In series Nos. 10 and 11, where Hammett's equation is not followed, it is possible that the extent of bond-making and bond-breaking in the transition state no longer remains constant along the series. The *p*-nitro substituent might well induce a much greater degree of extension in the C–Hg bond in the transition state and so provide some "S_E1-character" to the substitution.

3.2 MORE GENERAL INFLUENCES IN THE SUBSTRATE

3.2.1 Substitution by mechanism S_E2

The allyl–metal, cyclopropyl–metal, and phenyl–metal bonds are all generally cleaved by electrophilic reagents much more readily than are bonds between the simple alkyl groups and metals. Aralkyl–metal bonds, such as benzyl–metal bonds, on the other hand do not seem to be especially reactive in S_E2 substitutions compared with alkyl–metal bonds[29].

Patterns of reactivity amongst the simple alkyl groups themselves seem inextricably connected to solvent influences and are discussed in Section 5.1, p. 224.

3.2.2 Substitution by mechanism S_E1

Constitutional effects in the formation of carbanions from hydrocarbons are well known (see Chapter 2, Section 2, p. 7) and suggest that in the formation of carbanions by ionisation of the R–M bond, groups such as R = $Ph_2CH–$, $PhCH_2–$, CH_2CO_2Et, etc. will yield the corresponding carbanions much more easily than will the simple alkyl groups. This seems to be confirmed by experiment; the only compounds of type RHgX that are known to react by mechanism S_E1 are those

for which R = $ArCHCO_2Et$, $p-NO_2–C_6H_4CH_2$, and $\overset{+}{H}N\!\!\diagup\!\!\diagdown\!\!-CH_2$. Eaborn

and co-workers[21] have shown that in the base-catalysed (S_E1-OR⁻) cleavage of compounds $RSiMe_3$, the rate of formation of the carbanion R⁻ increases through the series R = $PhCH_2$ (0.34) < $Ph_2CH(480)$ < $Ph_3C(630)$.

3.3 REACTIVITIES OF SUBSTRATES RMX_n WITH RESPECT TO THE METAL ATOM M

3.3.1. Substitution by mechanism S_E2

The more electropositive is the metal atom in a substrate RMX_n, the more polar will be the R–M bond in the sense $R^{\delta-}–M^{\delta+}$, the better will be the group MX_n

TABLE 4

THE EFFECT OF THE METAL ATOM IN SUBSTRATES R_nM ON THE REACTIVITY OF THE SUBSTRATE TOWARDS ELECTROPHILIC REAGENTS

Acetolysis in acetic acid		Iododemetallation in CCl_4		Substitution by $HgBr_2$ in MeCN		Substitution by $HgCl_2$ in MeOH	
Substrate	$10^6 k_1$ (sec^{-1})	Substrate	k_2 (l.mole^{-1}.sec^{-1})	Substrate	Rel. rate[h]	Substrate	k_2 (l.mole^{-1}.sec^{-1})
Me$_4$Sn	0.3(20 °C)[a]	Me$_4$Sn	10^{-6}(25 °C)[e]	Et$_3$Si·allyl	1	Me$_4$Sn	3.1(36 °C)[i]
Me$_4$Pb	5.5(20 °C)[b]	Me$_4$Pb	0.035(31 °C)[f]	Et$_3$Ge·allyl	167	Me$_2$Hg	1.05(36 °C)[j]
Bu$^n_4$Pb	3.1(25 °C)[c]	Me$_2$Hg	1.2×10^{-3}(28 °C)[g]	Et$_3$Sn·allyl	≫167		
Bu$^n_2$Hg	2.4(25 °C)[d]						

[a] Ref. 30.
[b] Ref. 31, extrapolated using the given activation energy.
[c] Ref. 31.
[d] Ref. 32.
[e] Estimated value from data given in Table 31, Chapter 8, Section 5, p. 172.
[f] Ref. 33.
[g] Ref. 34.
[h] Ref. 35.
[i] Ref. 8.
[j] Ref. 16.

as a leaving group, and the more reactive will be the substrate towards electrophilic reagents. In Table 4, details are given of the relatively few comparisons that may be made. The series for acidolyses may be extended by incorporation of the relative reactivities of diethylcadmium and diethylmercury $(166 : 1)^{36}$ towards hydrogen chloride in solvent dioxan/dimethylformamide, and final results are given in Table

TABLE 5

RELATIVE REACTIVITIES OF ALKYLS OF DIFFERENT METALS IN ELECTROPHILIC SUB-
STITUTIONS

Alkyl	X^a	Iodometallation	Acidolysis	Substitution by HgX_2
R_2Cd	1.7		2×10^4	
R_4Pb	1.8	3×10^4	165	
R_2Hg	1.9	10^3	130	0.3
R_4Sn	1.9	1	1	1
R_4Ge	1.8			10^{-1}
R_4Si	1.8			10^{-2}
R_4C	2.5	0	0	0

^{a} Pauling-scale electronegativity^{37} of the metal atom in the substrate.

5*. It is evident that the relative reactivities of the metal alkyls are strongly dependent on the type of electrophilic reagent used. There is a partial correlation of reactivity with the Pauling-scale electronegativity^{37} of the metal atom in the substrate, but in any case, a rigorous correlation of reactivity with a parameter such as the electronegativity of the metal atom in the substrate would not be expected, if for no other reason than because changes in mechanism are quite probable along any extended series.

The influence of the metal atom on mechanism is particularly well illustrated by the pronounced tendency of the alkyls of boron and aluminium to react via co-ordination complexes in which the metal atom is four-co-ordinate. It is no coincidence that the prime examples of substitution by mechanism S_E2(co-ord) occur in reactions of alkyl–boron compounds.

3.3.2 Substitution by mechanism S_E1

The calculations in Chapter 2, Section 1, p. 5, indicate that carbanion formation from various metal alkyls would be facilitated in the sense R–H < R–HgR < R–ZnR < R–Li, *i.e.* that the more electropositive is the metal atom in the leaving

---

* The sequence for acidolyses given in Table 5 is a little different from the one given on p. 107; the latter sequence was constructed using data on metal alkyls and aryls, whilst the sequence in Table 5 refers only to metal alkyls.

group, the more is carbanion formation aided. This seems to be the case for un-catalysed S_E1 reactions, in so far as PMR studies yield[38] the sequence R_2Hg, $R_3Al < R_2Zn < R_2Mg < RLi$ for inversion via mechanism S_E1 (see Chapter 4, Section 2.3, p. 32).

4. Salt effects and co-solvent effects in electrophilic substitution at saturated carbon

4.1 SUBSTITUTION BY MECHANISM S_E2

Bimolecular reactions between neutral molecules proceeding through transition states in which charge separations occur would be expected to undergo increases in rate on addition of inert salts[39]. Hence such reactions taking place by mechanism S_E2(open) should exhibit positive kinetic salt effects; if, however, mechanism S_E2(cyclic) obtains then either no kinetic salt effects or only slight positive kinetic salt effects would be expected, since charge separation in a cyclic transition state will probably be only slightly greater than charge separation in the reactants. Kinetic salt effects should therefore be a valuable criterion of mechanism. In Table 6 are collected data on kinetic salt effects for which quantitative results have been reported.

Abraham and Behbahany[7] have discussed kinetic salt effects on reactions in-volving neutral molecules in terms of the equation

$$\left[\frac{\log(k/k_0)}{I}\right]_{I\to0} = \frac{4\pi Ne^4}{2303(\varepsilon kT)^2}Z^2d$$

Here, k and k_0 are the rate coefficients in the presence of salt (at an ionic strength, I) and in the absence of salt respectively; other quantities are as defined on p. 85. The transition state is assumed to behave as a point dipole consisting of charges $+Z$ and $-Z$ separated by a distance d. The dielectric constant of the solvent is ε, and it can be seen that the magnitude of the kinetic salt effect, $\log(k/k_0)$ is inversely proportional to ε^2. It is therefore difficult, if not impossible, to compare kinetic salt effects observed using solvents of different dielectric constants unless the salt effects are "corrected" for the variation in solvent dielectric constant. One method of correction is clearly to tabulate the salt effects in terms of the quantity Z^2d, and in Table 6 are given these values together with the corresponding values of Z. In many cases, the given values of Z^2d and of Z are very approximate because the necessary data required for extrapolation to $I \to 0$ are not available. Never-theless, the Z^2d values do represent the kinetic salt effects reduced to the same scale and corrected for differences in solvent dielectric constant and for differences in temperature.

Most of the substitutions in Table 6 refer to metal-for-metal exchanges (Nos. 1–15). Of these fifteen reactions, fourteen are subject to positive kinetic salt effects;

TABLE 6

KINETIC SALT EFFECTS IN SOME BIMOLECULAR ELECTROPHILIC SUBSTITUTIONS IN-
VOLVING NEUTRAL REACTANTS

No.	Reactants	Solvent	Additive	Effect of Additive[a]	Z^2d	Z	Refs.
1	$MeHgBr+HgBr_2$	EtOH	$LiNO_3$	+	$\simeq1.01^b$	$\simeq0.58^b$	2
2	$Bu^sHgOAc+Hg(OAc)_2$	EtOH	$LiNO_3$	+	$\simeq0.90^b$	$\simeq0.55^b$	2
3	$Bu^s_2Hg+Bu^sHgBr$	EtOH	$LiNO_3$, $LiClO_4$	+			11
4	$Bu^s_2Hg+Bu^sHgNO_3$	EtOH	$LiNO_3$	+	$\simeq1.00^b$	$\simeq0.57^b$	11
5	$MeAuPPh_3+MeHgBr$	Aq. dioxan[c]	$LiClO_4$, NaOAc	+	$\simeq0.96^d$	$\simeq0.57^d$	19
6	$Et_4Sn+HgCl_2$	Aq. MeOH[e]	$LiClO_4$, $Bu^n_4NClO_4$	+	1.73	0.75	7, 8
7	$Et_4Sn+HgCl_2$	MeOH	$LiClO_4$, $Bu^n_4NClO_4$	+	2.21	0.84	7, 8
8	$Et_4Sn+HgCl_2$	t-BuOH	$Bu^n_4NClO_4$	+	2.56	0.91	7
9	$Et_4Sn+HgCl_2$	MeCN	$LiClO_4$, $Bu^n_4NClO_4$	+	0.66	0.47	9
10	$R_4Sn^f+HgI_2$	Aq. MeOH[e]	$LiClO_4$	+	0.92^g	0.54^g	40
11	$Et_4Sn+HgI_2$	MeOH	$Bu^n_4NClO_4$	+	3.03^g	0.99^g	7
12	$Et_4Sn+HgI_2$	t-BuOH	$Bu^n_4NClO_4$	+	2.56^g	0.91^g	7
13	$Et_4Sn+HgI_2$	MeCN	$Bu^n_4NClO_4$	+			9
14	$Et_4Sn+Hg(OAc)_2$	MeOH	$NaClO_4$, $Bu^n_4NClO_4$	+			7
15	$Et_4Sn+Hg(OAc)_2$	t-BuOH	$Bu^n_4NClO_4$	None	(0)	(0)	7
16	Me_4Sn+I_2	MeOH	$NaClO_4$	+	0.63^h	0.32^h	10
17	$Pr^i_4Sn+Br_2$	HOAc	$NaClO_4$	+	$\simeq0.28^h$	$\simeq0.22^h$	30
18	Et_4Pb+Br_2	MeOH	$NaClO_4$	+			42
19	$Neophyl_2Hg+HOAc$	HOAc	NaOAc	None	(0)	(0)	32

[a] Positive kinetic salt effects indicated by +.
[b] Calculated by the author, with d taken as 3.0 A.
[c] Dioxan (4)–water (1) (v/v).
[d] Calculated by the author, see p. 103.
[e] Methanol (96)–water (4) (v/v).
[f] R = Et, Pr^n, Bu^n.
[g] Calculated by the author, with d taken as 3.1 A.
[h] Calculated by the author, see also p. 148.

for those reactions for which it is possible to calculate values of Z^2d, the corresponding values of Z are consistent with transition states that carry substantial
separations of charge. The range of solvents used varies from aqueous methanol
($\varepsilon^{40} = 32.2$) and acetonitrile ($\varepsilon^{25} = 36.0$) down to tert.-butanol ($\varepsilon^{30} = 11.6$) and
aqueous dioxan ($\varepsilon^{25} = 10.7$), so that it seems quite possible for these metal-for-
metal substitutions to proceed by mechanism S_E2(open) even in solvents that are
considered to be of only moderate polarity. Reaction No. 14 (Table 6) exhibits
positive salt effects of such magnitude that the possibility of anion exchange
between the reagent, $Hg(OAc)_2$, and the added salt was considered[7]. The only

reaction in the metal-for-metal class in Table 6 that does not exhibit a positive kinetic salt effect is that between tetraethyltin and mercuric acetate in solvent *tert.*-butanol; perhaps in this case reaction proceeds by mechanism S_E2(cyclic).

Table 6 includes data on three halogenations of tetraalkylmetals. The iododemetallation of tetramethyltin in solvent methanol no doubt proceeds by mechanism S_E2(open)[10], and the calculated value of Z ($\simeq 0.32$) is compatible with such an assignment. However, the positive salt effect on the bromodemetallation of tetra-isopropyltin in solvent acetic acid ($\varepsilon^{20} = 6.15$) is quite small, and leads to a value of only $\simeq 0.28$ for Z^2d and a corresponding value of $\simeq 0.22$ for Z, assuming (see p. 148) that d is approximately 6 A. Here is a possible case, therefore, of a mechanism that is in the intermediate S_E2(open)–S_E2(cyclic) region.

The observation that added sodium acetate gives rise to no kinetic salt effect at all in the acetolysis of dineophylmercury has already been put forward[32] as evidence that the acetolysis proceeds by mechanism S_E2(cyclic).

Kinetic salt effects therefore represent an extremely valuable method of distinguishing mechanism S_E2(open) from mechanism S_E2(cyclic) for the case in which the reactants are neutral molecules. Naturally, since these two mechanisms may shade into each other, it is possible that in some cases only small positive kinetic salt effects will be observed if the mechanism is in the S_E2(open)–S_E2(cyclic) border line region.

Co-solvent effects are generally more difficult to interpret than are kinetic salt effects, largely because the addition of a new solvent produces much more violent changes in the reaction medium than does addition of an inert salt. The mechanism of substitution may therefore be affected by addition of a co-solvent; this is espicially true of reactions proceeding by mechanism S_E2, where the S_E2(open) and S_E2(cyclic) transition states may be regarded as the two extremes of a whole range of transition states.

As an example of the effect of a co-solvent, the substitution of methyl-(triphenylphosphine)gold(I) by mercuric bromide[19] may be considered, *viz.*

Solvent	Relative k_2
Dioxan	1
Dioxan (9 vol.) : water (1 vol.)	3.0
Dioxan (8 vol.) : water (2 vol.)	5.1

In the absence of other information, these results could be interpreted in at least two ways, (*i*) mechanism S_E2(open) obtains in all three reaction media, and the aqueous co-solvent stabilises the "open" transition state with respect to the reactants, or (*ii*) mechanism S_E2(cyclic) obtains in solvent dioxan, and the effect of the co-solvent is to shift the mechanism of reaction towards the S_E2(open) mechanism.

Similar interpretations could be placed on the effect of water as a co-solvent on the substitution of tetraethyltin by mercuric chloride[8] in solvent methanol, *viz.*

Solvent	Relative k_2
Methanol	1
Methanol (9 vol.) : water (1 vol.)	4.2
Methanol (8 vol.) : water (2 vol.)	11.8
Methanol (7 vol.) : water (3 vol.)	28.0

In this particular case, it is known that the substitution in methanol itself is accelerated by the addition of inert salts, and hence that mechanism S_E2(open) probably obtains in solvent methanol. The most reasonable interpretation of the co-solvent effect is, therefore, that mechanism S_E2(open) obtains in all four reaction media, and that the transition state is stabilised relative to the reactants in the more aqueous media.

Because of the difficulties mentioned above, little use has been made of co-solvent effects in the deduction of mechanism.

4.2 SUBSTITUTION BY MECHANISM S_E1

It would be expected[39] that in the ionisation of a neutral substrate, the addition of inert salts should produce considerable rate accelerations. In the racemisation of $(-)$-α-carbethoxybenzylmercuric bromide by mechanism S_E1 in solvent dimethylsulphoxide, however, the addition of inert salts was found to result in either rate acceleration (*e.g.* with $LiNO_3$) or retardation (*e.g.* with $LiClO_4$)[43].

5. Solvent effects and constitutional effects of the substrate, in electrophilic substitution at saturated carbon

5.1 RELATIVE REACTIVITIES OF SUBSTRATES RMX_n WITH RESPECT TO THE ALKYL GROUP R

5.1.1 Substitution by mechanism S_E2

Following the work of Dessy *et al.*[12, 13, 41] on mechanisms of electrophilic substitution in which internal nucleophilic assistance by the electrophilic reagent was important, Gielen and Nasielski[44] made a notable contribution in their correlation of the observed sequences of reactivity in 11 series of substrates of type* RMX_n (where R is varied through Me, Et, Pr", etc. in each series) with the polarity of the solvent. Abraham and Hill[45] later interpreted the reactivity sequences of some 31 series of substitutions on the basis of the possible mechanisms of substitution; they suggested that there was a continuous gradation from mechanism S_E2(open) through S_E2(cyclic) to S_E2(co-ord). For the cases in which the electrophile may be denoted as E–N, this gradation may be illustrated as follows.

$$(2)$$

The observed sequences of reactivity (with respect to the alkyl group) of substrates in electrophilic substitutions were then correlated with the relative dominance of electrophilic attack and nucleophilic attack of the electrophile at the substrate. Tables 7–9 contain data on the relative reactivities of substrates in over 50 different series of electrophilic substitutions, and we shall use essentially the procedure of Abraham and Hill[45] in the interpretation of these series. All of the tabulated series are thought to proceed by mechanism S_E2, and the type of transition state postulated by the various investigators is also given. Only reactions which have been studied in homogeneous solution have been included in these tables**. Since most of the data in Tables 7–9 has been taken from tabulations given in previous chapters, it seems more convenient to refer back to these tabulations rather than to give the original sources again. The appropriate statistical corrections to observed rate coefficients have been applied where necessary. For example, in a comparison of the rate of cleavage of a methyl group from compounds Me_4Sn and $MeSnEt_3$ the observed rate coefficient for substitution of Me_4Sn is divided by a factor of 4 and the corrected rate coefficient then compared with that for the cleavage of the methyl group in compound $MeSnEt_3$.

In the interpretation of the data in Tables 7–9 we shall discuss first reactions proceeding by mechanism S_E2(open), then those proceeding by mechanism S_E2(co-ord), and finally discuss the rather complicated situation that arises in the intermediate region of S_E2(cyclic) mechanisms.

It is clear from the first eight series listed in Table 8 that in substitutions pro-

---

* As usual, the substrate is denoted as RMX_n, where X may be an electronegative group (as in RHgBr) or may itself be an alkyl group (as in R_4Sn or $RSnMe_3$).
** Results[46] on the rate of gas evolution on stirring dialkylmercurys with concentrated aqueous hydrochloric acid are hence not included, since in this case the reaction medium is probably heterogeneous.

TABLE 7

RELATIVE RATES OF ELECTROPHILIC SUBSTITUTIONS PROCEEDING BY MECHANISM S_E2

No.	Reactants	Temp. (°C)	Solvent	R						Transition state[a]	See page
				Me	Et	Pr^n	Bu^n	Pr^i	Bu^t		
1	$R_4Sn+HgCl_2$	25	MeOH	100	0.21	0.04	0.04	$<10^{-7}$		Open	77
2	$R_4Sn+HgI_2$	25	96% MeOH +4% H_2O	100	0.67	0.10	0.11	$<10^{-5}$		Open	70
3	R_4Sn+I_2	20	DMSO	100	4.4					Open	152
4	$R_4Pb+HClO_4$	25	AcOH	100	11	4.0	3.1			Open	125
5	R_4Sn+I_2	20	MeOH	100	12	1.4	0.57	0.06		Open	148
6	R_4Pb+I_2	20	DMSO	100	24					Open	167
7	R_4Sn+I_2	20	AcOH	100	33	3.9	3.3	0.03		Open	151
8	R_4Pb+I_2	35	MeCOMe	100	38					Open	167
9	R_4Pb+I_2	25	MeCN	100	39	8.3				Open	167
10	R_4Pb+I_2	25	MeOH	100	39	13				Open	167
11	$RHgI+H_3O^+$	110	H_2O	100	40	22	18	13		Open	114
12	$R_2Hg+HCl^b$	70	90% dioxan +10% H_2O	100	40				0.81	Open	118
13	$RHgX+HgX_2$	60–100	EtOH	100	42			6^c		Open	41
14	R_4Pb+I_2	25	Pr^nOH	100	45	6.2				Open	167
15	R_4Sn+Br_2	20	DMF	100	46	6.1				Open	153
16	R_4Pb+I_2	25	EtOH	100	48	7.2				Open	167

No.	Reaction	Temp. (°C)	Solvent								Ref.
19	$R_4Pb + AcOH$	25	AcOH	100	69	19	27		2.5		125
20	$R_4Sn + Br_2$	20	AcOH	100	83	12	10			Open	154
21	$R_4Pb + Br_2$	25	MeOH	100	85					Open	171
22	$R_3SnBr + Br_2$	20	AcOH	100	146	18					154
23	$R_2Zn +$ *p*-toluidine	35	Et_2O	100	244	102	65		149	Cyclic	112
24	$R_2Zn +$ *p*-toluidine	68	Pr^i_2O	100	244	81			163	Cyclic	112
25	$R_4Sn + I_2$	20	PhCl	100	600	78	60		560	Cyclic	156
26	$R_2Hg + AcOH$	25	AcOH	100	630	390	60		580[c]	Cyclic	115
27	$R_2Hg + HCl$	50	DMSO/dioxan	100	750	300			430	Cyclic	117
28	$R_4Sn + HCl$		Benzene	100	450	1700			300		128
29	$R_2Zn + PhHgCl$	35	Et_2O	100	1200	370			2200	Cyclic	64
30	$R_4Pb + I_2$	35	Benzene	100	1230	470					171
31	$R_4Sn + Br_2$	20	PhCl	100	1670	980	450		310	Co-ord	157
32	"$RMgBr$" + 1-hexyne	35	Et_2O	100	1670	980			3500	Cyclic	109
33	$R_2Mg + $ 1-hexyne	35	Et_2O	100	2080	2260			3000	Cyclic	111
34	$R_2Hg + HgI_2$	25	Dioxan	100	3040	1200			1920	Cyclic	51
35	$R_4Pb + I_2$	31	CCl_4	100		1200					171
36	$R_4Sn + Br_2$	20	CCl_4	100	9500	4400	5400		8.6×10^4	Cyclic	164
37	$R_4Sn + CrO_3$	20	AcOH	100	13500	9000	7000		2.0×10^4	Cyclic	189
38	$RB(OH)_2 + HOOH_2^+$ [d]	25	H_2O	100			3020	3.9×10^4		Co-ord	183
39	$RB(OH)_2 + HOO^-$ [d]	25	H_2O	100			3800	5.6×10^4		Co-ord	183
40	$RB(OH)_2 + HCrO_4^-$ [d]	30	H_2O	100	3×10^5			3×10^7	1.8×10^4[c]	Co-ord	184

[a] As suggested by the various investigators.
[b] The nature of the electrophile is not known; possibly H_3O^+ could be involved as well as HCl.
[c] These values are for $R = Bu^s$.
[d] Overall relative rate coefficients are given.

TABLE 8

STATISTICALLY CORRECTED RELATIVE RATES OF ELECTROPHILIC SUBSTITUTION AT R-M BONDS FOR SERIES OF SUBSTRATES IN WHICH THE LEAVING GROUP REMAINS CONSTANT

No.	Reactants[a]	Temp. (°C)	Solvent	R						Transition state[b]	See page
				Me	Et	Pr^n	Bu^n	Pr^i	Bu^t		
1	$R\text{-}SnBu^n_3 + HgI_2$	25	Aq. MeOH	100			0.1			Open	71
2	$R\text{-}SnBu^n_3 + I_2$	20	MeOH	100			0.7			Open	149
3	$R\text{-}SnEt_3 + I_2$	20	MeOH	100	6.2	1.8	1.7	0.1		Open	149
4	$R\text{-}SnMe_3 + I_2$	20	MeOH	100	14	3.1	7.5	0.6	$\simeq 0$	Open	149
5	$R\text{-}SnMe_3 + I_2$	20	AcOH	100	16	2.7	5.2	0.8	$\simeq 0.1$	Open	152
6	$R\text{-}SnMe_3 + Br_2$	20	AcOH	100	41	12	19	1.0		Open	155
7	$R\text{-}HgI + H_3O^+$	110	H_2O	100	40	22		13	0.81	Open	114
8	$R\text{-}HgX + HgX_2$	60–100	EtOH	100	42			6^c		Open	41
9	$R\text{-}SnMe_3 + Br_2$	20	PhCl	100	73	32	32	72	$\simeq 190$	Co-ord	158
10	$R\text{-}B(OH)_2 + HOOH_2^+$	25	H_2O	100			3020		3.9×10^4	Co-ord	183
11	$R\text{-}B(OH)_2 + HOO^-$	25	H_2O	100			3800	1.8×10^{4c}	5.6×10^4	Co-ord	183
12	$R\text{-}B(OH)_2 + HCrO_4^-$	30	H_2O	100	3×10^5				3×10^7	Co-ord	184

[a] The relative rates of cleavage refer only to cleavage of the group R.

[b] Type of S_E2 transition state as suggested by the various investigators.

[c] These values are for $R = Bu^s$.

ceeding by mechanism S_E2(open) the moving group (that is the alkyl group undergoing substitution) has a profound effect on the reactivity of the substrate. The leaving group, however, has but little effect (see Nos. 1–7, 9, and 10, in Table 9) unless it contains very bulky substituents, when the reactivity is reduced (Nos. 5 and 6, Table 9). To some extent, an increase in size of the leaving group actually increases the rate of cleavage of the methyl group (Nos. 1–3, 5–7, and 10, Table 9) by mechanism S_E2(open), but it is not clear whether this is a genuine effect or an artifact arising out of the strict application of the statistical factor correction*. In any case, a reasonable conclusion is that the moving group exerts a steric effect in the transition state through non-bonded interactions with the incoming electrophile and with the leaving group. These interactions result in a sequence of reactivity, with respect to the moving group, *viz.*

$$R = Me > Et > Pr^n \simeq Bu^n > Pr^i > Bu^t \tag{3}$$

If the leaving group also contains alkyl groups, then the larger alkyl groups might also contribute to the overall steric effect**. Hence for alkyls RMX_n (and R_nM), mechanism S_E2(open) will normally result in the reactivity sequence (3). Naturally, the quantitative aspect will depend on such factors as the size and shape of the electrophile, the size of the metal atom M, and the number of alkyl groups, R_{n-1}, attached to the leaving metal atom. Although, quantitatively, steric effects will also depend on whether the S_E2(open) substitution takes place with inversion or with retention of configuration, the steric sequence (3) seems to be qualitatively characteristic of both of these variants of the S_E2(open) mechanism and no attempt will be made to distinguish between them.

The effect of the moving group in substitutions proceeding by mechanism S_E2(co-ord) is completely the reverse to its effect in mechanism S_E2(open). Series Nos. 10–12 (Table 8) suggest that in mechanism S_E2(co-ord) the effect of the moving group is to accelerate reaction in the sense

$$R = Me < n\text{-alkyl} < Pr^i < Bu^t \tag{4}$$

Unfortunately, there is a lack of data on leaving group effects in mechanism S_E2(co-ord), but the analysis on p. 161 indicates that leaving groups of the type SnR_3 accelerate reaction in the sense $SnMe_3 < Sn(n\text{-alkyl})_3 < SnPr^i_3$. Thus for alkyls of type RMX_n (and R_nM), the sequence of reactivity in substitution by mechanism S_E2(co-ord) should approach that of sequence (4).

For substitutions proceeding by mechanism S_E2(cyclic), it might be expected that substituent effects would reflect, at least in part, the relative importance of electrophilic attack at the carbon atom undergoing substitution compared with nucleophilic attack at the metal atom in the leaving group, *i.e.* the position of

---

* See ref. 47, and the discussion on p. 72.
** For a detailed analysis see the discussion on p. 72 of the work of Abraham and Spalding[47].

TABLE

STATISTICALLY CORRECTED RELATIVE RATES OF ELECTROPHILIC SUBSTITUTION AT
SUBSTITUTION

No.	Reactants	Temp. (°C)	Solvent	R_3 in leaving group SnR_3				
				Me_3	Et_3	$Pr^n{}_3$	$Bu^n{}_3$	$Pr^i{}_3$
1	$Me–SnR_3+HgI_2$	25	Aq. MeOH	100			120	
2	$Me–SnR_3+Me_2Sn^{2+}$	25	MeOH	100	173	133	108	64
3	$Me–SnR_3+I_2$	20	MeOH	100	200	95	80	10
4	$Et–SnR_3+I_2$	20	MeOH	100	86			
5	$Me–SnR_3+I_2$	20	MeOH	100				
6	$Me–SnR_3+I_2$	20	AcOH	100				
7	$Me–SnR_3+Br_2$	20	AcOH	100				
8	$Me–SnR_3+Br_2$	20	PhCl	100				
9	$Et–SnR_3+I_2$	20	MeOH	100	86			
10	$Me–SnR_3+Me_2Sn^{2+}$	25	MeOH	100	173			

[a] Type of S_E2 transition state as suggested by the various investigators.

the transition state in series (2) p. 225. Hence the reactivity sequence (3) might be expected to obtain for compounds of type RMX_n (and R_nM) if in the cyclic transition state electrophilic attack is dominant, and the sequence (4) would then correspond to cyclic transition states in which nucleophilic attack is dominant. There is also likely to be a region in which reactivity sequences that are "mixtures" of (3) and (4) are observed.

The data in Table 7 illustrate these points quite well. Series Nos. 1–21 correspond, by and large, to sequence (3), and of these 21 series of substitutions no less than 16 have been suggested to proceed by mechanism S_E2(open)*. The remaining series, to which no detailed mechanisms have yet been assigned, are also quite likely to proceed by mechanism S_E2(open).

Series Nos. 22–37 (Table 7) show the gradual change-over from sequence (3) to sequence (4). Of these 16 series, 11 have been suggested to proceed via the S_E2-(cyclic) mechanism, 1 by the S_E2(co-ord) mechanism, and 4 have not been assigned a detailed mechanism. It seems to the author that these latter four series (Nos. 22, 28, 30, and 35 in Table 7) might well proceed by mechanism S_E2(cyclic). At first sight, it would appear unreasonable to assign this mechanism to reaction No. 22 (Table 7) — the brominolysis of trialkyltin bromides in solvent acetic acid — whereas the brominolysis of tetraalkyltins in the same solvent (No. 20,

* From the kinetic salt effect data in Table 6, mechanism S_E2(open) would be assigned to the series (Nos. 1, 2, 5, and 13, Table 7) since quite high values are calculated for the charge separation in the transition states. The transition state in series No. 20 (R = Pr^i, Table 7) has a much lower calculated charge separation and on this basis, series No. 20 (Table 7) could well follow a mechanism in the S_E2(open)–S_E2(cyclic) borderline region. From the general trend in Table 7 it seems as though a gradual change from mechanism S_E2(open) to S_E2(cyclic) is indeed taking place at around series No. 20 (Table 7).

9

Me–Sn AND Et–Sn BONDS FOR A SERIES OF SUBSTRATES IN WHICH THE GROUP UNDERGOING
REMAINS CONSTANT

Me_2Et	Me_2Pr^n	Me_2Bu^n	Me_2Pr^i	Me_2Bu^t	Et_2Pr^n	Et_2Bu^n	Et_2Pr^i	$MeEt_2$	Transition state[a]	See page
									Open	71
									Open	89
									Open	149
									Open	149
132	110	110	85	0.6					Open	149
143	104	104	95	≈8.2					Open	152
148	118	110	117						Open	155
193	210	216	197	≈140					Co-ord	155
					96	97	100		Open	149
112								173	Open	89

Table 7) has been suggested[30] to proceed via mechanism S_E2(open)*. The trialkyltin halides, though, are very much better Lewis acids than are tetraalkyltins, and hence the tin atom in transition state (VI) might well be able to coordinate to a bromine atom more easily than can the tin atom in (VII).

(VI)   (VII)

It is noticeable that the substitution of tetraalkyltins by bromine in solvent chlorobenzene (No. 31, Table 7) follows a much more "steric" sequence than does the substitution of the same series of substrates by bromine in solvent carbon tetrachloride (No. 36, Table 7) or by chromium trioxide in solvent acetic acid (No. 37, Table 7), and on this basis it could be suggested that the substitutions in series No. 31 (Table 7) follow mechanism S_E2(cyclic) rather than mechanism S_E2(co-ord).

It is apparent, however, that the various series of substitutions detailed in Tables 7–9 show sequences of reactivity, with respect to the alkyl group R, that can be interpreted rather well using the hypotheses of Gielen and Nasielski[44] and of Abraham and Hill[45]. On these theories, mechanism S_E2(open) is characterised by the steric sequence (3); this sequence arises[45, 47] mainly as a result of

---

* But see footnote on p. 230, indicating that the brominolysis of tetraalkyltins in solvent acetic acid proceeds by a mechanism on the S_E2(open)–S_E2(cyclic) borderline.

steric compression in the transition state between the alkyl group undergoing substitution and (*i*) the incoming electrophile and (*ii*) the leaving group. In contrast, mechanism S_E2(co-ord) is characterised by the polar sequence (4), and mechanism S_E2(cyclic) by various combinations or "mixtures" of the steric and polar sequences.

Other interpretations of sequence (3) have, however, been advanced. In connection with the substitution of tetraalkyltins in solvent methanol, Tagliavini and co-workers[48] suggest that such a sequence arises because the $+I$ effect of the alkyl groups attached to the metal atom reduces solvation of the initial state in the step

$$R_4Sn + S \rightleftharpoons R_4Sn \leftarrow S \tag{5}$$

The interaction of the solvent, S, with the substrate is supposed to take place prior to attack by the electrophilic reagent on the solvated species $R_4Sn \leftarrow S$. Since the $+I$ inductive effect of alkyl groups increases in the sense $R = Me < Et < Pr^n < Bu^n < Pr^i$, solvation of the solvent to the metal atom should be reduced along the same sequence. According to Tagliavini and co-workers[48], reaction should proceed with the most solvated species reacting the fastest, so that the sequence of reactivity $R = Me > Et > Pr^n > Bu^n > Pr^i$ is thus explained.

There are a number of difficulties with this hypothesis, however. First of all, solvation of the initial state will always[49] result in a reduction in the free energy of the initial state, and hence lead to an increase in $\Delta G^{\ddagger}$ and to a decrease in the rate coefficient. Thus if solvation of the initial state is reduced along the sequence $R = Me, Et, Pr^n, Bu^n, Pr^i$, this sequence should be that of increasing reactivity (and not, as observed, that of decreasing reactivity). The only way in which solvation can result in an increase in reactivity of a species is if the transition state is more solvated than the initial state (*i.e.* if the free energy of the transition state is lowered even more than is the free energy of the initial state). Secondly, the equilibrium (5) is set up independently of the attacking electrophile, so that if the effects of the alkyl groups were felt only with regard to solvation of the tetraalkyltin, then the overall pattern of steric effects should be independent of the nature of the electrophile. This seems inherently improbable, and data in Tables 7–9 on substitution of a given series of tetraalkyltins in a given solvent by different electrophiles suggest that the nature of the electrophile is a factor influencing the steric sequence. Thirdly, the hypothesis of Tagliavini and co-workers[48] does not easily explain the reactivity sequences observed when mixed alkyls (*i.e.* alkyls of type RMR'_3 or $R_2MR'_2$) are used. For example, data in Table 9 show that for reactions in the nucleophilic solvent methanol, variation in the leaving-group alkyl groups has little influence* on the statistically corrected rate coefficient, yet these alkyl groups should have a considerable influence on the equilibrium (5).

---

* Only when the alkyls in the leaving group are large (*e.g.* $SnPr^i_3$ or $SnMe_2Bu^t$) do they affect the rate coefficient greatly, yet even replacement of Et for Me in the moving group (Table 8) produces a decrease in the rate coefficient.

Bade and Huber[50] have put forward another interpretation of reactivity sequences with reference to the acidolysis of tetraalkyltins and tetraalkylleads in solvent methanol (Table 10). In contrast to Tagliavini and co-workers, these workers suggest that the sequences for acidolysis of the compounds R_4Sb and R_4Pb arise through solvation of the products. (Presumably, the argument implies that solvation of the transition state parallels solvation of the products.) Then inhibition of the solvation of the products, R_3M^+ or R_3MCl, by alkyl groups in the order of increasing inhibition $R = Me < Et < Pr^n < Bu^n < Pr^i$ will lead to stabilisation of the transition state (i.e. a reduction in free energy of the transition state) in the sense $R = Me > Et > Pr^n > Bu^n > Pr^i$ and a consequent reactivity sequence in the same sense. This suggestion of Bade and Huber[50] is clearly quite reasonable and it is a difficult problem to distinguish steric inhibition of solvation in the transition state from the steric interaction in the transition state, suggested by Abraham et al.[45,47]. Bade and Huber[50] restricted their discussion to the reactivity sequences in Table 10 (see also p. 127); extension to the sequences shown in Tables 7–9 again leads to difficulties in the interpretation of reactivities of mixed alkyls. It seems evident that the steric effect of the alkyl group undergoing substitution is very much greater than the steric effects of the alkyl groups on the leaving metal atom (see Tables 8 and 9), but there is no obvious reason why the moving group should inhibit solvation so much more than the leaving group. Indeed, if the SnR_3 or PbR_3 groups leave as $^+MR_3$, then the alkyl groups on the leaving group are nearer to the site of solvation (the positively charged metal atom) than is the moving group.

TABLE 10

RELATIVE RATES OF ACIDOLYSIS BY HCl IN SOLVENT METHANOL AT 50 °C[50]

	R				
Substrate	Me	Et	Pr^n	Bu^n	Pr^i
R_4Sn	100			35	30
R_4Pb	100	22	12	11	
R_3SnCl	100			230	110
R_3PbCl	100	116	591	356	

Tables 7–10 contain data only for alkyls of non-transition metals (Zn, Cd, and Hg have completely filled d-shells), but Gregory and Ingold[19] have determined rate coefficients for cleavage of Me–Au and Et–Au bonds by mercuric salts and methylmercuric salts. Relative rate coefficients are given in Table 11. Gregory and Ingold suggested that the reactions of $RAuPPh_3$ with mercuric salts in solvent dioxan ($\varepsilon = 2.1$) proceeded by mechanism S_E2(open), a conclusion which if correct would be in conflict with the hypotheses of Gielen and Nasielski[44] and of Abraham

TABLE 11

RELATIVE RATES OF CLEAVAGE OF R–Au BONDS AT 25 °C[19]

Reactants	Solvent	R	
		Me	*Et*
$RAuPPh_3+HgCl_2$	Dioxan	100	340
$RAuPPh_3+HgBr_2$	Dioxan	100	410
$RAuPPh_3+HgI_2$	Dioxan	100	440
$RAuPPh_3+MeHgOAc$	DMSO	100	19
$RAuPPh_3+MeHgCl$	DMSO	100	25
$RAuPPh_3+MeHgBr$	DMSO	100	67
$RAuPPh_3+MeHgI$	DMSO	100	93

and Hill[45]. No experiments on kinetic salt effects were carried out by Gregory and Ingold[19] and it seems to the author that, especially in view of the very low dielectric constant of the solvent, mechanism S_E2(cyclic) could well obtain for these substitutions in solvent dioxan.

On the basis of the various reactions listed in Tables 7 and 8, it would seem reasonable to assign mechanism S_E2(cyclic) to the metal-for-metal exchange (Table 11) in solvent dioxan, and to suggest that the exchanges in solvent DMSO are more likely to proceed by mechanism S_E2(open).

5.2 SOLVENT EFFECTS ON THE RATES OF ELECTROPHILIC SUBSTITUTIONS

5.2.1 Substitution by mechanism S_E2

Most of the information on this subject refers to solvent effects on the substitution of tetraalkyltins by mercuric chloride (see Chapter 6, Section 1.5, p. 70) and to the iododemetallation of tetraalkyltins and tetraalkylleads (see Chapter 8, Section 6, p. 173). Data on solvent effects for a number of reactions are summarised in Table 12. There is only a partial correlation of the rate coefficients with any of the usual parameters[51] of solvent polarity; in particular, the solvents acetonitrile and acetone increase the value of the various rate coefficients to a much greater extent than would be predicted on the basis of solvent dielectric constant (ε), or Z values, or E_T values, etc.[51]. This may be seen especially for the iododemetallation of tetramethyllead, reaction (6) (R = Me)

$$R_4Pb+I_2 \rightarrow RI+R_3PbI \tag{6}$$

Gielen and Nasielski[44] have defined a solvent parameter, X, as

$$\log (k/k_0) = pX \tag{7}$$

where k is the rate coefficient for electrophilic substitution of a given substrate in a solvent, k_0 is the rate coefficient for the same electrophilic substitution in solvent acetic acid, and p is some parameter that depends on the particular reaction. For the bromodemetallation of tetramethyltin, p was defined as unity. By definition, the value of X for solvent acetic acid is 0. Using equation (7), Gielen and Nasielski[44] calculated values of X to be

Solvent	CCl$_4$	PhCl	AcOH	MeOH	
X	-4.8	-1.9	0	0.91	(8)

They also observed[44] that for the above solvents there was a linear relation between X and $\log [k(\text{Me})/k(\text{Et})]$, where $k(\text{Me})/k(\text{Et})$ denotes the relative reactivities of tetramethyltin to tetraethyltin towards a given electrophile. For a number of solvents, the absolute values of rate coefficients were not determined* and values of X were calculated from the above linear relationship[44, 52]. These indirectly determined (X^i) values are

Solvent	MeCN	DMF	DMSO	
X^i	0.04	0.8	1.6	(9)

Where absolute values of rate coefficients are known for reactions run in the solvents listed in sequence (9) it is apparent that the X^i values do not correlate well with the observed rate coefficients; for example the rate of reaction (6) in solvent acetonitrile is much greater than predicted by $X(\text{MeCN}) = 0.04$, and the rate in solvent DMSO less than predicted by $X(\text{DMSO}) = 1.6$.

Two factors were put forward by Gielen and Nasielski[44] to account for solvent effects on the rate of halogenodemetallation of tetraalkyltins, viz. (a) a general stabilisation of charge-separation in the transition state, this stabilisation increasing with increasing solvent dielectric constant, ε, and (b) a specific solvation of the leaving tin atom in the transition state by nucleophilic solvents, e.g. as in (VIII).

(VIII)   (IX)

---

* The rate coefficients for halogenodemetallation of tetraalkyltins and tetraalkylleads in DMSO are actually available, but Gielen and Nasielski[52] preferred to use the indirect method to determine X^i. The direct method would lead to a value of $\simeq 1.1$ for X for solvent DMSO.

TABLE 12

SOLVENT EFFECTS ON THE RATE OF SOME BIMOLECULAR ELECTROPHILIC SUBSTITUTIONS

Solvent	ε	$J(^{117}Sn-H)$*	Second-order rate coefficients at 25 °C ($l.mole^{-1}.sec^{-1}$)[a]			
			1	2	3	4
MeCN	36.0	61.2	2.4×10^5	39	1.9×10^{-2}	
MeCOMe	20.5	62.0	3.8×10^4			24
MeOH	32.6	64.5	1.0×10^4		3.3×10^{-3}	
DMSO	46.7	66.9	1×10^4			
EtOH	24.3		2.8×10^3	3.5×10^{-2}		3.9
Pr^{n}OH	20.5		1.6×10^3			
MeCO$_2$H	6.2	58.8	3.0×10^4			
PhCl	5.6	55.0	$\simeq 17^b$			
C$_6$H$_6$	2.3		1.1			
CCl$_4$	2.2	55.0	2.6×10^{-2}			

[a] 1, For reactants Me$_4$Pb+I$_2$, data taken from Table 32, p. 173; 2, for reactants CH$_2$=CHCH$_2$ GeEt$_3$+HgBr$_2$, ref. 35; 3, for reactants Et$_4$Sn+HgCl$_2$, refs. 8 and 9; 4, for reactants Bu$^s_2$Hg+ HgBr$_2$, ref. 15.
[b] Estimated value, using the quoted X-values.
* $J(^{117}Sn-H)$ values taken from refs. 44 and 52.

The coupling constants $J(^{117}Sn-H)$ and $J(^{119}Sn-H)$ in trimethyltin halides are solvent-dependent and appear to reflect the nucleophilic power of the solvent towards the tin atom; the larger are the coupling constants the nearer[53] does the trimethyltin halide approach the trigonal bypyramid (IX). Since the environments of the tin atoms in (VIII) and (IX) are quite similar, the variation of $J(^{117}Sn-H)$ or $J(^{119}Sn-H)$ with solvent might be expected to provide also an estimate of the power of the solvent to solvate the leaving tin atom in the transition state. Table 12 contains data on solvent dielectric constants and on $J(^{117}Sn-H)$ constants for trimethyltin bromide, but it is evident that the dipolar aprotic solvents do not conform to the fairly regular pattern observed for the non-polar solvents and the hydroxylic solvents. Acetonitrile and acetone and also acetic acid are more powerful solvents than expected from their ε and J values, and dimethylsulphoxide a much weaker solvent than expected.

Undoubtedly the factors suggested by Gielen and Nasielski are important; nucleophilic co-ordination of solvent to tin is probably the reason why compound (X) suffers restricted rotation in solvent carbon tetrachloride (when the intra-molecularly-bridged species (Xa) is present) but not in solvents such as alcohols, ethers, or pyridine/carbon tetrachloride (when the species present is now (Xb), with the tin atom five-co-ordinate through external coordination)[54].

(Ⅹa)   (Ⅹb)

The difficulty in dealing with solvent influences on reaction rates is that the free energy of activation, $\Delta G^{\ddagger}$, depends not only on the free energy of the transition state but also on the free energy of the initial state. It is therefore of considerable interest to dissect solvent influences on $\Delta G^{\ddagger}$ into initial-state and transition-state contributions. As far as electrophilic substitution at saturated carbon is concerned, the only cases for which such a dissection has been carried out are (*a*) for the substitution of tetraalkyltins by mercuric chloride in the methanol–water solvent system (see page 79), and (*b*) for the iododemetallation of tetraalkylleads in a number of solvents (see p. 173). Data on the latter reaction (6) are more useful from the point of view of the correlation of transition-state effects with solvent properties, and in Table 13 are listed values of $\Delta G_t^0(Tr)$, the free energy of transfer (on the mole fraction scale) of the tetraalkyllead/iodine transition states from methanol to other solvents.

TABLE 13

FREE ENERGIES OF TRANSFER, ON THE MOLE FRACTION SCALE, FROM METHANOL TO OTHER SOLVENTS OF TETRAALKYLLEAD/IODINE TRANSITION STATES (kcal.mole^{-1}) AT 298 °K^a

Solvent	ΔG_t^0		ε	$J(^{117}Sn-H)$
	$[Me_4Pb/I_2]^{\ddagger}$	$[Et_4Pb/I_2]^{\ddagger}$		
MeCN	−1.3	−1.1	36.0	61.2
MeCOMe	0.0	−0.1	20.5	62.0
MeOH	0	0	32.6	64.5
EtOH	0.2	−0.2	24.3	
Pr^{n}OH	0.3	−0.1	20.5	
MeCO$_2$H	−0.1	−0.2	6.2	58.8
C$_6$H$_6$	3.7	1.1	2.3	55.0^b
CCl$_4$	6.5	3.5	2.2	55.0

^{a} Taken from Table 34, p. 174.
^{b} Value for PhCl.

There is little correlation of the $\Delta G_t^0(Tr)$ values with the parameters ε and $J(^{117}Sn-H)$, although the free energy of the transition states tends to be lower when the values of $J(^{117}Sn-H)$ are higher (especially when aprotic solvents alone are considered). If, indeed, $J(^{117}Sn-H)$ does measure the extent of solvation at the leaving tin atom, then there should be such a correlation. Neither ε nor $J(^{117}Sn-H)$ correlates well with $\Delta G^0_t(Tr)$ values when both hydroxylic and aprotic solvents are considered. Abraham[55] has suggested that electrically neutral transition states in which there is a high separation of charge (say 0.8 units) are stabilised by hydroxylic solvents in comparison with aprotic solvents. On the other hand, polarisable transition states in which there is a lower separation of charge (say 0.3–0.4 units) are more stabilised by dipolar aprotic solvents than by hydroxylic solvents. From the data in Table 13, it appears as though the tetraalkyllead/iodine transition states (in solvents methanol, acetonitrile, and acetone) behave as if they also carried a charge separation in the range $\simeq$ 0.3–0.4 units of charge, a conclusion in agreement with the value of $\simeq 0.3$ as obtained from results of kinetic salt effects on the Me_4Sn/I_2 transition state using solvent methanol. Such values of charge separation are compatible with the assignment of mechanism $S_E2(open)$ to halogenodemetallations in polar solvents; as the solvent becomes less polar, the mechanism of halogenodemetallation would be expected gradually to change to mechanism $S_E2(cyclic)$, with a concomitant reduction in the degree of charge separation. Unfortunately, the data in Table 13 are not extensive enough to warrant speculation on this point.

5.2.2 Substitution by mechanism S_E1

Except for the S_E1 reactions of pyridiomethyl–metal compounds in aqueous solution, nearly all the S_E1 reactions amenable to kinetic study have proceeded in dipolar aprotic solvents such as dimethylsulphoxide (DMSO), DMSO/dioxan, and probably also in acetonitrile. There is no obvious reason why such solvents should stabilise the produced carbanions to a much greater extent than hydroxylic solvents, and so the accelerating influence of these aprotic solvents on S_E1 reactions is not easily accounted for. The substrates involved have, to date, been either alkylmercury or alkylgold compounds. It is possible, therefore, that the influence of solvents such as DMSO is to stabilise the counter cation produced in reactions such as

$$PhCH(CO_2Et)HgBr \rightleftharpoons PhC\overset{-}{H}(CO_2Et) + HgBr^+ \qquad (10)$$

$$BuC(CN)(CO_2Et)AuPPh_3 \rightleftharpoons Bu\overset{-}{C}(CN)CO_2Et + AuPPh_3^+ \qquad (11)$$

(In other words, the developing positive charges in the transition states are stabilised by such solvents.) Dipolar aprotic solvents, especially DMSO and DMF, dissolve mercuric halides very readily, and also co-ordinate strongly to the $PhHg^+$ cation. Reutov and co-workers[57] have determined the equilibrium constants for the reaction

$$PhHg^+ + L \underset{}{\overset{K}{\rightleftharpoons}} PhHgL^+ \tag{12}$$

using solvent dichloromethane; some values for various ligands, L, are

L	Pyridine	HMPT	DMSO	DMF	MeOH	MeCN	THF	MeCOMe
log K	8.1	6.2	3.3	2.8	1.4	1.1	0.6	small

The efficacy of DMSO in promoting S$_E$1 heterolyses of alkyl–metal bonds could thus well be due to a strong tendency to solvate the partial positive charge at the leaving metal centre.

6. Metal-metal bonding in S$_E$2 transition states

Matteson[58] has suggested that transition states for the generalised electrophilic substitution

$$R–M + E–N \rightarrow R–E + M–N \tag{13}$$

may be represented as in (XI)

(XI)

with a three-centre bond between the carbon atom undergoing displacement, the leaving metal atom, M, and the entering electrophile. If E is itself a metal atom, as in the metal-for-metal exchange reactions, then the M – – – E bond is a metal–metal bond. As Matteson himself observes "Simple direct evidence for metal–metal bonding in transition states for electrophilic displacement is not yet available . . .", so that it is difficult at the moment to discuss usefully the concept, especially in the absence of any operational definition of metal–metal bonding in transition states.

ACKNOWLEDGEMENTS

I thank Dr. J. I. Bullock and Dr. L. F. Larkworthy for their helpful comments on parts of Chapters 6 and 7, Mrs. S. Manasseh for so efficiently typing the manuscript, and my wife for her help and encouragement during the writing of the manuscript.

References pp. 240–241

REFERENCES

1 H. WEINGARTEN AND J. R. VAN WAZER, *J. Am. Chem. Soc.*, 88 (1966) 2700.
2 E. D. HUGHES, C. K. INGOLD, F. G. THORPE AND H. C. VOLGER, *J. Chem. Soc.*, (1961) 1133.
3 O. A. REUTOV, T. A. SMOLINA AND V. A. KALYAVIN, *Dokl. Akad. Nauk SSSR*, 139 (1961) 389; *Proc. Acad. Sci. USSR*, 139 (1961) 697; *Zh. Fiz. Khim.*, 36 (1962) 119; *Russ. J. Phys. Chem.*, *Engl. Transl.*, 36 (1962) 59.
4 R. E. DESSY AND Y. K. LEE, *J. Am. Chem. Soc.*, 82 (1960) 689.
5 R. E. DESSY, Y. K. LEE AND J.-Y. KIM, *J. Am. Chem. Soc.*, 83 (1961) 1163.
6 O. A. REUTOV, T. P. KARPOV, É. V. UGLOVA AND V. A. MALYANOV, *Izv. Akad. Nauk SSSR, Ser. Khim.*, (1964) 1580; *Bull. Acad. Sci. USSR*, (1964) 1492.
7 M. H. ABRAHAM AND F. BEHBAHANY, *J. Chem. Soc. A*, (1971) 1469.
8 M. H. ABRAHAM AND G. F. JOHNSTON, *J. Chem. Soc. A*, (1970) 188.
9 M. H. ABRAHAM AND M. J. HOGARTH, *J. Chem. Soc. A*, (1971) 1474.
10 M. GIELEN AND J. NASIELSKI, *Bull. Soc. Chim. Belges*, 71 (1962) 32.
11 H. B. CHARMAN, E. D. HUGHES, C. K. INGOLD AND F. G. THORPE, *J. Chem. Soc.*, (1961) 1121.
12 R. E. DESSY AND F. E. PAULIK, *J. Chem. Educ.*, 40 (1963) 185.
13 L. H. TOPORCER, R. E. DESSY AND S. I. E. GREEN, *J. Am. Chem. Soc.*, 87 (1965) 1236.
14 A. JUBIER, É. HENRY-BASCH AND P. FRÉON, *Compt. Rend. Ser. C*, 267 (1968) 842; A. JUBIER, G. EMPTOZ, É. HENRY-BASCH AND P. FREON, *Bull. Soc. Chim. France*, (1969) 2032.
15 H. B. CHARMAN, E. D. HUGHES AND C. K. INGOLD, *J. Chem. Soc.*, (1959) 2530.
16 M. D. RAUSCH AND J. R. VAN WAZER, *Inorg. Chem.*, 3 (1964) 761.
17 R. G. COOMBES AND M. D. JOHNSON, *J. Chem. Soc. A*, (1966) 1805.
18 I. P. BELETSKAYA, I. L. ZHURAVLEVA AND O. A. REUTOV, *Zh. Org. Khim.*, 4 (1968) 729; *J. Org. Chem. USSR*, 4 (1968) 711.
19 B. J. GREGORY AND C. K. INGOLD, *J. Chem. Soc. B*, (1969) 276.
20 C. EABORN AND S. H. PARKER, *J. Chem. Soc.*, (1955) 126.
21 R. W. BOTT, C. EABORN AND T. W. SWADDLE, *J. Chem. Soc.*, (1963) 2342.
22 O. A. REUTOV, B. PRAISNER, I. P. BELETSKAYA AND V. I. SOKOLOV, *Izv. Akad. Nauk SSSR, Otd. Khim. Nauk*, (1963) 970; *Bull. Acad. Sci. USSR, Div. Chem. Sci.*, (1963) 884.
23 O. A. MAKSIMENKO, I. P. BELETSKAYA AND O. A. REUTOV, *Izv. Akad. Nauk SSSR, Ser. Khim.*, (1966) 662; *Bull. Acad. Sci. USSR*, (1966) 627.
24 O. A. REUTOV, T. A. SMOLINA AND V. A. KALYAVIN, *Dokl. Akad. Nauk SSSR*, 155 (1964) 596; *Proc. Acad. Sci. USSR*, 155 (1964) 273.
25 O. A. REUTOV, T. A. SMOLINA AND V. A. KALYAVIN, *Dokl. Akad. Nauk SSSR*, 139 (1961) 389; *Proc. Acad. Sci. USSR*, 139 (1961) 697.
26 O. A. REUTOV, I. P. BELETSKAYA AND G. A. ARTAMKINA, *Zh. Fiz. Khim.*, 36 (1962) 3582; *Russ. J. Phys. Chem., Engl. Transl.*, 36 (1962) 1407; *Izv. Akad. Nauk SSSR, Ser. Khim.*, (1964) 1737; *Bull. Acad. Sci. USSR*, (1964) 1651.
27 O. A. REUTOV, G. A. ARTAMKINA AND I. P. BELETSKAYA, *Dokl. Akad. Nauk SSSR*, 153 (1963) 588; *Proc. Acad. Sci. USSR*, 153 (1963) 939.
28 O. A. REUTOV, I. P. BELETSKAYA AND T. P. FETISOVA, *Dokl. Akad. Nauk SSSR*, 155 (1964) 1095; *Proc. Acad. Sci. USSR*, 155 (1964) 347.
29 A. N. NESMEYANOV, A. E. BORISOV AND I. S. SAVEL'EVA, *Dokl. Akad. Nauk SSSR*, 155 (1964) 603; *Proc. Acad. Sci. USSR*, 155 (1964) 280.
30 M. GIELEN AND J. NASIELSKI, *Bull. Soc. Chim. Belges*, 71 (1962) 601.
31 G. C. ROBINSON, *J. Org. Chem.*, 28 (1963) 843.
32 S. WINSTEIN AND T. G. TRAYLOR, *J. Am. Chem. Soc.*, 77 (1955) 3747.
33 G. PILLONI AND G. TAGLIAVINI, *J. Organometal. Chem.*, 11 (1968) 557.
34 A. LORD AND H. O. PRITCHARD, *J. Phys. Chem.*, 70 (1966) 1689.
35 R. M. G. ROBERTS, *J. Organometal. Chem.*, 12 (1968) 97.
36 R. E. DESSY AND J.-Y. KIM, *J. Am. Chem. Soc.*, 83 (1961) 1167.
37 L. PAULING, *The Nature of the Chemical Bond*, Cornell University Press, New York, 1960, 3rd edn., p. 93.
38 M. WITANOWSKI AND J. D. ROBERTS, *J. Am. Chem. Soc.*, 88 (1966) 737.
39 C. K. INGOLD, *Structure and Mechanism in Organic Chemistry*, Bell, London, 1953.

40 M. H. ABRAHAM AND T. R. SPALDING, *Chem. Commun.*, (1968) 46; *J. Chem. Soc. A*, (1969) 784.

41 R. E. DESSY, G. F. REYNOLDS AND J.-Y. KIM, *J. Am. Chem. Soc.*, 81 (1959) 2683.

42 M. GIELEN, J. NASIELSKI, J. E. DUBOIS AND P. FRESNET, *Bull. Soc. Chim. Belges*, 73 (1964) 293.

43 E. D. HUGHES, C. K. INGOLD AND R. M. G. ROBERTS, *J. Chem. Soc.*, (1964) 3900.

44 M. GIELEN AND J. NASIELSKI, *J. Organometal. Chem.*, 1 (1963) 173.

45 M. H. ABRAHAM AND J. A. HILL, *Chem. Ind. (London)*, (1965) 561; *J. Organometal. Chem.*, 7 (1967) 11.

46 C. S. MARVEL AND H. O. CALVERY, *J. Am. Chem. Soc.*, 45 (1923) 820.

47 M. H. ABRAHAM AND T. R. SPALDING, *J. Chem. Soc. A*, (1969) 399.

48 G. PLAZZOGNA, S. BRESADOLA AND G. TAGLIAVINI, *Inorg. Chim. Acta*, 2 (1968) 333.

49 J. E. LEFFLER AND E. GRUNWALD, *Rates and Equilibria of Organic Reactions*, Wiley, New York, 1963, p. 34.

50 V. BADE AND F. HUBER, *J. Organometal. Chem.*, 24 (1970) 387.

51 C. REICHARDT, *Angew. Chem. Intern. Ed. Engl.*, 4 (1965) 29.

52 M. GIELEN AND J. NASIELSKI, *J. Organometal. Chem.*, 7 (1967) 273.

53 G. MATSUBAYASHI, Y. KAWASAKI, T. TANAKA AND R. OKAWARA, *Bull. Chem. Soc. Japan*, 40 (1967) 1566.

54 F. B. BOER, J. J. FLYNN, H. H. FREEDMAN, S. V. MCKINLEY AND V. R. SANDEL, *J. Am. Chem. Soc.*, 89 (1967) 5068.

55 M. H. ABRAHAM, *J. Chem. Soc. B*, (1971) 299.

56 F. R. JENSEN AND B. RICKBORN, *Electrophilic Substitution of Organomercurials*, McGraw-Hill, New York, 1968.

57 K. P. BUTIN, A. N. RYABTSEV, V. S. PETROSYAN, I. P. BELETSKAYA, AND O. A. REUTOV, *Dokl. Akad. Nauk SSSR*, 183 (1968) 1328; *Proc. Acad. Sci. USSR*, 183 (1968) 1107.

58 D. S. MATTESON, *Organometal. Chem. Rev.*, 4A (1969) 263.

Index

A

acetate ion, and PhCH$_2$Cr(III)+acid, 132
—, and RB(OH)$_2$+H$_2$O$_2$, 183
—, effect on boronic esters+HgCl$_2$, 66, 68, 69
—, — on MeHgBr+HgBr$_2$, 43
acetic acid, and Me(Me$_3$)AuPPh$_3$+HgX$_2$, 100–102, 104
—, Br$_2$/Br$^-$ equilibrium in, 139
—, effect on boronic esters+HgCl$_2$, 66, 68, 69
—, MeCH=CHCH$_2$HgBr+HClO$_4$ in, 204
—, PbR$_4$+I$_2$ in, 166, 167, 173, 174, 227, 236, 237
—, pK of, 7
—, RB(OH)$_2$+H$_2$O$_2$ in, 183
—, reaction+BEt$_3$, 123
—, —+HgR$_2$, 13, 115, 116, 219, 222, 223, 227
—, —+Me$_3$PbOAc, PbR$_4$, 124, 125, 219, 226, 227
—, —+PhCH$_2$Cr(III), 132
—, —+SnR$_4$, 219
—, RHgBr+Br$_2$ in, 136
—, SnR$_4$+Br$_2$ in, 154, 155, 222, 223, 228, 230, 235
—, SnR$_4$+CrO$_3$ in, 189, 190, 227, 231
—, SnR$_4$+I$_2$ in, 151, 152, 173, 174, 226, 228, 230
acetolysis, 1
acetone, addition to PhHg$^+$, 239
—, alkyl exchange in RHgX in, 43
—, Br$_2$/Br$^-$, I$_2$/I$^-$ equilibrium in, 139, 209
—, HgCl$_2$+RHgCl in, 40
—, HgR$_2$+HgBr$_2$ in, 48, 49, 234, 236
—, MeAuPPh$_3$+HgX$_2$ in, 100, 101, 213
—, Me$_3$AuPPh$_3$+HgBr$_2$ in, 104
—, Me$_2$SnCl$_2$+SnMe$_4$ in, 89
—, PbR$_4$+I$_2$ in, 166, 167, 169, 173–175, 226, 234, 236
—, PhCHMeCl racemisation in, 53
—, Ph$_3$SnCH$_2$CH=CHR+I$_2$ in, 208–210
—, pK of, 7
—, RB(OH)$_2$+HgCl$_2$ in, 65
—, R$_3$SnBr/I$^-$ equilibrium in, 160
—, Sn allyl$_4$+I$_2$ in, 206–208
acetonitrile, addition to PhHg$^+$, 239
—, Br$_2$/Br$^-$, I$_2$/I$^-$ equilibrium in, 139, 208, 209
—, Et$_3$Ge(Si, Sn)CH$_2$CH=CH$_2$+HgX$_2$ in, 198, 200, 219, 234, 236
—, heat of solution of SnEt$_4$ in, 151

—, HgBz$_2$+HCl in, 121
—, Me$_3$SiCH$_2$CH=CH$_2$+HgX$_2$ in, 197, 198
—, Me$_2$SnCl$_2$+SnMe$_4$ in, 89
—, PbR$_4$+Br$_2$ in, 235
—, PbR$_4$+I$_2$ in, 166–170, 173–175, 226, 234, 236, 237
—, PhCH$_2$HgCl+I$_2$ in, 142
—, Ph$_3$SnCH$_2$CH=CHR+I$_2$ in, 208–210
—, pK of, 7
—, SnEt$_4$+HgX$_2$ in, 86, 87, 211, 213, 214, 222, 234, 236
acetophenone, pK of, 7
acid dissociation constant, of boronic esters, 67
—, of carbon compounds, 67
—, of HCl in MeOH, 126
—, of RCOOH, 123
—, —, and rate of BEt$_3$+HgPh$_2$, 212
activation energy, *see also* enthalpy of activation
—, of CH$_2$=CHCH$_2$HgCl+HCl, 203
—, of HgBu$_2$+HgX$_2$, 49
—, of HgBu$_2$+RHgBr, 58
—, of HgMe$_2$+I$_2$, 145
—, of HgR$_2$+HCl, 117, 119
—, of hydrolysis of NC$_5$H$_4$CH$_2$Mn(CO)$_5$, 32
—, of inversion of metal alkyls, 33
—, of Me$_3$SiCH$_2$CH=CH$_2$+HgCl$_2$, 198
—, of Me$_3$SiCH$_2$Ph+OH$^-$, 34
—, of Me$_3$SiEt disproportionation, 188
—, of Me$_2$SnCl$_2$+SnR$_4$, 89
—, of p-NO$_2$C$_6$H$_4$CH$_2$HgBr+HgBr$_2$, 25
—, of PbEt$_4$+AcOH, 125, 126
—, of PhCH=CHCH$_2$Si(Sn)R$_3$+base, 196
—, of PhCH(CO$_2$Et)HgBr+HgBr$_2$, 24, 47
—, of PhCH$_2$HgBr+HgBr$_2$, 44
—, of PhCH$_2$HgCl+Br$_2$, 144
—, of PhCH$_2$HgCl+HCl, 121
—, of RHgBr+HgBr$_2$, 40, 41
—, of R$_3$SnCH$_2$CH=CHR′+HgI$_2$, 200
—, of SnMe$_4$+I$_2$, 147
—, of XC$_6$H$_4$CH$_2$HgCl+I$_2$, 140, 142, 143
activity coefficients, of SnR$_4$ and transition states with HgI$_2$, 74
allylmercuric chloride, reaction+HCl, 203
allylmercuric iodide, reaction+HClO$_4$, 201–203
aluminium bromide, catalysis of Me$_3$SiR disproportionation, 188, 189
aluminium tributyl, reaction+Ph$_2$CO, 188

aluminium triethyl, reaction+PhCN, XC$_6$H$_4$CO$_2$Et, 185
aluminium tri-iso-hexyl, inversion of, 33, 221
aluminium trimethyl, reaction+PhCN, 185, 186
—, —+Ph$_2$CO, 186–188
ammonia, reaction+RHgBr, 53–57, 216
ammonium bromide, and PhCH$_2$HgCl+Br$_2$, 143
antimony trichloride, reaction+SbMe$_3$, 90, 91, 211
antimony trimethyl, reaction+SbCl$_3$, 90, 91, 211
association factor, for R$_2$Mg, 111
—, for RMgX, 64, 109
autoperturbation, in SnR$_4$+Br$_2$, 163

B

benzaldehyde, reaction+ZnR$_2$, 180
benzene, AlMe$_3$+Ph$_2$CO in, 186, 187
—, ArSiR$_3$+Br$_2$ in, 162
—, HgR$_2$+HgI$_2$ in, 51, 211
—, Me$_3$SiR disproportionation in, 188, 189
—, PbR$_4$+HCl in, 126, 227
—, PbR$_4$+I$_2$ in, 170, 171, 173–175, 227, 236, 237
—, PhCH$_2$HgCl+I$_2$ in, 142
—, pK of, 7
—, SnBu$_4$+I$_2$ in, 165
—, SnR$_4$+HCl in, 124, 128
—, ZnR$_2$+Ph$_2$NH in, 113
benzoic acid, reaction+BEt$_3$, 123
benzonitrile, reaction+AlR$_3$, 185, 186
benzophenone, reaction+AlR$_3$, 186–188
—, —+MeMgBr, 180
benzyl alcohols, reaction+CdEt$_2$, 112–114, 212
benzylboronic acids, reaction+HgCl$_2$, 66–68
—, —+H$_2$O$_2$, 183
benzylchromium ion, reaction+acids, HgCl$_2$, 132
benzyl cyanide, reaction+MeAlX$_2$, 185
benzylmercuric bromides, Hg exchange+ HgBr$_2$, 44–46, 211, 216, 217
—, reaction+I$_2$, 143, 216
benzylmercuric chlorides, reaction+Br$_2$, 143, 144
—, —+HCl, 121
—, —+I$_2$, 139–142
1-bicycloheptylboronic acid, reaction+H$_2$O$_2$, 183
bipyridyl, effect on ZnR$_2$+Ph$_2$NH, 112, 113
bis(triethylphosphine)halomethylplatinum (II), reaction+acid, 129–132
bond dissociation energy, of PbR$_4$, SnR$_4$, 127

bond length, in PbMe$_4$ and SnMe$_4$, 170
boronic acids, reaction+H$_2$O$_2$, 182–184, 227
boronic esters, reaction+acids, 108
—, —+HgCl$_2$, 66–69
boron triethyl, reaction+RCOOH, 122–124, 212
boron tri-exo-norbornyl, reaction+acid, 108
bromide ion, and BzHgCl+Br$_2$, 143, 144
—, catalysis by, 19, 26, 27
—, effect on BzHgBr+HgBr$_2$, 46
—, — on HgBr$_2$+MeAuPPh$_3$, 101
—, — on HgBr$_2$+NC$_5$H$_4$CH$_2$Cr(III), 92–94
—, — on MeHgBr+HgBr$_2$, 42
—, — on SnR$_4$+Br$_2$, 153–155
—, equilibrium+Br$_2$, 139, 153, 154
bromine, equilibrium+Br$^-$, 139, 153, 154
—, reaction+ArSiR$_3$, 162
—, —+organometallics, 136, 137
—, —+PbR$_4$, 171, 222, 227
—, —+PhCH$_2$HgCl, 143, 144
—, —+RHgBr, 135, 136
—, —+R$_3$SnBr, 154, 227
—, —+SnR$_4$, 18, 153–165, 171, 222, 223, 227, 228, 230, 231, 235
butane, pK of, 7
butanol, heat of solution of SnEt$_4$ in, 151
—, HgBz$_2$+HCl in, 121
—, PhCH$_2$HgCl+I$_2$ in, 142
—, RHgCl+HgCl$_2$ in, 40
—, SnEt$_4$+HgX$_2$ in, 84–86, 211, 222
butylboronic acid, reaction+CrO$_3$, 184, 227, 228
—, —+HgCl$_2$, 67
—, —+H$_2$O$_2$, 183, 227, 228
n-butyl bromide, SnBu$_4$+I$_2$ in, 165
butyl chloride, ΔG and ΔS for transfer from MeOH, 81, 82, 84
sec-butyl cobaltoximine, reaction+Br$_2$, I$_2$, 137
butylmercuric acetate, Hg exchange+ Hg(OAc)$_2$, 40, 41, 215
—, reaction+HgBu$_2$, 58, 59
butylmercuric bromide, reaction+Br$_2$, 135, 136, 144, 145
—, —+HgBu$_2$, 58, 59, 222, 228
butylmercuric iodide, reaction+H$_2$SO$_4$, 114, 115, 226, 228
—, —+I$_2$, 137
butylmercuric nitrate, reaction+HgBu$_2$, 58, 59, 215, 222

C

cadmium diethyl, acidolysis of, 220
—, reaction+XC$_6$H$_4$CH$_2$OH, 112–114, 212
cadmium iodide, effect on ArHgBr(Cl)+I$_2$,

139, 140, 142, 143

4-camphylmercuric iodide, reaction+I_2, 137

α-carbalkoxybenzylmercuric bromides, reaction+NH_3, 53–57

α-carbethoxybenzylmercuric bromide (Chloride), racemisation of, 24–26

—, reaction+Ar_3CBr, 180–182, 216

—, —+$HClO_4$/Cl^-, 121, 122

—, —+HgX_2, 23, 24, 40, 47, 48, 216

—, —+$HgPh_2$, 57

—, —+I_2, 143, 216, 217

—, —+NH_3, 53–57, 216

carbon disulphide, $RHgBr$+Br_2 in, 135, 136

carbon tetrachloride, $ArSiR_3$+Br_2 in, 162

—, $BuHgBr$+Br_2 in, 136, 144

—, heat of solution of $SnEt_4$ in, 151

—, HgR_2+I_2 in, 145, 219

—, Me_3SnR+acid in, 107

—, PbR_4+I_2 in, 170, 171, 173–175, 219, 227, 236, 237

—, SnR_4+Br_2 in, 160, 161, 163–165, 171, 172, 227, 231, 235

—, SnR_4+I_2 in, 165, 219

—, SnR_4+IBr in, 172

charge separation, in transition state, of bimolecular electrophilic substitution, 85, 221–223, 235

—, —, of Me_3AuPPh_3+HgX_2, 104

—, —, of $MeAuPPh_3$+$MeHgBr$, 103

—, —, of $PbEt_4$+I_2, 175, 238

—, —, of $RAuPPh_3$+HgX_2, 101

—, —, of SnR_4+$HgCl_2$, 76, 81–86

—, —, of SnR_4+I_2, 148, 238

chloride ion, effect on $MeHgBr$+$HgBr_2$, 43

—, — on $MeSnCl_2$+$SnMe_4$, 88

—, — on $NC_5H_4CH_2Co(III)$+$HgCl_2$, $TlCl_3$, 97–99

—, — on $NC_5H_4CH_2Co(III)$+HNO_2, 191

—, — on $NC_5H_4CH_2Cr(III)$+$HgCl_2$, $TlCl_3$, 93–95

—, — on $NC_5H_4CH_2Fe(CO)_2\pi$-C_5H_5, 96

—, — on $NC_5H_4CH_2$ metal hydrolysis, 32

—, — on $Pt(PEt_3)_2MeX$+acid, 129–131

—, — on $RB(OH)_2$+$HgCl_2$, 65

—, reaction+$ClHgCH_2C_5H_4N$, 27–29

—, —+$PhCH(CO_2Et)HgCl$, 121, 122

chlorine, reaction+$RHgCl$, 136, 137

chloroacetic acid, reaction+BEt_3, 123

chlorobenzene, $PbMe_4$+I_2 in, 236

—, SnR_4+Br_2 in, 156–163, 171, 227, 228, 230, 231, 235

—, SnR_4+I_2 in, 155, 156, 173, 174, 227

chloroform, $RHgBr$+Br_2 in, 136

—, $RHgBr$+NH_3 in, 53–57, 216

chromic acid, reaction+$RB(OH)_2$, 184, 227, 228

—, —+SnR_4, 189, 190, 226, 231

cinnamylmercuric acetate (bromide), reaction+acid, 204

coupling constant, in Me_3SnX, solvent effect, 236, 237

crotyl anions, reactions of, 16

crotylmercuric acetate(bromide), reaction+acid, 204

1-cyano-1-carbethoxypentyl-(triphenylphosphine) gold(I), reaction+HgX_2, 100–102, 213

—, —+$RHgX$, 31, 102, 103

cyclic transition state, 12, 13, 17, 20, 214, 226, 227, 231, 232

—, in $AlMe_3$+Ph_2CO, 187

—, in $ArCH_2HgBr$+$HgBr_2$, 217

—, in $CH_2[B(OMe)_2]_2$+$HgCl_2$, 69

—, in Hg exchange, 42, 43, 45, 46, 48, 51, 52, 54

—, in HgR_2+HCl, 118, 120

—, in LiR, $RMgBr$+CO_2, 179

—, in $MeMgX$+R_3SiX, 64

—, in $NC_5H_4CH_2Cr(III)$+HgX_2, 93

—, in $PhCH_2HgCl$+I_2, 140

—, in $RAuPPh_3$+HgX_2, 101

—, in $RMgBr$, R_2Mg+hexyne-1, 109, 111

—, in $SbCl_3$+$SbMe_3$, 91

—, in SnR_4+Br_2, 164

—, in SnR_4+CrO_3, 190

—, in SnR_4+I_2, 156

—, in ZnR_2+$R'NH_2$, 112

cyclohexane, Me_3SiEt disproportionation in, 188

—, SnR_4+br_2, BrI in, 172

—, SnR_4+I_2 in, 165

cyclohexylamine, reaction+$ZnEt_2$, 111, 112

cyclopentadiene, pK of, 7

D

deuterium chloride, reaction+$HgBz_2$, 121

deuterium oxide, OD^-+boronic ester in, 9, 108

dichloroacetic acid, reaction+BEt_3, 123

1,2-dichloroethane, Ar_3CBr+$RHgBr$ in, 180, 181, 216

dielectric constant, and charge separation in transition state, 235, 236

—, and kinetic salt effect, 85, 221

—, effect on SnR_4+$HgCl_2$, 76, 78, 85, 234

—, of solvents and mechanism, 234–238

diethylether, see ether

diglyme, BEt_3+$RCOOH$ in, 122, 123, 212

1,2-dimethoxyethane, effect on ZnR_2+Ph_2NH, 112, 113

dimethyl(allenyl)tin bromide, structure in

solution, 236, 237

dimethylaluminium chloride, reaction+RCN, 185

8-dimethylaminoquinoline, effect on ZnEt₂+ Ph₂NH, 112, 113

2,2′-dimethylbutyl-dicarboxyl ion cyclopentadienyl, reaction+Br₂, 137

3,3-dimethylbutylmagnesium chloride, inversion of, 33

dimethylformamide, addition to PhHg⁺, 239

—, alkyl group transfer in, 69

—, and ionisation of HgBu₂, 25

—, Br₂/Br⁻, I₂/I⁻ equilibria in, 139, 153

—, BzHgCl+Br₂ in, 144

—, BzHgCl+I₂ in, 140–142, 216

—, HgBz₂+HCl in, 121

—, HgEt₂+HCl in, 118, 220

—, MeAuPPh₃+HgBr₂ in, 100, 101

—, PhCH(CO₂Et)HgBr+HgBr₂ in, 47, 48

—, SbCl₃+SbMe₃ in, 90, 91, 211

—, SnR₄+Br₂ in, 153, 235

1,5-dimethylhexylmercuric bromide, reaction +HgBu₂, 58, 211

dimethylsulphoxide, addition to PhHg⁺, 239

—, ArHgBr+HgBr₂ in, 23–27, 216

—, Bu(CN)(CO₂Et)CAuPPh₃+RHgX in, 31

—, BzHgCl+I₂ in, 141, 142

—, HgR₂+HCl in, 117, 121

—, I₂/I⁻ equilibrium in, 139, 208, 209

—, PbR₄+I₂ in, 166, 167, 169, 173, 226, 236

—, PhCH(CO₂Et)HgBr racemisation in, 25, 26

—, Ph₃SnCH₂CH=CHR+I₂ in, 208–210

—, RAuPPh₃+MeHgX in, 102, 103, 213, 234

—, SnR₄+Br₂ in, 235

—, SnR₄+I₂ in, 152, 173, 226

—, XC₆H₄CH₂HgBr+HgBr₂ in, 45, 46, 216

dimethyltin dichloride, reaction+SnR₄, 87–90, 230, 231

dioxan, Bu(CN)(CO₂Et)CAuPPh₃+RHgX in, 31

—, BzHgCl+Br₂ in, 144

—, BzHgCl+HCl in, 121

—, BzHgCl+I₂ in, 139–142

—, CH₂=CHCH₂HgCl+HCl in, 203

—, HgR₂+HCl in, 117–120, 212, 220

—, HgR₂+HgI₂ in, 51, 211, 226

—, Me₃AuPPh₃+HgX₂ in, 104, 213

—, Me₂SnCl₂+SnMe₄ in, 89

—, MgR₂ inversion in, 33

—, PhCH(CO₂Et)HgBr+HgBr₂ in, 23, 47

—, RAuPPh₃+HgX₂ in, 100–102, 213, 223, 233

—, RAuPPh₃+MeHgX in, 102, 103, 213, 222, 234

—, RHgCl+HgCl₂ in, 40

—, RHgI+I₂ in, 137

—, SnR₄+HCl in, 124

diphenylacetic acid, reaction+BEt₃, 123

diphenylamine, reaction+ZnR₂, 112, 113

diphenylmethane, pK of, 7

di-iso-propylether, ZnR₂+RNH₂ in, 111, 112, 227

E

electron affinity, of CH₃, 5

energy of inversion, of CH₃⁻, 8

energy profile, for HgR₂+HgX₂, 50

—, for R′HgX+HgR₂, 59

—, for RHgX+HgX₂, 44

enthalpy of activation, see also activation energy

—, of boronic esters+HgCl₂, 65, 69

—, of HgR₂+AcOH, 116

—, of HgR₂+HgI₂, 51

—, of MeHgBr+HgBr₂, 41

—, of MeMgX+R₃SiX, 64

—, of NC₅H₄CH₂Cr(III)+HgCl₂, 91

—, of PbR₄+AcOH, 125

—, of PhCH(CO₂Et)HgBr+HgBr₂, 47

—, of PhCH₂HgBr+HgBr₂, 44

—, of Ph₃SnCH₂CH=CHR+I₂, 209, 210

—, of RHgI+H₂SO₄, 114

—, of SbMe₃+SbCl₃, 91

—, of SnR₄+HgX₂, 70, 71, 76, 77, 84, 87

enthalpy of formation, of CH₃⁻, 5, 6

—, of I₃⁻, 147, 209

—, of Me₃SnCl · base, 150

enthalpy of solution, of CH₄, CH₃⁻ and H⁺, 5

—, of HgMe₂ and ZnMe₂, 6

—, of SnEt₄, 151

enthalpy of transfer, from MeOH of HgCl₂, KBr, SnEt₄ and transition state, 79

entropy change, for acid dissociation, 5, 6

entropy of activation, see also pre-exponential factor

—, of boronic esters+HgCl₂, 65, 69

—, of BzHgBr+HgBr₂, 44, 211

—, of BzHgCl+Br₂, 144

—, of BzHgCl+HCl, 121

—, of CH₂=CHCH₂HgI+HI, 201

—, of EtSiMe₃ disproportionation, 188

—, of HgBu₂+RHgBr, 58

—, of HgR₂+AcOH, 116

—, of HgR₂+HCl, 117, 119

—, of HgR₂+HgI₂, 51, 211

—, of MeMgX+R₃SiX, 64

—, of Me₂SnCl₂+SnR₄, 89

—, of NC₅H₄CH₂Cr(III)+HgCl₂, 91

—, of NC₅H₄CH₂Mn(CO)₅ hydrolysis, 32

—, of PbR₄+AcOH, 125

—, of PhCH(CO₂Et)HgBr+HgBr₂, 47

—, of RHgI+H₂SO₄, 114

—, of R₃SnCH₂CH=CHR′+HgI₂, 200

—, of SbCl₃+SbMe₃, 91, 211

—, of SnR₄+HgX₂, 70, 71, 76, 77, 84, 87, 210

—, of SnR₄+I₂, 211

—, of XC₆H₄CH₂HgCl+I₂, 140, 142, 143

entropy of transfer, from MeOH of RX, SnR₄, HgCl₂ and transition states, 83, 84

entropy of translation, for formation of SbCl₃/SbMe₃ transition state, 91

equilibrium constant, for Br₂/Br⁻, I₂/I⁻, 139 147, 153, 154, 166, 169, 208, 209

—, for 2EtMgBr, 110

—, for EtSnCl in MeOH, 128

—, for HgX₂+X⁻, 27, 52, 76

—, for PhHg⁺+solvents, 239

—, for R₃SnBr+I⁻, 160

—, for SnR₄+Br₂, 160

ethane, from CdEt₂+BzOH, 113

—, pK of, 7

ethanol, ArB(OH)₂+HgCl₂ in, 66

—, HgBu₂+HgX₂ in, 48, 49, 213, 215, 226

—, HgBu₂+RHgX in, 58, 59, 211, 215, 222

—, HgR₂+HCl in, 119, 120

—, I₂/I⁻ equilibrium in, 139

—, PbR₄+I₂ in, 166, 167, 169, 170, 173–175, 226, 236, 237

—, PhCH=CHCH₂SiMe₃+NaOH in, 196

—, PhCH(CO₂Et)HgBr+HgBr₂ in, 47

—, PhCH₂Cr(III)+acid in, 132

—, PhCH₂HgCl+I₂ in, 140–142

—, Ph₃Sn allyl+I₂ in, 208

—, R₃Ge(Sn)CH₂CH=CH₂+HgX₂ in, 198–200, 236

—, RHgX+HgX₂ in, 40–43, 211, 222, 228

ether, AlR₃+Ph₂CO in, 187, 188

—, CdEt₂+XC₆H₄CH₂OH in, 112

—, EtMgBr+hexyne-1 in, 109, 227

—, HgR₂+HgCl₂ in, 48

—, inversion of metal alkyls in, 33

—, LiR+Br₂, I₂ in, 136

—, LiR+MeOH in, 107

—, MeMgX+R₃SiX in, 64, 65

—, MgR₂+hexyne-1 in, 110, 111, 227

—, PhCH=CHCH₂HgX+HCl in, 204

—, ZnR₂+MeC₆H₄NH₂ in, 112, 227

—, ZnR₂+PhCHO in, 180

—, ZnR₂+PhHgCl in, 64, 227

ethyl acetate, MeCH=CH₂HgX+HCl in, 204

—, pK of, 7

ethyl benzene, SnBu₄+I₂ in, 165

ethyl benzoates, reaction+AlEt₃, 185

ethylboronic acid, reaction+CrO₃, 184, 227, 228

ethylene, pK of, 7

ethylmagnesium bromide, reaction+hexyne-1, 108–110, 227

ethylmercuric bromide, Hg exchange+HgBr₂, 40, 41, 226, 228

ethylmercuric chloride, reaction+Bu(CN)(CO₂Et)CAu PPh₃, 31

ethylmercuric iodide, reaction +H₂SO₄, 114, 201, 226, 228

ethyl(triphenylphosphine)gold(I), reaction + HgX₂, 100–102, 213, 234

—, reaction+MeHgX, 102, 103, 213, 234

F

fluorosulphonic acid, reaction+Me₃SiCH₂Cl, SiMe₄, 124

free energy change, for acid dissociation of CH₄, 6

free energy of activation, of MeMgX+R₃SiX, 64

—, of RB(OH)₂+HgCl₂, 65

—, of SnR₄+HgX₂, 70, 71, 76, 77, 84, 87

free energy of transfer, from MeOH to aq. MeOH of solutes and transition states, 79–82, 131

—, from MeOH to other solvents of I₂, PbR₄, SnR₄ and transition states, 173, 174, 237

G

gallium trichloride, reaction + NC₅H₄CH₂Co(III), 97

germanium triethyl(allyl), reaction+HgX₂, 198–200

glycerol, and ArB(OH)₂+HgCl₂, 66–68

Grignard reagents, reaction+alkynes, 108–110

H

Hammett correlation, and ArB(OH)₂+HgCl₂, 67

—, and CdEt₂+XC₆H₄CH₂OH, 113

—, and cleavage of benzyl metal compounds, 215–218

—, and XC₆H₄CH₂HgCl+I₂, 142

heat of solution, see enthalpy of solution

hexane, ΔG of transfer from MeOH to aq. MeOH, 82

—, ZnPr₂+MeC₆H₄NH₂ in, 112

hexanoic acid, reaction+boronic ester, 108

hexyne-1, reaction+EtMgBr, 108–110, 227

—, —+MgR₂, 110, 111, 227

hydration, see enthalpy of solution

hydrochloric acid, hydrolysis of NC₅H₄CH₂ metal in, 32

—, reaction+HgR₂, 225

—, —+Me₃PbCl, PbR₄, R₃SnCl, SnR₄,

126–128
—, —+Me$_3$SnR, 107
hydrogen bonding, and SnR$_4$+HgI$_2$, 73
hydrogen bromide, reaction+Me$_3$SnR, 107
hydrogen chloride, reaction+CdEt$_2$, 220
—, —+HgR$_2$, 117–120, 203, 220
—, —+PhCH$_2$HgCl, 121
—, —+RCH–CHCH$_2$HgX, 203, 204
—, —+R$_3$Sn allyl(allenyl), 204, 205
—, —+SnR$_4$, 124
hydrogen cyanide, entropy of acid dissociation
   of, 6
hydrogen ion, effect on NC$_5$H$_4$CH$_2$Cr(III),
   Co(III)+HNO$_2$, 191
—, — on PhCH(CO$_2$Et)HgCl+Cl$^-$, 122
—, — on Pt(PEt$_3$)$_2$ MeX+acid, 129–131
—, — on RB(OH)$_2$+CrO$_3$, 184
—, from CH$_4$, 5
—, reaction+BEt$_3$, 124
hydrogen peroxide, reaction+RB(OH)$_2$, 182–
   184, 227, 228
hydroxide ion, reaction+ArSi(Sn)Me$_3$, 216
—, —+boronic ester, 108
—, —+Ge, Si and Sn compounds, 34, 35

I

indene, reaction+RMgX, 109
indium trichloride, reaction + NC$_5$H$_4$CH$_2$Co
   (III), 97
inversion, of alkyl metal compounds, 32, 221
—, of CH$_3^-$, 8
iodide ion, effect on CH$_2$=CHCH$_2$HgI+
   HClO$_4$, 201–203
—, — on MeHgBr+HgBr$_2$, 43
—, — on PbR$_4$+I$_2$, 166–169
—, — on Ph$_3$Sn allyl+I$_2$, 208, 209
—, — on RHgI+I$_2$, 137–139
—, — on Sn allyl$_4$+I$_2$, 206, 207
—, — on SnR$_4$+I$_2$, 146
—, — on XC$_6$H$_4$CH$_2$HgCl+I$_2$, 139–142
—, equilibrium+I$_2$, 139, 147, 166, 208, 209
—, —+R$_3$SnBr, 160
iodine, ΔG of transfer from MeOH to other
   solvents, 174
—, equilibrium+I$^-$, 139, 147, 166, 208, 209
—, reaction+HgMe$_2$, 219, 220
—, —+LiR, Me$_3$SnR, 136
—, —+PbR$_4$, 166–171, 219, 220, 226, 227,
   234, 236, 237
—, —+Ph$_3$Sn allyl, 208–210
—, —+RHgBr, 143, 216
—, —+RHgI, 137
—, —+Sn allyl$_4$, 206–208
—, —+SnR$_4$, 146–152, 155, 156, 165, 166, 211,
   219, 220, 222, 223, 226–228, 230, 231, 238

—, —+XC$_6$H$_4$CH$_2$HgCl, 139–142
iodine bromide, reaction+SnR$_4$, 172
ionic strength, effect on CH$_2$=CHCH$_2$HgI+
   HI, 201
—, — on PbEt$_4$+Br$_2$, 171
—, — on Pt(PEt$_3$)$_2$MeCl+acid, 131
—, — on SnMe$_4$+Br$_2$, 153
—, — on SnMe$_4$+I$_2$, 147
ion pairs, ΔG and ΔS of transfer from MeOH
   to aq. MeOH, 81, 82, 84
iridium alkyl, reaction+Br$_2$, 137
isotope effect, *see* kinetic isotope effect

K

kinetic isotope effect, in acidolysis of Grignard
   reagents, 110
—, in Ar(Me$_3$)Si and Sn+OH$^-$, 35, 36
—, in CdEt$_2$+BzOH, 113
—, in CH$_2$=CHCH$_2$HgI+HI, 201
—, in EtMgBr+hexyne-1, 109
—, in Me$_3$SiCH$_2$CH–CHPh+base, 196
—, in MgEt$_2$+hexyne-1, 110
—, in RB(OH)$_2$+HgCl$_2$, 65
—, in RHgI+H$_2$SO$_4$, 115
kinetic salt effect, and charge separation in
   transition state, 85, 221
—, on BuHgBr+HgBu$_2$, 211, 215
—, on BuHgOAc+Hg(OAc)$_2$, 215
—, on PbEt$_4$+I$_2$, 169
—, on SnMe$_4$+I$_2$, 147, 148
—, on SnMe$_4$+HgX$_2$, 73–75, 85–87, 213–215

L

lead tetra-iso-amyl, reaction+AcOH, 125
lead tetrabutyl, reaction+acid, 125, 127, 219,
   226, 233
lead tetraethyl, bond dissociation energy, 127
—, ΔG of transfer from MeOH to other solvents, 174
—, reaction+acid, 125, 127, 226, 233
—, —+Br$_2$, 171, 222, 226
—, —+I$_2$, 166, 167, 170, 171, 173–175, 226,
   227, 237
lead tetramethyl, bond dissociation energy, 127
—, ΔG for transfer from MeOH to other solvents, 174
—, reaction+acid, 125–127, 219, 226, 233
—, —+Br$_2$, 171, 227
—, —+I$_2$, 166–171, 173–175, 219, 226, 227,
   234, 236, 237
lead tetraphenyl, reaction+HCl, 126
lead tetrapropyl, reaction+acid, 125, 127, 226,
   233
—, —+Br$_2$, 166, 167, 170, 171, 226, 227

lithium acetate, effect on $HgBu_2+BuHgBr$, 58
lithium alkyls, carbonation of, 179
—, reaction+Br_2, 136
lithium bromide, effect on $HgBu_2+BuHgBr$, 58
—, — on $HgBu_2+HgBr_2$, 49
—, — on $Me_3AuPPh_3+HgBr_2$, 104
—, — on $MeAuPPh_3+MeHgBr$, 103
—, — on $MeHgBr+HgBr_2$, 42
—, — on $NC_5H_4CH_2Cr(III)+HgBr_2$, 94
—, — on SnR_4+Br_2, 155
lithium butyl, reaction+CO_2, 179
lithium chloride, effect on $NC_5H_4CH_2Cr(III)$ $+HgCl_2$, 94
lithium 3,3-dimethylbutyl, inversion of, 32, 33, 221
lithium methyl, formation of CH_3^- from, 6
lithium 2-methylbutyl, inversion of, 33, 34, 221
lithium 1-methyl-2,2-diphenylcyclopropyl, carbonation of, 179
—, reaction+acids, 107
lithium nitrate, effect on $BuHgOAc+$ $Hg(OAc)_2$, 215, 222
—, — on $HgBu_2+BuHgX$, 58, 215, 222
—, — on $MeHgBr+HgBr_2$, 41, 222
—, — on $PhCH(CO_2Et)$ HgBr racemisation, 25, 26
lithium octyl, reaction+CO_2, 179
lithium perchlorate, effect on $HgBu_2+BuHgBr$, 58, 222
—, — on $MeAuPPh_3+HgBr_2$, 101
—, — on $MeAuPPh_3+MeHgBr$, 103
—, — on $PhCH(CO_2Et)HgBr$ racemisation, 25, 26
—, — on SnR_4+HgX_2, 73–76, 86, 222

M

magnesium di(3,3-dimethylbutyl), inversion of, 32, 33, 221
magnesium diethyl, reaction+hexyne-1, 110, 111, 227
magnesium dimethyl, association factor, 111
—, reaction+ArCOMe, 180
magnesium di(2-methylbutyl), inversion of, 33, 221
magnesium dipropyl, reaction+hexyne-1, 110, 111, 227
magnesium halides, effect on $ZnR_2+PhCHO$, 180
manganese alkyl, reaction+Br_2, 137
Menschutkin reactions, transition states, 82, 175
mercuric acetate, Hg exchange+RHgOAc, 40, 41, 215, 222
—, reaction+HgR_2, 48, 49, 51, 52, 213
—, —+Me_3AuPPh_3, 104

—, —+$Me_3SiCH_2CH=CH_2$, 198
—, —+$NC_5H_4CH_2Cr(III)$, 92, 93
—, —+$RAuPPh_3$, 100–102, 213
—, —+SnR_4, 86, 87, 222
mercuric bromide, catalysis of PhCHMeCl racemisation by, 52, 53
—, effect on $Ar_3CBr+RHgBr$, 180, 181
—, Hg exchange+RHgBr, 40–42, 210, 216, 222
—, — —+XC_6H_4HgBr, 44–46, 210
—, reaction+HgR_2, 48, 49, 52, 213, 236
—, —+$NC_5H_4CH_2Cr(III)$, 91–94, 213
—, —+p-$NO_2C_6H_4CH_2HgBr$, 24, 25, 27
—, —+$PhCH(CO_2Et)HgBr$, 23, 24, 47, 48
—, —+$RAuPPh_3$, 100–102, 213, 223, 233
—, —+R_3AuPPh_3, 104, 213
—, —+$R_3Ge(Sn)CH_2CH=CHR$, 200, 219, 235
—, —+$RMgCl$, 66
—, —+$R_3SiCH_2CH=CH_2$, 198, 219
mercuric chloride, catalysis of PhCHMeCl racemisation by, 52, 53
—, ΔG, ΔH and ΔS for transfer from MeOH to aq. MeOH, 79, 80, 83
—, reaction+boronic esters, 65–69
—, —+$ClHgCH_2C_6H_4NH^+$, 29, 30
—, —+HgR_2, 48, 50, 52, 213, 219
—, —+$Me_3SiCH_2CH=CH_2$, 197, 198
—, —+$NC_5H_5CH_2Co(III)$, 97–99, 190, 191
—, —+$NC_5H_5CH_2Cr(III)$, 91–94, 190, 191, 213
—, —+$NC_5H_5CH_2Fe(CO)_2\pi$-C_5H_5, 96
—, —+$PhCH(CO_2Et)HgCl$, 24
—, —+$PhCH_2Cr(III)$, 132, 213
—, —+$RAuPPh_3$, 100–102, 213, 233
—, —+$R_3Ge(Sn)CH_2CH=CHR'$, 200
—, —+SnR_4, 75–78, 84–87, 128, 211, 213, 214, 219, 222, 223, 226, 236
mercuric iodide, catalysis of PhCHMeCl racemisation, 52, 53
—, Hg exchange+HgR_2, 51, 52, 211, 213
—, — —+MeHgI, 41
—, reaction+$Me_3SiCH_2CH=CH_2$, 198
—, —+$RAuPPh_3$, 100–102, 213, 233
—, —+$R_3SnCH_2CH=CHR'$, 200
—, —+SnR_4, 70–75, 78, 86, 87, 211, 213, 214, 222, 226, 228, 230
mercuric nitrate, Hg exchange+$MeHgNO_3$, 41
—, reaction+HgR_2, 48, 49, 51, 213
—, —+$NC_5H_5CH_2Cr(III)$, 92, 93, 213
—, —+$RAuPPh_3$, 100, 101, 213
mercury(I), reaction+$NC_5H_4CH_2Cr(III)$, 99
mercury butyl phenyl, reaction+$HgBr_2$, 50
mercury dialkyls, acetolysis of, 1, 115–117
—, reaction+HCl, 117–120
—, —+HgX_2, 48–53
—, —+RHgX, 58, 59

mercury diallyl, reaction+HCl, 203

mercury dibenzyl, protolysis of, 25

—, reaction+HCl, 120, 121

mercury dibutyl, ionisation of, 25

—, reaction+acids, 13, 107, 115, 116, 118, 219, 226, 227

—, —+HgX$_2$, 48, 49, 213, 222, 236

—, —RHgX, 58, 59, 215, 222

mercury di(4-camphyl), reaction+AcOH, 115–117

mercury di(α-carbethoxybenzyl), reaction+HgBr$_2$, 48

mercury di(3,3'-dimethylbutyl), inversion of, 32, 33, 221

mercury di(1,5-dimethylhexyl), reaction+RHgBr, 58

mercury di(1,4-dimethylpentyl), reaction+HgBr$_2$, 48

mercury diethyl, reaction+HCl, 117, 118, 203, 226, 227

—, —+HgI$_2$, 51, 227

mercury dimethyl, formation of CH$_3^-$ from, 6

—, reaction+HCl, 117

—, —+HgX$_2$, 51, 52, 213, 219, 227

—, —+I$_2$, 219

mercury di(4-methylcyclohexyl), reaction+acids, 107

mercury dineophyl, reaction+AcOH, 115–117, 222, 223

mercury diphenyl, reaction+HCl, 117, 212

—, —+HgI$_2$, 51

—, —+PhCH(CO$_2$Et)HgBr, 57

mercury dipropyl, reaction+HCl, 117

—, —+HgI$_2$, 51, 227

mercury ethyl phenyl, reaction+HgCl$_2$, 50

mesitylene, SnBu$_4$+I$_2$ in, 165

methane, ΔG of transfer from MeOH to aq. MeOH, 82

—, formation of CH$_3^-$ from, 5, 6

methanol, addition to PhHg$^+$, 239

—, boronic esters+HgCl$_2$ in, 68, 69

—, Br$_2$/Br$^-$ and I$_2$/I$^-$ equilibria in, 139

—, BzHgCl+Br$_2$ on, 144

—, BzHgCl+I$_2$ in, 140–142

—, effect on BuHgBr+Br$_2$, 144

—, OH$^-$+Ge, Si and Sn organics in, 34–36

—, PbR$_4$+I$_2$ in, 166, 167, 169, 170, 173–175, **236, 237**

—, PbR$_4$, R$_3$PbCl, R$_3$SnCl, SnR$_4$+HCl in, 126–128, 233

—, PhCH=CHCH$_2$Si(Sn)R$_3$+NaOH in, 196

—, Ph$_3$SnCH$_2$CH=CHR+I$_2$ in, 208–210

—, Pt(PEt$_3$)$_2$MeX+acid in, 129, 131

—, reaction+LiR, 107

—, RHgI+H$_2$SO$_4$ in, 114

—, R$_3$Sn allyl(allenyl)+HCl in, 204, 205

—, SnEt$_4$+HgCl$_2$ in, 222, 224

—, SnR$_4$+Br$_2$ in, 18, 235

—, SnR$_4$+HgI$_2$ in, 70, 71, 73–86, 211, 213, 214, 222, 226

—, SnR$_4$+I$_2$ in, 146–151, 173, 174, 208, 211, 222, 223, 228, 230, 238

—, SnR$_4$+IBr in, 172

—, SnR$_4$+Me$_2$SnCl$_2$ in, 88–90, 230

methoxide ion, and boronic esters+HgCl$_2$, 69

—, reaction+Ge, Si and Sn compounds, 34, 35

2-methoxycyclohexylmercuric chloride, Hg exchange+HgCl$_2$, 39, 40

p-methoxyphenyl-methyl-1-naphthyl-phenyl-silane, reaction+Br$_2$, 162

methylaluminium dichloride, reaction+RCN, 185

methyl anion, energy of inversion, 8

—, enthalpy of formation, 5, 6

α-methylbenzyl chloride, racemisation of catalysed by HgX$_2$, 52, 53

methylboronic acid, reaction+CrO$_3$, 184, 227, 228

—, —+H$_2$O$_2$, 183, 227, 228

2-methylbutylmagnesium bromide, inversion of, 33

4-methylcyclohexylmercuric bromide, reaction+Br$_2$, 135, 136

1-methyl-2,2-diphenylcyclopropylmagnesium bromide, reaction+CO$_2$, 179

methylene chloride, RHgBr+Br$_2$ in, 136

methylmagnesium halides, reaction+hexyne-1, 109, 227

—, —+Ph$_2$CO, 180

—, —+R$_3$SiX, 64, 65

4-methylmercaptoacetophenone, reaction+MgMe$_2$, 180

methylmercuric acetate, Hg exchange+Hg(OAc)$_2$, 41

—, reaction+Bu(CN)(CO$_2$Et)CAuPPh$_3$, 31

—, —+RAuPPh$_3$, 213, 233

methylmercuric bromide, Hg exchange+HgBr$_2$, 40–42, 211, 222, 228

—, reaction+Bu(CN)(CO$_2$Et)CAuPPh$_3$, 31

—, —+RAuPPh$_3$, 213, 222, 233

methylmercuric chloride, reaction+Bu(CN)(CO$_2$Et)CAuPPh$_3$, 31

—, —+RAuPPh$_3$, 213, 233

methylmercuric iodide, Hg exchange+HgI$_2$, 41

—, reaction+H$_2$SO$_4$, 114, 228

—, —+RAuPPh$_3$, 213, 233

methylmercuric nitrate, Hg exchange+Hg(NO$_3$)$_2$, 41

—, reaction+RAuPPh$_3$, 213

α-methylnaphthalene, SuBu$_4$+I$_2$ in, 165

methyl radical, electron affinity of, 5

methyl(triphenylphosphine)gold(I), reaction +HgX₂, 100–102, 213, 223, 233
—, —+MeHgX, 102, 103, 213, 222, 233
microscopic reversibility, and RHgX+HgX₂, 43

N

neophylmercuric iodide, reaction+I₂, 137
nephelometry, and RHgBr+NH₃, 53
nitrate ion, effect on boronic esters+HgCl₂, 69
—, — on MeHgBr+HgBr₂, 43
nitrobenzene, PbMe₄, Me₃PbOAc+AcOD in, 126
—, PhCHMeCl racemisation in, 53
p-nitrobenzylchloride, reaction+Me₃N, 175
p-nitrobenzylmercuric bromide, reaction + HgBr₂, 24–27
nitromethane, pK of, 7
nitrous acid, reaction+NC₅H₄CH₂Cr(III), Co(III), 190–192
nitrosyl chloride, reaction + NC₅H₄CH₂Cr (III), 191, 192
5-norbornene-2-boronic acid, reaction+HgCl₂, 65, 66
norbornylmagnesium bromide, reaction+ CO₂, 179
norbornylmagnesium chloride, reaction+ HgBr₂, 66
nuclear magnetic resonance, see also proton magnetic resonance
—, and inversion of metal alkyls, 32–34
—, and Me₃SnBr+Br₂, 158
—, and RHgBr+NH₃, 54

O

octanoic acid, reaction+BEt₃, 123
order of reaction, see also rate law
—, and mechanism, 11–15
—, of AlR₃+PhCN, 185, 186
—, of Ar₃CBr+RHgBr, 181
—, of ArHgBr+HgBr₂, 23, 24, 44, 45, 47
—, of BEt₃+RCOOH, 122
—, of Bu(CN)(CO₂Et)CAuPPh₃+RHgX, 31
—, of BuHgBr+Br₂, 144, 145
—, of BzHgCl+Br₂, 143, 144
—, of BzHgCl+I₂, 140
—, of CH₂=CHCH₂HgCl+HCl, 203
—, of CH₂=CHCH₂HgI+HI, 201
—, of EtMgBr+hexyne-1, 109, 110
—, of Hg exchange in HgBu₂+HgX₂, 49
—, of Hg exchange in RHgX+HgX₂, 40
—, of HgR₂+AcOH/HClO₄, 115, 116
—, of HgR₂+HCl, 120, 121
—, of HgR₂+hexyne-1, 110

—, of HgR₂+I₂, 145
—, of hydrolysis of NC₅H₄CH₂ metal, 31
—, of inversion of metal alkyls, 33, 34
—, of Me₃Si(CH₂)₂COOH+H₂SO₄, 124
—, of PbR₄+AcOH, 124, 126
—, of PbR₄+Br₂, 156, 164
—, of PbR₄+I₂, 170
—, of PhCH(Me)Cl racemisation catalysed by HgX₂, 52
—, of RAuPPh₃+HgX₂, 100
—, of RB(OH)₂+CrO₃, 184
—, of RB(OH)₂+H₂O₂, 182
—, of RHgI+I₂, 137
—, of RHgX+NH₃, 53
—, of R'HgX+R₂Hg, 58
—, of R₃Sn allyl(allenyl)+acid, 204
—, of SbCl₃+SbMe₃, 91
—, of SnEt₄+HgCl₂, 75
—, of SnR₄+I₂, 156
—, of ZnR₂+PhCHO, 180
—, of ZnR₂+PhHgCl, 64

P

pentaaquopyridiomethylchromium(III), reaction+Hg(1), 99
—, —+HgX₂, 91–94, 98, 190, 191,213
—, —+HNO₂, 190, 191
—, —+TlCl₃, 95, 98, 190, 191
pentane, LiR+Br₂ in, 136
pentylmercuric bromide, Hg exchange+ HgBr₂, 41
perchloric acid, and ClHgCH₂C₅H₄NH⁺+ Cl⁻, 27–29
—, and NC₅H₄CH₂Cr(III)+TlCl₃, 95
—, and PhCH(CO₂Et)HgCl+Cl⁻, 121
—, effect on HgR₂+AcOH, 116, 117
—, RB(OH)₂+H₂O₂ in, 182, 183
—, reaction+CH₂=CHCH₂HgI, 201–203
—, —+MeCH=CHCH₂HgBr, 204
—, —+MeHgI, 114
—, —+Me₃SnCH₂CH=CH₂, 204
—, —+PbR₄, 125, 126, 226
perturbers, in SnR₄+Br₂, 163
phenylacetic acid, reaction+BEt₃, 123
phenylboronic acid, reaction+H₂O₂, 183
phenylethylboronic acid, esters of, reaction+ acid, 108
—, —, —+OD⁻, 9
phenylmercuric chloride, reaction+ZnR₂, 64, 227
phenylmercuric ion, equilibrium+solvents, 239
phenylpentylmagnesium chloride, inversion of, 33
phthalate ion, effect on RB(OH)₂+HgCl₂, 65
pi-complex, of Et₃ metal allyl+HgX₂, 199

—, of Ph$_3$SnCH$_2$CH=CHR+I$_2$, 209
pK, *see* acid dissociation constant
platinum (IV), in Pt(PEt$_3$)$_2$MeX+acid, 130, 131
polarography, and pK of carbon acids, 7
potassium bromide, ΔH for transfer from MeOH to aq. MeOH, 79
—, effect on BzHgBr+HgBr$_2$, 46
—, — on p-NO$_2$C$_6$H$_4$CH$_2$HgBr+HgBr$_2$, 27
potassium perchlorate, effect on PhCH(CO$_2$Et)HgBr racemisation, 25, 26
pre-exponential factor, of BzHgBr+HgBr$_2$, 44
—, of BzHgCl+Br$_2$, 144
—, of BzHgCl+I$_2$, 140
—, of CH$_2$=CHCH$_2$HgCl+HCl, 203
—, of HgMe$_2$+I$_2$, 45
—, of MeHgBr+HgBr$_2$, 41
—, of Me$_3$SiBz+OH$^-$, 34
—, of Me$_3$SiCH$_2$CH=CH$_2$+HgCl$_2$, 198
—, of p-NO$_2$C$_6$H$_4$CH$_2$HgBr+HgBr$_2$, 25
—, of PhCH=CHCH$_2$Si(Sn)R$_3$+base, 196
—, of PhCH(CO$_2$Et)HgBr+HgBr$_2$, 24, 47
propane, pK of, 7
propanol, I$_2$/I$^-$ equilibrium in, 139
—, PbR$_4$+I$_2$ in, 166, 167, 169, 170, 173–175, 226, 236, 237
propionic acid, reaction+BEt$_3$, 123
propylmagnesium bromide, reaction+ hexyne-1, 109, 227
propylmercuric bromide, Hg exchange+ HgBr$_2$, 40, 226, 228
propylmercuric iodide, reaction+H$_2$SO$_4$, 114, 115, 226, 228
proton magnetic resonance, *see also* nuclear magnetic resonance
—, and exchange of R groups in metal alkyls, 105
—, and inversion of metal alkyls, 32–34, 221
pyridine, addition to PhHg$^+$, 239
—, effect on ZnEt$_2$+Ph$_2$NH, 112, 113
—, PhCH(CO$_2$Et)HgBr+HgBr$_2$ in, 47, 48
—, RHgBr+Br$_2$ in, 135, 136
pyridiomethyldicarbonyl-π-cyclopentadienyl iron, reaction+HgCl$_2$, TlCl$_3$, 96, 98
4-pyridiomethylmanganese,-molybdenum, tungsten carbonyls, hydrolysis of, 31, 32
4-pyridiomethylmercuric chloride ion, reaction +Cl$^-$, 27–29
—, —+HgCl$_2$, 30
pyridiomethylpentaaquochromium(III), *see* penta-aquopyridiomethylchromium(III)
pyridiomethylpentacyanocobalt(III), reaction +Ga(III), Hg(II), In(III), Tl(III), 97–99
—, —+HNO$_2$, 190–192

Q

quinoline, XC$_6$H$_4$CH$_2$HgBr+HgBr$_2$ in, 44, 45, 211, 216

R

racemisation, of PhCH(CO$_2$Et)HgBr, 24–26
—, of PhCH(Me)Cl catalysed by HgX$_2$, 52
radioactive isotopes, in electrophilic substitution, 23–27, 30, 39 *et seq.*
rate coefficient, for electrophilic substitution, 14
—, of AlMe$_3$+PhCN, 185
—, of AlMe$_3$+Ph$_2$CO, 188
—, of Ar$_3$CBr+RHgBr, 181
—, of BEt$_3$+RCOOH, 123
—, of boronic esters+HgCl$_2$, 65, 69
—, of Bu(CN)(CO$_2$Et)CAuPPh$_3$+RHgX, 31
—, of CdEt$_2$+XC$_6$H$_4$CH$_2$OH, 113
—, of CH$_2$=CHCH$_2$HgCl+HCl, 203
—, of CH$_2$=CHCH$_2$HgI+HI, 201, 202
—, of ClHgCH$_2$C$_5$H$_4$NH$^+$+Cl$^-$, 28–30
—, of EtMgBr+hexyne-1, 109
—, of HgMe$_2$+I$_2$, 45, 219
—, of HgR$_2$+AcOH, 116, 219
—, of HgR$_2$+HCl, 117, 118, 120, 203
—, of HgR$_2$+HgX$_2$, 49, 51, 52, 219, 236
—, of hydrolysis of NC$_5$H$_4$CH$_2$ metal, 32
—, of Me$_3$AuPPh$_3$+HgX$_2$, 104
—, of Me$_3$SiCH$_2$CH=CH$_2$+HgX$_2$, 198
—, of Me$_3$SiEt disproportionation, 188, 189
—, of MgR$_2$+hexyne-1, 111
—, of NC$_5$H$_4$CH$_2$Co(III), Cr(III)+HNO$_2$, 190, 191
—, of NC$_5$H$_4$CH$_2$Co(III)+HgCl$_2$, TlCl$_3$, 97, 98, 100
—, of NC$_5$H$_4$CH$_2$Cr(III)+Hg(I), 99
—, of NC$_5$H$_4$CH$_2$Cr(III)+HgX$_2$, 91, 92, 98
—, of NC$_5$H$_4$CH$_2$Cr(III)+TlCl$_3$, 95, 98, 190
—, of p-NO$_2$C$_6$H$_4$CH$_2$HgBr+HgBr$_2$, 25, 27
—, of OH$^-$+Ge, Si and Sn compounds, 34, 35
—, of PbR$_4$+Br$_2$, 171
—, of PbR$_4$+I$_2$, 166–171, 173, 219, 236
—, of PbR$_4$ and R$_3$PbCl+acid, 125–127, 219
—, of PhCH(CO$_2$Et)HgBr(Cl)+HgX$_2$, 23, 24, 47
—, of PhCH(CO$_2$Et)HgBr racemisation, 25, 26
—, of PhCH$_2$HgCl+Br$_2$, 144
—, of PhCH$_2$HgCl+HCl, 121
—, of PhCH$_2$HgCl+I$_2$, 140, 142
—, of Ph$_3$SnCH$_2$CH=CHR+I$_2$, 218, 219
—, of Pt(PEt$_3$)MeX+acid, 129–131
—, of RAuPPh$_3$+HgX$_2$, MeHgX, 100, 102, 103

—, of $RB(OH)_2+CrO_3$, 184
—, of $RB(OH)_2+H_2O_2$, 183
—, of $R_3Ge(Sn)CH_2CH=CHR'+HgX_2$, 200, 236
—, of $RHgBr+NH_3$, 54–56
—, of $RHgI+H_2SO_4$, 114, 201
—, of $RHgI+I_2$, 137
—, of $RHgX+HgBu_2$, 59
—, of $RHgX+HgX_2$, 40–42
—, of $R_3Si(Sn)CH_2CH=CHPl+$ base, 196, 200
—, of R_3Sn allyl(allenyl)$+HCl$, 205
—, of $SbCl_3+SbMe_3$, 91
—, of Sn allyl$_4+I_2$, 207
—, of SnR_4+Br_2, 153–155, 157–159, 162–164, 172
—, of SnR_4+CrO_3, 189
—, of SnR_4+HgX_2, 70, 71, 73, 75–78, 84, 86, 87, 198, 219, 236
—, of SnR_4+I_2, 146–149, 151, 152, 156, 165, 173, 219
—, of SnR_4+IBr, 172
—, of $SnR_4+Me_2SnCl_2$, 87–90
—, of $XC_6H_4HgBr+HgBr_2$, 44–46
—, of $ZnR_2+PhCHO$, 180
—, of $ZnR_2+PhHgCl$, 64
—, of $ZnR_2+R'NH_2$, 111–113
rate determining step, in $AlMe_3+PhCN$, 185, 186
—, in AlR_3+Ph_2CO, 186
—, in $BEt_3+RCOOH$, 123
—, in $CdEt_2+XC_6H_4CH_2OH$, 113
—, in $CH_2=CHCH_2HgI+HClO_4$, 202, 203
—, in Grignard reagents$+$acid, 110
—, in $R_3SnCH_2CH=CHR'+HgX_2$, 199
rate law, see also order of reaction
—, for $AlMe_3+Ph_2CO$, 187, 188
—, for $BzHgCl+I_2$, 141
—, for $CdEt_2+XC_6H_4CH_2OH$, 113
—, for $CH_2=CHCH_2HgI+HI$, 202
—, for electrophilic substitution, 14, 15
—, for $HgCl_2+$boronic esters, 65, 66, 68
—, for $Me_3SiCH_2CH=CH_2+HgCl_2$, 197
—, for Me_3SiEt disproportionation, 188
—, for $NC_5H_4CH_2Co(III)$, $Cr(III)+HNO_2$, 191
—, for PbR_4+Br_2, 171
—, for PbR_4+I_2, 167, 168
—, for $PhCH(CO_2Et)HgCl+$acid, 122
—, for $Ph_3SnCH_2CH=CHR+I_2$, 208
—, for $Pt(PEt_3)_2MeX+$acid, 129
—, for $RHgBr+NH_3$, 53–55
—, for RMX_n+I_2, 138
—, for Sn allyl$_4+I_2$, 206, 207
—, for $SnEt_4+HgI_2$, 70
—, for SnR_4+Br_2, 153, 154, 157

—, for SnR_4+CrO_3, 183
—, for SnR_4+I_2, 146, 152, 165
—, for SnR_4+IBr, 172

S

salt effect, see kinetic salt effect
sodium acetate, effect on boronic esters$+$ $HgCl_2$, 66
—, — on HgR_2+AcOH, 115, 116, 222
—, — on $MeAuPPh_3+MeHgBr$, 103, 222
—, — on $RB(OH)_2+H_2O_2$, 183
sodium chloride, effect on HgR_2+HCl, 117
—, — on $SnEt_4+HgCl_2$, 75, 76
—, — on $SnMe_4+Me_2SnCl_2$, 88
sodium hydroxide, reaction$+$ $PhCH=CHCH_2Si(Sn)R_3$, 196
sodium nitrate, effect on boronic esters$+$ $HgCl_2$, 69
sodium perchlorate, effect on $PbEt_4$, SnR_4+ Br_2I_2, $Hg(OAc)_2$, 222
—, — on $PbMe_4+I_2$, 169
—, — on $SnMe_4+Me_2SnCl_2$, 88
sodium sulphate, effect on HgR_2+HCl, 117
statistical factor, and SnR_4+Br_2, 155, 158, 228
—, and SnR_4+HgI_2, 71, 72, 228
—, and SnR_4+I_2, 148, 152, 228
steric effects, in $HgR_2+R'HgBr$, 58
—, in OH^-+organometallics, 34
—, in PbR_4+HCl, 233
—, in PbR_4+I_2, 169
—, in $RB(OH)_2+HgCl_2$, 68
—, in $RB(OH)_2+H_2O_2$, 183, 184
—, in $RHgX+HgX_2$, 41
—, in SnR_4+Br_2, 160, 231
—, in SnR_4+HCl, 233
—, in SnR_4+HgX_2, 71, 78
—, in SnR_4+I_2, 150, 151
—, in $ZnR_2+PhHgCl$, 65
sulpholane, equilibria Br_2/Br^- and I_2/I^- in, 139
sulphur dioxide, and Me_3SiCH_2Cl, $SiMe_4+$ FSO_3H, 124
sulphuric acid, effect on HgR_2+HCl, 117
—, reaction$+Me_3Si(CH_2)_2COOH$, 124
—, —$+RHgI$, 114

T

tetrabutylammonium halides, effect on ZnR_2+ $PhCHO$, 180
tetrabutylammonium perchlorate, effect on $SnEt_4+HgCl_2$, 75, 76, 85–87, 222
—, — on $SnEt_4+Hg(OAc)_2$, 215, 222
tetraethylammonium bromide, effect on

PhCH(CO$_2$Et)HgBr racemisation, 25, 26

tetraethylammonium iodide, ΔG of transfer from MeOH to aq. MeOH, 82

—, ΔG of transfer from MeOH to other solvents, 175

tetrahydrofuran, addition to PhHg$^+$, 239

—, Br$_2$+Ir, Mn alkyls in, 137

—, EtMgBr+hexyne-1 in, 110

—, HgBz$_2$+HCl in, 121

—, HgEt$_2$+HCl in, 118

—, inversion magnesium alkyls in, 33

—, ZnR$_2$+PhHgCl in, 64, 65

tetramethylammonium salts, ΔG and ΔS for transfer from MeOH to aq. MeOH, 81, 82, 84

N,N,N',N'-tetramethylethylenediamine, effect on ZnR$_2$+Ph$_2$NH, 112, 113

tetramethylsilane, reaction+FSO$_3$H, 124

tetrapropylammonium perchlorate, ΔG of transfer from MeOH to aq. MeOH, 81, 82

thallium trichloride, reaction+ NC$_5$H$_4$CH$_2$Co(CN)$_5$(III), 97–99, 190, 191

—, —+NC$_5$H$_4$CH$_2$Cr(III), 95, 190, 191

—, —+NC$_5$H$_4$CH$_2$Fe(CO)$_2$-π-C$_5$H$_5$, 96

tin dimethyl diethyl, reaction+Me$_2$SnCl$_2$, 89, 90

tin tetraalkyl, bond dissociation energy, 127

—, reaction+acids, 107, 124, 127, 128, 227, 233

—, —+CrO$_3$, 189, 190, 227

—, —+halogens, 18, 136, 137, 145 et seq., 222, 226–228, 230

—, —+HgX$_2$, 70–87, 222, 226

—, —+Me$_2$SnCl$_2$, 87–90

tin tetraallyl, reaction+I$_2$, 148, 206–208

tin tetrabutyl, bond dissociation energy, 127

—, ΔG for transfer from MeOH to aq. MeOH, 80

—, reaction+Br$_2$, 154, 155, 157–159, 163, 164, 227, 228

—, —+CrO$_3$, 189, 190, 227

—, —+HCl, 127, 128, 233

—, —+HgCl$_2$, 77, 78, 128, 226

—, —+HgI$_2$, 70–74, 78, 222, 226, 228

—, —+I$_2$, 148, 149, 151, 152, 156, 165, 226, 228

tin tetradodecyl, reaction+Br$_2$, 157, 164

tin tetraethyl, bond dissociation energy, 127

—, ΔG, ΔH, ΔS for transfer from MeOH to other solvents, 78–84, 174

—, heat of solution, 151

—, reaction+Br$_2$, 153–155, 157–159, 163, 164, 171, 227, 228, 235

—, —+CrO$_3$, 189, 190, 227

—, —+HCl, 124, 128, 227

—, —+HgCl$_2$, 75–79, 84–87, 128, 198, 211, 213, 214, 222, 224, 226, 236

—, —+HgI$_2$, 70–74, 86, 87, 211, 213, 214, 222, 226, 228

—, —+Hg(OAc)$_2$, 86, 87, 213, 222

—, —+I$_2$, 148, 149, 151, 152, 156, 173–175, 208, 222, 226, 230

—, —+Me$_2$SnCl$_2$, 89

tin tetramethyl, bond dissociation energy, 127

—, ΔG, ΔS of transfer from MeOH to other solvents, 80, 84, 174

—, reaction+acid, 124, 127, 128, 219, 233

—, —+Br$_2$, 153–155, 157–159, 162–164, 171, 172, 227, 228, 235

—, —+CrO$_3$, 189, 190, 227

—, —+HgCl$_2$, 77, 78, 128, 219, 226

—, —+HgI$_2$, 70–73, 78, 226, 228

—, —+I$_2$, 146–149, 151, 152, 156, 173–175, 211, 219, 222, 223, 226, 228, 230, 238

—, —+IBr, 172

—, —+Me$_2$SnCl$_2$, 88–90, 230

tin tetraoctyl, reaction+Br$_2$, 157, 164

tin tetrapropyl, bond dissociation energy, 127

—, ΔG of transfer from MeOH to aq. MeOH, 80, 81

—, reaction+Br$_2$, 153–155, 157, 158, 162–164, 171, 172, 222, 223, 227

—, —+CrO$_3$, 189, 190

—, —+HCl, 124, 127, 128, 227

—, —+HgCl$_2$, 77, 78, 128, 226

—, —+HgI$_2$, 70–74, 222, 226

—, —+I$_2$, 148, 149, 151, 152, 156, 226

—, —+IBr, 172

tin tributyl alkyl, reaction+Br$_2$, 159, 164, 165, 228

—, —+I$_2$, 149, 228

—, —+Me$_2$SnCl$_2$, 89, 90, 230

tin tributyl allyl, reaction+HgI$_2$, 200

tin tributyl phenallyl, reaction+base, 196

tin triethyl alkyl, reaction+Br$_2$, 155, 159, 164, 228

—, —+I$_2$, 149, 228

—, —+Me$_2$SnCl$_2$, 89, 90, 230

tin triethyl allenyl, reaction+HCl, 205

tin triethyl allyl, reaction+HgX$_2$, 198–200, 219

tin triethyl phenallyl, reaction+base, 196, 200

tin trimethyl allenyl(allyl), reaction+HCl, 204–206

tin trimethyl benzyls, reaction+OH$^-$, 216

tin trimethyl butyl, bond dissociation energy, 127

—, reaction+Br$_2$, 155, 159, 228

—, —+HgI$_2$, 70–73, 228

—, —+I$_2$, 149, 152, 228

tin trimethyl ethyl, bond dissociation energy, 127

—, reaction+Br$_2$, 155, 159, 163, 228

—, —+I_2, 149, 152, 228, 230

—, —+Me_2SnCl_2, 89, 90

tin trimethyl p-methoxyphenyl, reaction+I_2, 208

tin trimethyl 1-methyl-2,2-diphenylcyclopropyl, reaction+acids, 107

—, —+halogens, 136

tin trimethyl propyl, bond dissociation energy, 127

—, reaction+Br_2, 155, 159, 228

—, —+I_2, 149, 152, 228

tin triphenyl allenyl, reaction+HCl, 205

tin triphenyl allyl, reaction+HCl, 205

—, —+HgX_2, 200

—, —+I_2, 208, 209

tin tripropyl methyl, reaction+Br_2, 164

—, —+I_2, 149

—, —+Me_2SnCl_2, 89, 90

toluene, ArHgBr(Cl)+I_2 in, 143, 216

—, inversion of metal alkyls in, 33

—, $PbEt_4$+AcOH in, 125, 126

—, pK of, 7

p-toluene sulphonic acid, reaction+$Pt(PEt_3)_2$ MeX, 129

p-toluidine, reaction+ZnR_2, 111, 112, 227

transition state, for $AlMe_3$+Ph_2CO, 187

—, for AlR_3+$ArCO_2Et$, PhCN, 185, 186

—, for Ar_3CBr+RHgBr, 181, 182

—, for ArHgBr+Br^-, 26

—, for BEt_3+RCOOH, 212

—, for boronic esters+$HgCl_2$, 68, 69

—, for BzHgBr+$HgBr_2$, 45, 46, 211, 217, 218

—, for BzHgCl+I_2, 140, 142

—, for $CdEt_2$+$XC_6H_4CH_2OH$, 114, 213

—, for $ClHgCH_2C_5H_4NH^+$+Cl^-, 29

—, for electrophilic substitution, 12, 13, 17–20, 214, 221–224, 226–233, 235, 237–239

—, for HgR_2+AcOH, 116, 117

—, for HgR_2+HCl, 118, 120

—, for HgR_2+HgX_2, 49–52, 211

—, for HgR_2+R′HgX, 59

—, for LiR, RMgX+CO_2, 179

—, for Me_3AuPPh_3+HgX_2, 104

—, for $MeAuPPh_3$+MeHgBr, 103

—, for Me_3SiCH_2CH=CH_2+$HgCl_2$, 198

—, for Me_3SiR disproportionation, 189

—, for Me_3SnCH_2CH=$CHMe$+HCl, 206

—, for Me_2SnCl_2+$MeSnR_3$, 89

—, for MgR_2+hexyne-1, 111

—, for $NC_5H_4CH_2Co(III)$, Cr(III)+HNO_2, NOCl, 191, 192

—, for $NC_5H_4CH_2Cr(III)$+HgX_2, $TlCl_3$, 93–95

—, for $NC_5H_4CH_2Fe(CO)_2$ π-C_5H_5+$HgCl_2$, $TlCl_3$, 96

—, for OH^-+Me_3SnBz, 35

—, for PbR_4+AcOH, 125

—, for PbR_4+Br_2, 171

—, for PbR_4+I_2, 169, 170, 174, 175

—, for PhCH=CHCH_2Si(Sn)R_3+base, 196

—, for $Pt(PEt_3)_2MeX$+acid, 130–132

—, for $RB(OH)_2$+H_2O_2, 183

—, for RHgBr+NH_3, 54, 56, 57

—, for RHgI+acid, 114

—, for RHgI+I_2, 137

—, for RHgX+HgX_2, 41–44

—, for RMgBr+hexyne-1, 109

—, for R_3SnCH_2CH=CHR'+HgX_2, 200

—, for $SbCl_3$+$SbMe_3$, 91

—, for SnR_4+Br_2, 160, 162–164, 174, 175, 231, 235

—, for SnR_4+CrO_3, 190

—, for SnR_4+HgX_2, 71–73, 75, 76, 78–83, 85, 211

—, for SnR_4+I_2, 147, 148, 150–152, 156, 211, 235

—, for ZnR_2+PhHgCl, 64

—, for ZnR_2+$R'NH_2$, 112

translational entropy, see entropy of translation

tributyllead chloride, reaction+HCl, 127, 233

tributyltin bromide, reaction+I^-, 160

tributyltin chloride, reaction+HCl, 127, 128, 233

trichloroacetic acid, reaction+BEt_3, 123

triethyl allyl germane (silane), reaction+$HgBr_2$, 219, 236

triethylamine, effect on $ZnEt_2$+Ph_2NH, 112, 113

triethyl aryl germanes, reaction+OH^-, 35

triethyl diphenylmethyl silane, reaction+OH^-, 35

triethyllead acetate, reaction+AcOH, 125

triethyllead chloride, reaction+HCl, 127, 233

triethyltin bromide, reaction+Br_2, 154, 227

—, —+I^-, 160

trimethylacetic acid, reaction+BEt_3, 123

trimethyl alkyl silanes, disproportionation of, 188, 189

trimethyl allyl silane, reaction+HgX_2, 197, 198

trimethylamine, reaction+p-$NO_2C_6H_4CH_2Cl$, 175

trimethyl aryl germanes, reaction+OH^-, 34, 35

trimethyl aryl silanes, reaction+OH^-, 34–36, 216, 218

trimethyl aryl stannanes, reaction+OH^-, 35, 36

trimethyllead chloride, reaction+HCl, 127, 233

trimethyl phenallyl silane, reaction+base, 196

trimethylsilyl chloride, reaction+
  $Me_3SiCH_2CH=CH_2$, 197
trimethylsilyl chloromethane, reaction+
  FSO_3H, 124
trimethylsilyl halides, reaction+MeMgX, 63,
  64
trimethylsilyl propionic acid, reaction+
  H_2SO_4, 124
trimethyltin bromide, coupling constant
  (NMR), 236, 237
—, effect on SnR_4+Br_2, 162
—, reaction+Br_2, 154, 158, 227
—, —+I^-, 160
trimethyltin chloride, complexes with C_5H_5N,
  MeCN, Me_2CO, Me_2SO, 150
—, coupling constant (NMR), 236, 237
—, reaction+HCl, 127, 128, 233
trimethyl(triphenylphosphine)gold(III),   reac-
  tion+HgX_2, 104, 213
triphenylamine, effect on SnR_4+I_2, 156
triphenylbromomethane, reaction+RHgBr,
  180–182
triphenylmethane, pK of, 7
tripropyllead chloride, reaction+HCl, 127, 233
tripropyltin bromide, reaction+Br_2, 154, 227
—, —+I^-, 160
tripropyltin chloride, reaction+HCl, 127, 128
tritolylbromomethane, reaction+RHgBr, 181,
  182

V

vinylboronic acid, reaction+H_2O_2, 183

W

water, and $MeAuPPh_3+HgBr_2$, 100, 101
—, and $Me_3AuPPh_3+HgBr_2$, 104

—, and $NC_5H_4CH_2Cr(III)+HgX_2$, 93, 94
—, and $RHgI+H_2SO_4$, 115
—, Br_2/Br^- and I_2/I^- equilibria in, 139
—, effect on boronic esters+$HgCl_2$, 69
—, — — $BzHgCl+Br_2$, 144
—, — — $HgPh_2+HCl$, 117
—, — — $MeHgBr+HgBr_2$, 41
—, — — $Pt(PEt_3)_2MeCl+acid$, 131
—, — — SnR_4+HgX_2, 76–86
—, heat of solution of $SnEt_4$ in, 151
—, reaction+BEt_3, 124

X

Xylene, AlR_3+PhCN in, 185
—, $SnBu_4+I_2$ in, 165

Z

zinc bromide, and $RHgBr+Br_2$, 136
—, effect on $ZnR_2+PhCHO$, 180
zinc butyl ethyl(propyl), reaction+
  $MeC_6H_4NH_2$, 112
zinc dibutyl, reaction+RNH_2, 112, 113, 227
zinc di(3,3′-dimethylbutyl), inversion of, 32,
  33, 221
zinc diethyl, reaction+PhCHO, 180
—, —+PhHgCl, 64, 65, 227
—, —+RNH_2, 112, 113, 227
zinc dimethyl, formation of CH_3^- from, 6
—, reaction+PhHgCl, 64, 65, 227
zinc dipropyl, reaction+PhCHO, 180
—, —+PhHgCl, 64, 65, 227
—, —+RNH_2, 111, 112, 227